Model Identification and Data Analysis

Model Identification and Data Analysis

Sergio Bittanti
Politecnico di Milano
Milan, Italy

Registered Office
John Wiley & Sons, Inc., 111 River Street, Hoboken, NJ 07030, USA

Editorial Office
111 River Street, Hoboken, NJ 07030, USA

For details of our global editorial offices, customer services, and more information about Wiley products visit us at www.wiley.com.

Wiley also publishes its books in a variety of electronic formats and by print-on-demand. Some content that appears in standard print versions of this book may not be available in other formats.

Library of Congress Cataloging-in-Publication Data

Names: Bittanti, Sergio, author.
Title: Model identification and data analysis / Sergio Bittanti, Politecnico
  di Milano, Milan, Italy.
Description: Hoboken, NJ, USA : Wiley, [2019] | Includes bibliographical
  references and index. |
Identifiers: LCCN 2018046965 (print) | LCCN 2018047956 (ebook) | ISBN
  9781119546412 (Adobe PDF) | ISBN 9781119546313 (ePub) | ISBN 9781119546368
  (hardcover)
Subjects: LCSH: Mathematical models. | Quantitative research. | System
  identification.
Classification: LCC TA342 (ebook) | LCC TA342 .B58 2019 (print) | DDC
  511/.8–dc23
LC record available at https://lccn.loc.gov/2018046965

Cover design: Wiley
Cover image: © Oleksii Lishchyshyn/Shutterstock, © gremlin/Getty Images

Set in 10/12pt WarnockPro by SPi Global, Chennai, India

Contents

Introduction

Today, a deluge of information is available in a variety of formats. Industrial plants are equipped with distributed sensors and smart metering; huge data repositories are preserved in public and private institutions; computer networks spread bits in any corner of the world at unexpected speed. No doubt, we live in the *age of data*.

This new scenario in the history of humanity has made it possible to use new paradigms to deal with old problems and, at the same time, has led to challenging questions never addressed before. To reveal the information content hidden in observations, models have to be constructed and analyzed.

The purpose of this book is to present the first principles of model construction from data in a simple form, so as to make the treatment accessible to a wide audience. As R.E. Kalman (1930–2016) used to say "Let the data speak," this is precisely our objective.

Our path is organized as follows.

We begin by studying signals with stationary characteristics (Chapter 1). After a brief presentation of the basic notions of random variable and random vector, we come to the definition of white noise, a peculiar process through which one can construct a fairly general family of models suitable for describing random signals. Then we move on to the realm of frequency domain by introducing a spectral characterization of data. The final goal of this chapter is to identify a wise representation of a stationary process suitable for developing prediction theory.

In our presentation of random notions, we rely on elementary concepts: the mean, the covariance function, and the spectrum, without any assumption about the probability distribution of data. In Chapter 2, we briefly see how these features can be computed from data.

For the simple dynamic models introduced in Chapter 1, we present the corresponding prediction theory. Given the model, this theory, explained in Chapter 3, enables one to determine the predictor with elementary computations. Having been mainly developed by Andrey N. Kolmogorov and Norbert Wiener, we shall refer it as *Kolmogorov–Wiener theory* or simply *K–W theory*.

Then, in Chapter 4, we start studying the techniques for the construction of a model from data. This transcription of long sequences of apparently confusing numbers into a concise formula that can be scribbled into our notebook is the essence and the magic of identification science.

The methods for the parameter estimation of input–output models are the subject of Chapter 5. The features of the identified models when the number of snapshots tends to infinity is also investigated (asymptotic analysis). Next, the recursive version of the various methods, suitable for real-time implementation, are introduced.

In system modeling, one of the major topics that has attracted the attention of many scholars from different disciplines is the selection of the appropriate complexity. Here, the problem is that an overcomplex model, while offering better data fitting, may also fit the noise affecting measurements. So, one has to find a trade-off between accuracy and complexity. This is discussed in Chapter 6.

Considering that prediction theory is model-based, our readers might conclude that the identification methods should be presented prior to the prediction methods. The reason why we have done the opposite is that the concept of prediction is very much used in identification.

In Chapter 7, the problem of identifying a model in a state space form is dealt with. Here, the data are organized into certain arrays from the factorization of which the system matrices are eventually identified.

The use of the identified models for control is concisely outlined in Chapter 8. Again prediction is at the core of such techniques, since their basic principle is to ensure that the prediction supplied by the model is close to the desired target. This is why these techniques are known as *predictive control* methods.

Chapter 9 is devoted to Kalman theory (or simply K theory) for filtering and prediction. Here, the problem is to estimate the temporal evolution of the state of a system. In other words, instead of parameter estimation, we deal with signal estimation. A typical situation where such a problem is encountered is deep space navigation, where the position of a spacecraft has to be found in real time from available observations.

At the end of this chapter, we compare the two prediction theories introduced in the book, namely we compare K theory with K–W theory of Chapter 3.

We pass then to Chapter 10, where the problem of the estimation of an unknown parameter in a given model is treated.

Identification methods have had and continue to have a huge number of applications, in engineering, physics, biology, and economics, to mention only the main disciplines. To illustrate their applicability, a couple of case studies are discussed in Chapter 11. One of them deals with the analysis of Kobe earthquake of 1995. In this study, most facets of the estimation procedure of input–output models are involved, including parameter identification and model complexity selection. Second, we consider the problem of estimating the unknown frequency of a periodic signal corrupted by noise, by resorting to the input–output approach as well as with the state space approach by nonlinear Kalman techniques.

There are, moreover, many numerical examples to accompany and complement the presentation and development of the various methods.

In this book, we focus on the discrete time case. The basic pillars on which we rely are random notions, dynamic systems, and matrix theory.

Random variables and stationary processes are gradually introduced in the first sections of the initial Chapter 1. As already said, our concise treatment hinges on simple notions, culminating in the concept of white process, the elementary brick for the construction of the class of models we deal with. Going through these pages, the readers will become progressively familiar with stationary processes and ideas, as tools for the description of uncertain data.

The main concepts concerning linear discrete-time dynamical systems are outlined in Appendix A. They range from state space to transfer functions, including their interplay via realization theory.

In Appendix B, the readers who are not familiar with matrix analysis will find a comprehensive overview not only of eigenvalues and eigenvectors, determinant and basis, but also of the notion of rank and the basic tool for its practical determination, singular value decomposition.

Finally, a set of problems with their solution is proposed in Appendix C.

Most simulations presented in this volume have been performed with the aid of MATLAB® package, see https://it.mathworks.com/help/ident/.

The single guiding principle in writing this books has been to introduce and explain the subject to readers as clearly as possible.

Acknowledgments

The birth of a new book is an emotional moment, especially when it comes after years of research and teaching.

This text is indeed the outcome of my years of lecturing *model identification and data analysis* (MIDA) at the Politecnico di Milano, Italy. In its first years of existence, the course had a very limited number of students. Nowadays, there are various MIDA courses, offered to master students of automation and control engineering, electronic engineering, bio-engineering, computer engineering, aerospace engineering, and mathematical engineering.

In my decades of scientific activity, I have had the privilege of meeting and working with many scholars. Among them, focusing on the Italian community, are Paolo Bolzern, Claudio Bonivento, Marco Claudio Campi, Patrizio Colaneri, Antonio De Marco, Giuseppe De Nicolao, Marcello Farina, Simone Formentin, Giorgio Fronza, Simone Garatti, Roberto Guidorzi, Alberto Isidori, Antonio Lepschy, Diego Liberati, Arturo Locatelli, Marco Lovera, Claudio Maffezzoni, Gianantonio Magnani, Edoardo Mosca, Giorgio Picci, Luigi Piroddi, Maria Prandini, Fabio Previdi, Paolo Rocco, Sergio Matteo Savaresi, Riccardo Scattolini, Nicola Schiavoni, Silvia Carla Strada, Mara Tanelli, Roberto Tempo, and Antonio Vicino.

I am greatly indebted to Silvia Maria Canevese for her generous help in the manuscript editing, thank you Silvia. Joshua Burkholder, Luigi Folcini, Chiara Pasqualini, Grace Paulin Jeeva S, Marco Rapizza, Matteo Zovadelli, and Fausto Vezzaro also helped out with the editing in various phases of the work.

I also express my gratitude to Guido Guardabassi for all our exchanges of ideas on this or that topic and for his encouragement to move toward the subject of data analysis in my early university days.

Some of these persons, as well as other colleagues from around the world, are featured in the picture at the end of the book (taken at a workshop held in 2017 at Lake Como, Italy).

A last note of thanks goes to the multitude of students I met over the years in my class. Their interest has been an irreplaceable stimulus for my never ending struggle to explain the subject as clearly and intelligibly as possible.

e-mail: sergio.bittanti@polimi.it   *Sergio Bittanti*
website: home.deib.polimi.it/bittanti/

The support of the Politecnico di Milano and the National Research Council of Italy (Consiglio Nazionale delle Ricerche–CNR) is gratefully acknowledged.

1

Stationary Processes and Time Series

1.1 Introduction

Forecasting the evolution of a man-made system or a natural phenomenon is one of the most ancient problems of human kind. We develop here a prediction theory under the assumption that the variable under study can be considered as stationary process. The theory is easy to understand and simple to apply. Moreover, it lends itself to various generalizations, enabling to deal with non-stationary signals.

The organization is as follows. After an introduction to the prediction problem (Section 1.2), we concisely review the notions of random variable, random vector, and random (or stochastic) process in Sections 1.3–1.5, respectively. This leads to the definition of white process (Section 1.6), a key notion in the subsequent developments. The readers who are familiar with random concepts can skip Sections 1.3–1.5.

Then we introduce the *moving average* (MA) process and the *autoregressive* (AR) process (Sections 1.7 and 1.8). By combining them, we come to the family of *autoregressive and moving average* (ARMA) processes (Section 1.10). This is the family of stationary processes we focus on in this volume.

For such processes, in Chapter 3, we develop a prediction theory, thanks to which we can easily work out the optimal forecast given the model.

In our presentation, we make use of elementary concepts of linear dynamical systems such as transfer functions, poles, and zeros; the readers who are not familiar with such topics are cordially invited to first study Appendix A.

1.2 The Prediction Problem

Consider a real variable v depending on discrete time t. The variable is observed over the interval $1, 2, \ldots, t - 1$. The problem is to predict the value that will take the subsequent sample $v(t)$.

Model Identification and Data Analysis, First Edition. Sergio Bittanti.
© 2019 John Wiley & Sons, Inc. Published 2019 by John Wiley & Sons, Inc.

Various prediction rules may be conceived, providing a guess for $v(t)$ based on $v(t-1), v(t-2), \ldots, v(2), v(1)$. A generic predictor is denoted with the symbol $\hat{v}(t|t-1)$:

$$\hat{v}(t|t-1) = f(v(t-1), v(t-2), \ldots, v(2), v(1)).$$

The question is how to choose function f.

A possibility is to consider only a bunch of recent data, say $v(t-1), v(t-2), \ldots, v(t-n)$, and to construct the prediction as a linear combination of them with real coefficients $a_1, a_2, \ldots, a_n$:

$$\hat{v}(t|t-1) = a_1 v(t-1) + a_2 v(t-2) + \cdots + a_n v(t-n).$$

The problem then becomes that of selecting the integer n and the most appropriate values for parameter $a_1, a_2, \ldots, a_n$.

Suppose for a moment that n and $a_1, a_2, \ldots, a_n$ were selected. Then the prediction rule is fully specified and it can be applied to the past time points for which data are available to evaluate the prediction error:

$$\epsilon(k) = v(k) - \hat{v}(k|k-1), \quad k = t-1, t-2, \ldots.$$

Let's now consider this fundamental question: Which characteristics should the prediction error ϵ exhibit in order to conclude that we have constructed a "good predictor"? In principle, the best one can hope for is that the prediction error be null at any time point. However, in practice, this is Utopian. Hence, we have to investigate the properties that a non-null ϵ should exhibit in order to conclude that the prediction is fair.

For the sake of illustration, consider the case when ϵ has the time evolution shown in Figure 1.1a. As can be seen, the mean value $\bar{\epsilon}$ of ϵ is nonzero. Correspondingly, the rule

$$\hat{v}(t|t-1) = a_1 v(t-1) + a_2 v(t-2) + \cdots + a_n v(t-n) - \bar{\epsilon}$$

would be better than the original one. Indeed, with the new rule of prediction, one can get rid of the systematic error.

As a second option, consider the case when the prediction error is given by the diagram of Figure 1.1b. Then the mean value is zero. However, the sign of ϵ changes at each instant; precisely, $\epsilon(t) > 0$ for t even and $\epsilon(t) < 0$ for t odd. Hence, even in such a case, a better prediction rule than the initial one can be conceived. Indeed, one can formulate the new rule:

$$v(t) > a_1 v(t-1) + a_2 v(t-2) + \cdots + a_n v(t-n) \quad \text{for } t \text{ even}$$

and

$$v(t) < a_1 v(t-1) + a_2 v(t-2) + \cdots + a_n v(t-n) \quad \text{for } t \text{ odd}.$$

From these simple remarks, one can conclude that the best predictor should have the following property: besides a zero mean value, the prediction error should have no regularity, rather it should be fully unpredictable. In this way, the

Figure 1.1 Possible diagrams of the prediction error.

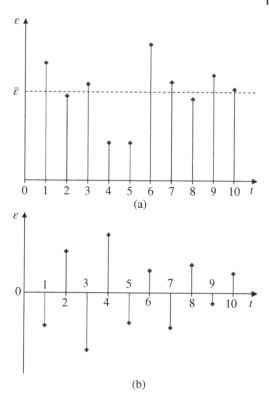

(a)

(b)

model captures the whole dynamic hidden in data and no useful information remains unveiled in the residual error, and no better predictor can be conceived. The intuitive concept of "unpredictable signal" has been formalized in the twentieth century, leading to the notion of white noise (WN) or white process, a concept we precisely introduce later in this chapter. For the moment, it is important to bear in mind the following conclusion: A prediction rule is appropriate if the corresponding prediction error is a white process.

In this connection, we make the following interesting observation. Assume that ϵ is indeed a white noise, then

$$v(t) = \hat{v}(t|t-1) + \epsilon(t) = a_1 v(t-1) + a_2 v(t-2) + \cdots + a_n v(t-n) + \epsilon(t).$$

Rewrite this difference equation by means of the delay operator z^{-1}, namely the operator such that

$$z^{-1}v(t) = v(t-1),$$
$$z^{-2}v(t) = v(t-2),$$
$$\vdots$$
$$z^{-k}v(t) = v(t-k).$$

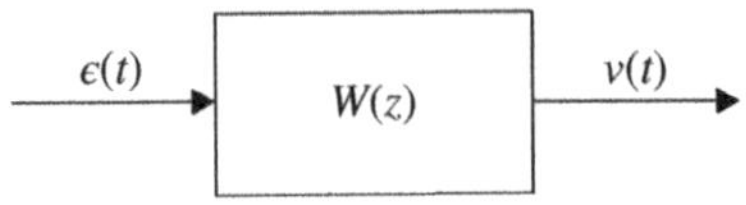

Figure 1.2 Interpreting a sequence of data as the output of a dynamic model fed by white noise.

Then

$$v(t) = a_1 z^{-1} v(t) + a_2 z^{-2} v(t) + \cdots + a_n z^{-n} v(t) + \epsilon(t)$$

from which

$$(a_1 z^{-1} + a_2 z^{-2} + \cdots + a_n z^{-n}) v(t) = \epsilon(t)$$

or

$$v(t) = W(z)\epsilon(t),$$

with

$$W(z) = \frac{1}{1 - a_1 z^{-1} - a_2 z^{-2} - \cdots - a_n z^{-n}} = \frac{z^n}{z^n - a_1 z^{n-1} - a_2 z^{n-2} - \cdots - a_n}.$$

By reinterpreting z as the complex variable, this relationship becomes the expression of a dynamical system with transfer function (from ϵ to v) given by $W(z)$.

Summing up, finding a good predictor is equivalent to determining a model supplying the given sequence of data as the output of a dynamical system fed by white noise (Figure 1.2).

This is why studying dynamical systems having a white noise at the input is a main preliminary step toward the study of prediction theory.

The road we follow toward this objective relies first on the definition of white noise, which we pursue in four stages: random variable → random vector → stochastic process → white noise.

1.3 Random Variable

A random (or stochastic) variable v is a real variable that depends upon the outcome of a random experiment. For example, the variable taking the value $+1$ or -1 depending on the result of the tossing of a coin is a random variable.

The outcome of the random experiment is denoted by s; hence, a random variable is a function of s: $v = v(s)$.

For our purposes, a random variable is described by means of its mean value (or expected value) and its variance, which we will denote by $E[v]$ and $Var[v]$, respectively.

The *mean value* is the real number around which the values taken by the variable fluctuate. Note that, given two random variables, v_1 and v_2 with mean values $E[v_1]$ and $E[v_2]$, the random variable

$$v = \alpha_1 v_1 + \alpha_2 v_2,$$

obtained as a linear combination of v_1 and v_2 via the real numbers α_1 and α_2, has a mean value:

$$E[v] = \alpha_1 E[v_1] + \alpha_2 E[v_2].$$

The variance captures the intensity of fluctuations around the mean value. To be precise, it is defined as

$$\text{Var}[v] = E[(v - m)^2],$$

where m denotes the mean value $E[v]$ of v. Obviously, $(v - m)^2$ being non-negative, the variance is a real non-negative number.

Often, the variance is denoted with symbols such as λ^2 or σ^2. When one deals with various random variables, the variance of the ith variable may be denoted as λ_{ii}^2 or σ_{ii}^2.

The square root of the variance is called *standard deviation*, denoted by λ or σ. If the random variable has a Gaussian distribution, then the mean value and the variance define completely the probability distribution of the variable. In particular, if a random variable is Gaussian, the probability that it takes value in the interval $m - 2\lambda$ and $m + 2\lambda$ is about 95%. So if v is Gaussian with mean value 10 and variance 100, then, in 95% cases, the values taken by v range from $10 - 2\sqrt{100} = -10$ to $10 + 2\sqrt{100} = 30$.

1.4 Random Vector

A random (or stochastic) vector is a vector whose elements are random variables. We focus for simplicity on the bi-dimensional case, namely, given two random variables v_1 and v_2,

$$v = \begin{bmatrix} v_1 \\ v_2 \end{bmatrix}$$

is a random vector (of dimension 2). The mean value of a random vector is defined as the vector of real numbers constituted by the mean values of the elements of the vector. Thus,

$$E[v] = \begin{bmatrix} m_1 \\ m_2 \end{bmatrix},$$

where $m_1 = E[v_1]$ and $m_2 = E[v_2]$ are the mean values of v_1 and v_2, respectively. The variance is a 2×2 matrix given by

$$\text{Var}[v] = \begin{bmatrix} \lambda_{11} & \lambda_{12} \\ \lambda_{21} & \lambda_{22} \end{bmatrix},$$

where

$$\lambda_{11} = \text{Var}[v_1],$$
$$\lambda_{22} = \text{Var}[v_2],$$
$$\lambda_{12} = E[(v_1 - m_1)(v_2 - m_2)],$$
$$\lambda_{21} = E[(v_2 - m_2)(v_1 - m_1)].$$

Here, besides variances λ_{11} and λ_{22} of the single random variables, the so-called "cross-variance" between v_1 and v_2, λ_{12}, and "cross-variance" between v_2 and v_1, λ_{21}, appear. Obviously, $\lambda_{12} = \lambda_{21}$, so that $\text{Var}[v]$ is a symmetric matrix.

It is easy to verify that the variance matrix can also be written in the form

$$\text{Var}(v) = E\left[\left\{ \begin{bmatrix} v_1 \\ v_2 \end{bmatrix} - \begin{bmatrix} m_1 \\ m_2 \end{bmatrix} \right\} \left\{ \begin{bmatrix} v_1 \\ v_2 \end{bmatrix} - \begin{bmatrix} m_1 \\ m_2 \end{bmatrix} \right\}' \right],$$

where $'$ denotes transpose.

In general, for a vector v of any dimension, the variance matrix is given by

$$\text{Var}(v) = E[(v - m)(v - m)'],$$

where $m = E[v]$ is the vector whose elements are the mean values of the random variables entering v.

If v is a vector with n entries, $\text{Var}[v]$ is a $n \times n$ matrix. In any case, $\text{Var}[v]$ is a symmetric matrix having the variances of the single variables composing vector v along the diagonal and all cross-variances as off-diagonal terms.

A remarkable feature of a variance matrix is that it is a positive semi-definite matrix.

Remark 1.1 (Positive semi-definiteness) The notions of positive semi-definite and positive definite matrix are explained in Appendix B. In a very concise way, given a $n \times n$ real symmetric matrix A, associate to it the scalar function $f(x)$ defined as $f(x) = x'Ax$, where x is an n-dimensional real vector. For example, if

$$A = \begin{bmatrix} a_{11} & a_{12} \\ a_{12} & a_{22} \end{bmatrix},$$

we take

$$x = \begin{bmatrix} x_1 \\ x_2 \end{bmatrix}.$$

Then

$$f(x) = x'\text{Var}[v]x = a_{11}x_1^2 + a_{22}x_2^2 + 2a_{12}x_1x_2.$$

Hence, $f(x)$ is quadratic in the entries of vector x. Matrix A is said to be

- positive semi-definite if $f(x) \geq 0$, $\quad \forall x$,
- positive definite if it is positive semi-definite and $f(x) = 0$ only for $x = 0$.

We write $A \geq 0$ and $A > 0$ to denote a positive semi-definite and a positive definite matrix, respectively.

We can now verify that, for any random vector v, $\mathrm{Var}[v]$ is positive semi-definite. Indeed, consider

$$x' \mathrm{Var}[v] x.$$

Then

$$
\begin{aligned}
x' \mathrm{Var}[v] x &= x' E[(v - m)(v - m)'] x \\
&= E[x'(v - m)(v - m)' x] \\
&= E[\{(v - m)' x\}' \{(v - m)' x\}].
\end{aligned}
$$

Here, we have used the property $(AB)' = B'A'$. Observe now that $(v - m)'x$, being the product of a row vector times a column vector, is a scalar. As such, it coincides with its transpose: $(v - m)'x = x'(v - m)$. Therefore,

$$x' \mathrm{Var}[v] x = E[\{(v - m)' x\}^2].$$

This is the expected value of a square, namely a non-negative real number. Therefore, this quantity is non-negative for any x. Hence, we come to the conclusion that any variance matrix is positive semi-definite. We simply write

$$\mathrm{Var}[v] \geq 0.$$

1.4.1 Covariance Coefficient

Among the remarkable properties of positive semi-definite matrices, there is the fact that their determinant is non-negative (see Appendix B). Hence, referring to the two-dimensional case,

$$\det \begin{bmatrix} \lambda_{11} & \lambda_{12} \\ \lambda_{21} & \lambda_{22} \end{bmatrix} = \lambda_{11}\lambda_{22} - \lambda_{12}^2 \geq 0.$$

Under the assumption that $\lambda_{11} \neq 0$ and $\lambda_{22} \neq 0$, this inequality suggests to define

$$\rho = \frac{\lambda_{12}}{\sqrt{\lambda_{11}}\sqrt{\lambda_{22}}}.$$

ρ is known as *covariance coefficient* between random variables v_1 and v_2. When v_1 and v_2 have zero mean value, ρ is also known as *correlation coefficient*. The previous inequality on the determinant of the variance matrix can be restated as follows:

$$|\rho| \leq 1.$$

One says that v_1 and v_2 are *uncorrelated* when $\rho = 0$. If instead $\rho = +1$ or $\rho = -1$, one says that they have maximal correlation.

Example 1.1 (Covariance coefficient for two random variables subject to a linear relation) Given a random variable v_1, with $E[v_1] = 0$, consider the variable

$$v_2 = \alpha v_1,$$

where α is a real number. To determine the covariance coefficient between v_1 and v_2, we compute the mean value and the variance of v_2 as well as the cross-covariance $v_1 - v_2$. The mean value of v_2 is

$$m_2 = E[v_2] = E[\alpha v_1] = \alpha E[v_1] = 0.$$

Its variance is easily computed as follows:

$$\lambda_{22} = E[(v_2 - m_2)^2] = E[v_2^2] = E[(\alpha v_1)^2] = E[\alpha^2 v_1^2]$$
$$= \alpha^2 E[v_1^2] = \alpha^2 \lambda_{11}.$$

As for the cross-variance, we have

$$\lambda_{12} = \lambda_{21} = E[(v_1 - m_1)(v_2 - m_2)]$$
$$= E[v_1 v_2] = E[v_1 \alpha v_1]$$
$$= E[\alpha v_1^2] = \alpha E[v_1^2] = \alpha \lambda_{11}.$$

Therefore,

$$\rho = \frac{\alpha \lambda_{11}}{\sqrt{\lambda_{11}}\sqrt{\alpha^2 \lambda_{11}}} = \frac{\alpha \lambda_{11}}{|\alpha| \lambda_{11}} = \begin{cases} 1, & \text{if } \alpha > 0, \\ -1, & \text{if } \alpha < 0. \end{cases}$$

Finally, if $\alpha = 0$, then $\rho = 0$. In conclusion,

$$\rho = \begin{cases} 0, & \text{if } \alpha = 0, \\ 1, & \text{if } \alpha > 0, \\ -1, & \text{if } \alpha < 0. \end{cases}$$

In particular, we see that, if $\alpha \neq 0$, the correlation is maximal in absolute value. This is expected, since, being $v_2 = \alpha v_1$, knowing the value taken by v_1, one can evaluate v_2 without any error.

1.5   Stationary Process

A random or stochastic process is a sequence of random variables ordered with an index t, referred to as time. We consider t as a discrete index $(t = 0, \pm 1, \pm 2, \ldots)$. The random variable associated with time t is denoted by $v(t)$. It is advisable to recall that a random variable is not a number, it is a real function of the outcome of a random experiment s. In other words,

$$v(t) = v(t, s).$$

Thus, a stochastic process is an infinite sequence of real variables, each of which depends upon two variables, time t and outcome s. Often, for simplicity in notation, the dependence upon s is omitted and one simply writes $v(t)$ to denote the process. However, one should always keep in mind that $v(t)$ depends also upon the outcome s of an underlying random experiment, $v(t, s)$.

   Once a particular outcome $s = \bar{s}$ is fixed, the set $v(t, \bar{s})$ defines a real function of time t: $v(\cdot, s)$. Such function is named *process realization*. To each outcome, a realization is associated. Hence, the set of realizations is the set of possible signals that the process can exhibit depending on the specific outcome of a random experiment. If, in the opposite, time $t = \bar{t}$ is fixed, then one obtains $v(\bar{t}, \cdot)$, the random variable at time $\bar{t}$ extracted from the process.

Example 1.2 (Random process determined by the tossing of a coin) Consider the following process: Toss a coin, if the outcome is heads, then we associate to it the function $\mathrm{sen}(\omega t)$, if the outcome is tails, we associate the function $-\mathrm{sen}(\omega t)$. The random process so defined has two sinusoidal signals as realizations, $\mathrm{sen}(\omega t)$ and $-\mathrm{sen}(\omega t)$. At a given time point $t = \bar{t}$, the process is a random variable $v(t)$, which can take two values, $\mathrm{sen}(\omega \bar{t})$ and $-\mathrm{sen}(\omega \bar{t})$.

   The simplest way to describe a stochastic process is to specify its mean function and its covariance function.

Mean function:
   The mean function is defined as

$$m(t) = E[v(t)].$$

Operator $E[\cdot]$ performs the average over all possible outcomes s of the underlying random experiment. Hence, we also write

$$m(t) = E_s[v(t)] = E_s[v(t, s)].$$

In such averaging, t is a fixed parameter. Therefore, $E[v(t)]$ does not depend upon s anymore; it depends on t only. $m(t)$ is the function of time around which the samples of the random variable $v(t)$ fluctuate.

Variance function:

The variance function of the process is

$$\text{Var}[t] = E[(v(t) - m(t))^2].$$

It provides the variances of the random variables $v(t)$ at each time point.

Covariance function:

The covariance function captures the mutual dependence of two random variables extracted from the process at different time points, say at times t_1 and t_2. It is defined as

$$\gamma(t_1, t_2) = E[(v(t_1) - m(t_1))(v(t_2) - m(t_2))].$$

It characterizes the interdependence between the deviation of $v(t_1)$ around its mean $m(t_1)$ and the deviation of $v(t_2)$ around its mean value $m(t_2)$. Note that, if we consider the same function with exchanged indexes, i.e. $\gamma(t_2, t_1)$, we have

$$\gamma(t_2, t_1) = E[(v(t_2) - m(t_2))(v(t_1) - m(t_1))].$$

Since $(v(t_2) - m(t_2))(v(t_1) - m(t_1)) = (v(t_1) - m(t_1))(v(t_2) - m(t_2))$, it follows that

$$\gamma(t_1, t_2) = \gamma(t_2, t_1). \tag{1.1}$$

Furthermore, by setting $t_2 = t_1 = t$, we obtain

$$\gamma(t, t) = E[(v(t) - m(t))^2]. \tag{1.2}$$

This is the variance of random variable $v(t)$. Hence, when the two time indexes coincide, the covariance function supplies the process variance at the given time point.

We are now in a position to introduce the concept of *stationary process*.

Definition 1.1 A stochastic process is said to be stationary when

- $m(t)$ is constant,
- $\text{Var}(t)$ is constant,
- $\gamma(t_1, t_2)$ depends upon $t_2 - t_1$ only.

Therefore, the mean value of a stationary process is simply indicated as

$$E[v(t)] = m,$$

and the covariance function can be denoted with the symbol $\gamma(\tau)$, where $\tau = t_2 - t_1$:

$$\gamma(\tau) = E[(v(t) - m)(v(t + \tau) - m)].$$

Note that, for $\tau = 0$, from this expression, we have $\gamma(0) = E[(v(t) - m)^2]$. In other words, $\gamma(0)$ is the variance of the process.

Summing up, a stationary stochastic process is described by its mean value (a real number) and its covariance function (a real function). The variance of the process is implicitly given by the covariance function at $\tau = 0$.

We now review the main properties of the covariance function of a stationary process.

- $\gamma(0) \geq 0$.
  Indeed, $\gamma(0)$ is a variance.
- $\gamma(-\tau) = \gamma(\tau)$.
  This is a consequence of (1.1) (taken $t_1 = t$ and $t_2 = t + \tau$).
- $|\gamma(\tau)| \leq \gamma(0)$.
  Indeed, consider any pair of random variables drawn for the process, say $v(t_1)$ and $v(t_2)$, with different time points. The covariance coefficient between such variables is

$$\rho = \frac{\lambda_{12}}{\sqrt{\lambda_{11}}\sqrt{\lambda_{22}}} = \frac{\gamma(\tau)}{\sqrt{\gamma(0)}\sqrt{\gamma(0)}} = \frac{\gamma(\tau)}{\gamma(0)}.$$

On the other hand, we know that $|\rho| < 1$, so that $|\gamma(\tau)|$ cannot oversize $\gamma(0)$.

This last property suggests the definition of the *normalized covariance function* as

$$\rho(\tau) = \frac{\gamma(\tau)}{\gamma(0)}.$$

Obviously, $\rho(0) = 1$, while $|\rho(\tau)| \leq 1, \forall \tau \neq 0$. Note that, for $\tau \neq 0$, $\gamma(\tau)$ and $\rho(\tau)$ may be both positive or negative.

Further properties of the covariance function are discussed in Section 2.2.

1.6 White Process

A white process is defined as the stationary stochastic process having the following covariance function:

$$\gamma(\tau) = 0, \ \forall \tau \neq 0.$$

This means that, if we take any pair of time points t_1 and t_2 with $t_2 \neq t_1$, the deviations of $v(t_1)$ and $v(t_2)$ from the process mean value are uncorrelated whatever t_1 and t_2 be. Thus, the knowledge of the value $v(t_1)$ of the process at time t_1 is of no use to predict the value of the process at time t_2, $t_2 \neq t_1$. The only prediction that can be formulated is the trivial one, the mean value. This is why the white process is a way to formalize the concept of *fully unpredictable signal*.

The white process is also named white noise (WN).

We will often use the compact notation

$$\eta(t) \sim WN(m, \lambda^2)$$

to mean that $\eta(t)$ is a white process with

- $E[\eta(t)] = m,$
- $\gamma(\tau) = 0, \quad \forall \tau \neq 0,$
- $\gamma(0) = \lambda^2.$

The white noise is the basic brick to construct the family of stationary stochastic processes that we work with.

1.7 MA Process

An MA process is a stochastic process generated as a linear combination of the current and past values of a white process $\eta(t)$:

$$v(t) = c_0\eta(t) + c_1\eta(t-1) + c_2\eta(t-2) + \cdots + c_n\eta(t-n), \quad \eta \sim WN(0, \lambda^2),$$

where $c_0, c_1, c_2, \ldots, c_n$ are real numbers.

We now determine the main features of $v(t)$. We start with the computation of the mean value and the variance of $v(t)$. As for the mean,

$$E[v(t)] = E[c_0\eta(t) + c_1\eta(t-1) + c_2\eta(t-2) + \cdots + c_n\eta(t-n)]$$
$$= c_0E[\eta(t)] + c_1E[\eta(t-1)] + c_2E[\eta(t-2)] + \cdots + c_nE[\eta(t-n)].$$

Since $E[\eta(t)] = 0, \forall t$, it follows that $E[v(t)] = 0, \forall t$.

Passing to the variance, we have

$$Var[v(t)] = E[v(t)^2]$$
$$= c_0^2E[\eta(t)^2] + c_1^2E[\eta(t-1)^2] + \cdots + c_n^2E[\eta(t-n)^2]$$
$$+ c_0c_1E[\eta(t)\eta(t-1)] + c_0c_2E[\eta(t)\eta(t-2)] + \cdots.$$

$\eta(\cdot)$ being white, all mean values of the cross-products of the type $E[\eta(i)\eta(j)]$ with $j \neq i$ are equal to zero. Hence,

$$Var[v(t)] = (c_0^2 + c_1^2 + \cdots + c_n^2)\lambda^2.$$

Turn now to the covariance function $\gamma(t_1, t_2) = E[v(t_1)v(t_2)]$. First, we consider the case when $t_2 = t_1 + 1$, and for simplicity, we set $t_1 = t$ and $t_2 = t + 1$. Then

$$E[v(t_1)v(t_2)]E[v(t)v(t+1)] =$$
$$= c_0c_1E[\eta(t)^2] + c_1c_2E[\eta(t-1)^2] + \cdots + c_{n-1}c_nE[\eta(t-n+1)^2]$$
$$= (c_0c_1 + c_1c_2 + \cdots + c_{n-1}c_n)\lambda^2.$$

It is easy to see that the same conclusion holds true if $t_2 = t_1 - 1$, so that

$$E[v(t)v(t \pm 1)] = (c_0 c_1 + c_1 c_2 + c_2 c_3 + \cdots + c_{n-1} c_n)\lambda^2.$$

Analogous computations can be performed for $t_2 = t_1 + 2$, obtaining

$$E[v(t)v(t \pm 2)] = (c_0 c_2 + c_1 c_3 + c_2 c_4 + \cdots + c_{n-2} c_n)\lambda^2.$$

We see that $E[v(t)v(t \pm 1)]$ and $E[v(t)v(t \pm 2)]$ do not depend on time t.

In general, we come to the conclusion that $\gamma(t_1, t_2) = E[v(t_1)v(t_2)]$ does not depend upon t_1 and t_2 separately; it depends upon $\tau = |t_2 - t_1|$ only. Precisely,

$$\gamma(1) = (c_0 c_1 + c_1 c_2 + c_2 c_3 + \ldots + c_{n-1} c_n)\lambda^2,$$

$$\gamma(2) = (c_0 c_2 + c_1 c_3 + \ldots + c_{n-2} c_n)\lambda^2,$$

$$\gamma(3) = (c_0 c_3 + + \ldots + c_{n-3} c_n)\lambda^2,$$

$$\vdots$$

$$\gamma(n) = (c_0 c_n)\lambda^2,$$

$$\gamma(n+1) = 0,$$

$$\gamma(n+2) = 0,$$

$$\vdots$$

Summing up, any MA process has

- constant mean value,
- constant variance,
- covariance function depending upon the distance between the two considered time points.

Therefore, it is a stationary process, whatever values parameters c_i, $i = 0, 1, 2, \ldots, n$, may take.

Observe that the expression of an MA process

$$v(t) = c_0 \eta(t) + c_1 \eta(t-1) + c_2 \eta(t-2) + \cdots + c_n \eta(t-n)$$

can be restated by means of the delay operator as

$$v(t) = (c_0 + c_1 z^{-1} + c_2 z^{-2} + \cdots + c_n z^{-n})\eta(t).$$

Then by introducing the operator

$$C(z) = c_0 + c_1 z^{-1} + c_2 z^{-2} + \cdots + c_n z^{-n},$$

one can write

$$v(t) = C(z)\eta(t).$$

From this expression, the transfer function $W(z)$ from $\eta(t)$ to $v(t)$ can be worked out

$$W(z) = c_0 + c_1 z^{-1} + c_2 z^{-2} + \cdots + c_n z^{-n} = \frac{c_0 z^n + c_1 z^{n-1} + c_2 z^{n-2} + \cdots + c_n}{z^n}.$$

Note that this transfer function has n poles in the origin of the complex plane, whereas the zeros, the roots of polynomial $c_0 z^n + c_1 z^{n-1} + c_2 z^{n-2} + \cdots + c_n$, may be located in various positions, depending on the values of the parameters.

Remark 1.2 (MA(∞) process) We extrapolate the above notion of MA(n) process and consider the MA(∞) case too:

$$v(t) = c_0 \eta(t) + c_1 \eta(t-1) + c_2 \eta(t-2) + \cdots + c_n \eta(t-n) + \cdots .$$

Of course, this definition requires some caution, as in any series of infinite terms. If the white process has zero mean value, then $v(t)$ also has a zero mean value. The variance can be obtained by extrapolating the expression of the variance for the MA(n) case, namely,

$$\mathrm{Var}[v(t)] = (c_0^2 + c_1^2 + \cdots + c_n^2 + \cdots)\lambda^2. \tag{1.3}$$

This infinite sum takes on a finite value if and only if

$$\sum_0^\infty c_i^2$$

is finite. If this condition is satisfied, then it is easy to verify that all infinite sums

$$c_0 c_1 + c_1 c_2 + \cdots + c_{n-1} c_n + \cdots$$
$$c_0 c_2 + c_1 c_3 + \cdots + c_{n-2} c_n + \cdots$$
$$c_0 c_3 + c_1 c_4 + \cdots + c_{n-3} c_n + \cdots$$
$$\vdots$$

exist as finite as well. In other words, under the condition $\sum_0^\infty c_i^2 < \infty$, the MA($\infty$) process makes sense as a stationary process and its covariance function is given by

$$\gamma(0) = (c_0^2 + c_1^2 + c_2^2 + \cdots + c_n^2 + \cdots)\lambda^2,$$
$$\gamma(1) = (c_0 c_1 + c_1 c_2 + \cdots + c_{n-1} c_n + \cdots)\lambda^2,$$
$$\gamma(2) = (c_0 c_2 + c_1 c_3 + \cdots + c_{n-2} c_n + \cdots)\lambda^2,$$
$$\gamma(3) = (c_0 c_3 + c_1 c_4 + \cdots + c_{n-3} c_n + \cdots)\lambda^2.$$
$$\vdots$$

Remark 1.3 (MA process with non-null mean value) The definition of MA process can be generalized as follows

$$v(t) = c_0 \eta(t) + c_1 \eta(t-1) + c_2 \eta(t-2) + \cdots + c_n \eta(t-n), \quad \eta \sim \mathrm{WN}(m, \lambda^2).$$

$v(t)$ has now a non-null mean value

$$E[v(t)] = (c_0 + c_1 + \cdots + c_n)\, m.$$

Note that the covariance function is the same as before, as it is easy to verify. Hence, we have again a stationary process.

Remark 1.4 (GMA process)   Turn now to the process

$$v(t) = c_0 \eta(t) + c_1 \eta(t-1) + c_2 \eta(t-2) + \cdots + c_n \eta(t-n),$$

where, instead of being a white noise, $\eta(\cdot)$ is any stationary process with zero mean and covariance function $\gamma_{\eta\eta}(\tau)$. Process $v(t)$ will be called *generalized moving average process* or simply *GMA process*.

The variance at time t of such process is given by

$$
\begin{aligned}
Var[v(t)] &= E[v(t)^2] \\
&= c_0^2 E[\eta(t)^2] + c_1^2 E[\eta(t-1)^2] + \cdots + c_n^2 E[\eta(t-n)^2] \\
&\quad + c_0 c_1 E[\eta(t)\eta(t-1)] + c_1 c_2 E[\eta(t-1)\eta(t-2)] + \cdots \\
&\quad + c_0 c_2 E[\eta(t)\eta(t-2)] + c_1 c_3 E[\eta(t)\eta(t-2)] \ldots
\end{aligned}
$$

Note that the various cross-terms $E[\eta(t)\eta(t-k)]$ are no more null since η is (in general) not white. Hence,

$$
\begin{aligned}
Var[v(t)] &= (c_0^2 + c_1^2 + \cdots + c_n^2)\, \gamma_{\eta\eta}(0) \\
&\quad + (c_0 c_1 + c_1 c_2 + \ldots)\gamma_{\eta\eta}(1) \\
&\quad + (c_0 c_2 + c_1 c_3 + \cdots)\gamma_{\eta\eta}(2) + \cdots .
\end{aligned}
\tag{1.4}
$$

As can be seen, this variance does not depend upon the considered time point t. Analogously, it can be shown that the covariance $\gamma_{vv}(t_1, t_2)$ is a function of $(t_2 - t_1)$ only.

In conclusion, *all processes generated by passing a stationary process through an MA model are stationary.*

We close this remark with a final observation on formula (1.4), which can be given an alternative expression. Let

$$
\bar{R}_{\eta\eta}^{(n+1)} = \begin{bmatrix} \gamma_{\eta\eta}(0) & \gamma_{\eta\eta}(1) & \cdots & \gamma_{\eta\eta}(n-1) \\ \gamma_{\eta\eta}(1) & \gamma_{\eta\eta}(0) & \cdots & \gamma_{uu}(n-2) \\ \cdots & \cdots & \cdots & \cdots \\ \gamma_{\eta\eta}(n) & \gamma_{\eta\eta}(n-1) & \cdots & \gamma_{\eta\eta}(0) \end{bmatrix}.
\tag{1.5}
$$

Observe that this matrix has equal elements along the diagonals, and it is therefore named *Toeplitz matrix*. Letting

$$c = [c_0 \; c_1 \; c_2 \cdots c_n]',$$

it is easy to verify that the variance of process y can be written as

$$\gamma_{vv}(0) = c' \bar{R}_{\eta\eta}^{(n+1)} c \, .$$

If, in particular, the stationary process η is white, then matrix (1.5) becomes diagonal and we return to expression (1.3).

1.8 AR Process

An AR process of order n (or simply an AR(n)) is defined starting from the model

$$v(t) = a_1 v(t-1) + a_2 v(t-2) + \cdots + a_n v(t-n) + \eta(t), \tag{1.6}$$

where $\eta(t) \sim \mathrm{WN}(0, \lambda^2)$.

By means of the operator $A(z) = 1 - az^{-1} - az^{-2} - \cdots - a^n z^{-n}$, Eq. (1.6) can be written as

$$A(z)v(t) = \eta(t).$$

The corresponding transfer function is

$$W(z) = \frac{1}{A(z)} = \frac{1}{1 - az^{-1} - az^{-2} - \cdots - a^n z^{-n}}$$

$$= \frac{z^n}{z^n - az^{n-1} - az^{n-2} - \cdots - a_n} \, .$$

We see that there are n zeros and n poles. The zeros are all located in the origin of the complex plane, whereas the poles location depends upon the values of parameters a_i.

The system is stable if and only if all poles have absolute values less than 1 (Appendix A). In such a case, whatever the initial condition be, the output $v(t)$ tends asymptotically to a stationary process.

Our study of these processes will start with the simplest case, the AR(1).

1.8.1 Study of the AR(1) Process

The AR(1) model is defined by equation

$$v(t) = av(t-1) + \eta(t).$$

We now show that this is equivalent to an MA(∞) model. Indeed, suppose that, at a certain time point, say t_0, the value of v is $v(t_0) = v_0$. Then,

$$v(t_0 + 1) = av_0 + \eta(t_0 + 1)$$

$$v(t_0 + 2) = av(t_0 + 1) + \eta(t_0 + 2) = a(av_0 + \eta(t_0 + 1)) + \eta(t_0 + 2)$$

$$= \eta(t_0 + 2) + a\eta(t_0 + 1) + a^2 v_0$$

and, in general,

$$v(t) = \eta(t) + a\eta(t-1) + a^2\eta(t-2) + \cdots + a^{t-t_0}v_0.$$

If $|a| < 1$, then the last term in this sum vanishes as $t_0 \to -\infty$, so that we obtain the MA(∞) model:

$$v(t) = \eta(t) + a\eta(t-1) + a^2\eta(t-2) + \cdots, \qquad (1.7)$$

with coefficients

$$c_0 = 1, \quad c_1 = a, \quad c_2 = a^2, \quad \ldots$$

The sum of the squares of such coefficients is

$$\sum_0^\infty c_i^2 = \sum_0^\infty a^{2i}.$$

This is just the well-known geometrical series, which, for $|a| < 1$, converges to

$$\sum_0^\infty c_i^2 = \sum_0^\infty a^{2i} = \frac{1}{1-a^2}.$$

Hence, we can conclude that, if $|a| < 1$, the AR(1) model generates a stationary process, equivalent to a well-defined MA(∞) process.

We note that, letting $A(z) = 1 - az^{-1}$, the model $v(t) = av(t-1) + \eta(t)$ can be written in operator form as $A(z)v(t) = \eta(t)$. The corresponding transfer function is

$$W(z) = \frac{1}{1 - az^{-1}} = \frac{z}{z-a}, \qquad (1.8)$$

characterized by a zero in the origin of the complex plane and a pole located in $z = a$.

Thus, the condition $|a| < 1$ imposes that the pole has absolute value less than 1. On the other hand, this is the well-known condition of stability for the systems with transfer function $W(z) = z/(z-a)$, see Section A.3 of Appendix A. The previous conclusion can be restated as follows: the output of a stable AR(1) model tends (as $t_0 \to -\infty$) to an MA(∞) stationary process.

This observation enables us to compute the properties of the output process. To be precise, we have already computed the variance of such process, namely

$$\text{Var}[v(t)] = \frac{1}{1-a^2}\lambda^2.$$

As for its covariance function, we can resort to the expression for $\gamma(\tau)$ of an MA(∞) process. In particular, since $c_0 = 1, c_1 = a, c_2 = a^2, \ldots$, we have for $\tau = 1$

$$\gamma(1) = (a + a^3 + a^5 + \cdots)\lambda^2$$
$$= a(1 + a^2 + a^4 + \cdots)\lambda^2$$
$$= a\frac{1}{1-a^2}\lambda^2.$$

In general, from the expression of the covariance function of the MA(∞) process, one can easily derive the covariance of an AR(1) process for any τ:

$$\gamma(\tau) = a^\tau \frac{1}{1-a^2}\lambda^2, \quad \tau \geq 1.$$

In particular, we see that

$$\gamma(\tau) = a\gamma(\tau - 1), \ \tau \geq 1. \tag{1.9}$$

Remark 1.5 (MA(∞) expression of AR(1) via long division) As above seen, the transfer function of an AR(1) model is given by (1.8). The MA(∞) expression (1.7) of this process can be worked out from this function by the so-called *long division* algorithm. This algorithm can be applied to the quotient of any pair of polynomials, the numerator polynomial and the denominator polynomial. In the division, we find a result polynomial plus a remainder polynomial so that we write

$$\frac{\text{Numerator}}{\text{Denominator}} = \text{Result} + \frac{\text{Remainder}}{\text{Denominator}}.$$

This is easily illustrated with the AR(1) case. The starting point is transfer function

$$W(z) = \frac{z}{z-a}$$

characterized by

$$\text{Numerator} = z,$$
$$\text{Denominator} = z - a.$$

If we develop $W(z)$ as

$$W(z) = w_0 + w_1 z^{-1} + w_2 z^{-2} + \cdots ,$$

Then process $v(t) = W(z)\eta(t)$ becomes

$$v(t) = W(z)\eta(t) = (w_0 + w_1 z^{-1} + w_2 z^{-2} + \cdots)\eta(t)$$
$$= w_0\eta(t) + w_1\eta(t - 1) + w_2\eta(t - 2) + \cdots$$

namely the MA expression of the process.

To determine w_0, we divide the numerator by the denominator for the first step

$$
\begin{array}{ll|ll}
z & & z & -a \\
z & -a & 1 & \\
\hline
& +a & &
\end{array}
$$

Hence,

$$\text{Result} = 1,$$
$$\text{Remainder} = a.$$

Therefore,

$$W(z) = \text{Result} + \frac{\text{Remainder}}{\text{Denominator}} = 1 + \frac{a}{z - a} \ .$$

Summing up,

$$w_0 = 1.$$

We now perform the second step of the long division:

```
z                          | z    -a
  z    -a                  | 1   +az⁻¹
     +a
     +a   -a²z⁻¹
        +a²z⁻¹
```

so that

$$\text{Result} = 1 + az^{-1},$$
$$\text{Remainder} = a^2 z^{-1}.$$

Therefore,

$$W(z) = \text{Result} + \frac{\text{Remainder}}{\text{Denominator}} = 1 + az^{-1} + \frac{a^2 z^{-1}}{z - a}.$$

In conclusion,

$$w_0 = 1,$$
$$w_1 = a.$$

The third step leads to

```
z                              | z    -a
   z    -a                     | 1   +az⁻¹   +a²z⁻²
      +a
       a    -a²z⁻¹
          +a²z⁻¹
          +a²z⁻¹    -a³z⁻²
              +a³z⁻²
```

so that

$$F(z) = \text{Result} + \frac{\text{Remainder}}{\text{Denominator}} = 1 + az^{-1} + a^2z^{-2} + \frac{a^3z^{-2}}{z-a}.$$

Consequently,

$$w_0 = 1,$$
$$w_1 = a,$$
$$w_2 = a^2,$$
$$\vdots$$

Thus,

$$W(z) = 1 + az^{-1} + a^2z^{-2} + a^3z^{-3} + \cdots$$

so that we re-obtain the MA(∞) model in the form (1.7).

1.9 Yule–Walker Equations

To compute the covariance function of an AR(1) process, we have previously adopted this rationale: first restate the AR(1) model as an MA(∞) one, and then resort to the expression of the covariance function of the MA(∞) process. In principle, the same rationale can be applied to an AR(2) model or to an AR model of higher order.

We will now see a simple alternative procedure, based on the so-called *Yule–Walker equations*. These equations relate the parameters of the AR model to the covariance function via an algebraic linear system. Such system can be used in two ways; from one side, it is possible to compute the parameters given the covariance function; from the other, one can compute the covariance function given the parameters.

1.9.1 Yule–Walker Equations for the AR(1) Process

Multiply both sides of the AR(1) difference equation

$$v(t) = av(t-1) + \eta(t)$$

by $v(t-\tau)$, and then apply the expectation operator to both sides:

$$E[v(t)v(t-\tau)] = aE[v(t-1)v(t-\tau)] + E[\eta(t)v(t-\tau)]. \tag{1.10}$$

Here, the left-hand side term, $E[v(t)v(t-\tau)]$, is $\gamma(\tau)$ while the first term appearing at the right-hand side, $E[v(t)v(t-\tau)]$, is $\gamma(\tau-1)$. Finally, from the structure of the AR(1) equation, it is apparent that $v(t-\tau)$ depends upon $\eta(t-\tau)$,

$\eta(t - \tau - 1)$, $\eta(t - \tau - 2)$, ..., namely the samples of the white process $\eta(\cdot)$ up to the instant $t - \tau$. Therefore, if $\tau \geq 1$, $v(t - \tau)$ does not convey useful information to predict $\eta(t)$, so that its only estimate is the trivial one, namely the process mean value, 0. Therefore, $E[\eta(t)v(t - \tau)] = 0$. These considerations lead from (1.10) to the equation

$$\gamma(\tau) = a\gamma(\tau - 1), \ \tau \geq 1,$$

which coincides with (1.9).

If instead $\tau = 0$, then $E[\eta(t)v(t - \tau)] = \lambda^2$, as it is easy to prove. Hence, Eq. (1.10) becomes

$$\gamma(0) = a\gamma(1) + \lambda^2.$$

In conclusion, we can write

$$\gamma(0) = a\gamma(1) + \lambda^2,$$
$$\gamma(\tau) = a\gamma(\tau - 1), \quad \tau \geq 1.$$

In particular,

$$\gamma(0) = a\gamma(1) + \lambda^2,$$
$$\gamma(1) = a\gamma(0) .$$

These are the so-called *Yule–Walker equations* (for the AR(1) process). They constitute an algebraic linear system relating samples $\gamma(0)$ and $\gamma(1)$ of the covariance function to the parameters a and λ^2 of the AR(1) process.

To be precise, given parameter a, the second equation allows the computation of $\gamma(1)$ from $\gamma(0)$. In turn, given a and λ^2, $\gamma(0)$ can be easily determined by combining equations $\gamma(1) = a\gamma(0)$ and $\gamma(0) = a\gamma(1) + \lambda^2$, so obtaining $\gamma(0) = (1 - a^2)^{-1}\lambda^2$.

Vice versa, given the first two samples of the covariance function, from the equation $\gamma(1) = a\gamma(0)$, parameter a can be computed; then from equation $\gamma(0) = a\gamma(1) + \lambda^2$, parameter λ^2 follows.

1.9.2 Yule–Walker Equations for the AR(2) and AR(n) Process

Pass now to the AR(2) model

$$v(t) - u_1 v(t - 1) + u_2 v(t - 2) + \eta(t), \quad \eta(\cdot) \sim WN(0, \lambda^2).$$

By multiplying the two members by $v(t - \tau)$ and applying the expectation operator,

$$E[v(t)v(t - \tau)]$$
$$= a_1 E[v(t - 1)v(t - \tau)] + a_2 E[v(t - 2)v(t - \tau)] + E[\eta(t)v(t - \tau)].$$

In this expression, $E[v(t)v(t - \tau)] = \gamma(\tau)$ and $E[v(t - 2)v(t - \tau)]\gamma(\tau - 1)$. Moreover, from the AR(2) equation, we see that $v(t - \tau)$ depends on

$$\eta(t - \tau), \eta(t - \tau - 1), \ldots$$

but does not depend on future values of $\eta(\cdot)$. $\eta(\cdot)$ being a white noise, we have

$$E[\eta(t)v(t - \tau)] = 0, \quad \forall \tau > 1.$$

Hence, for $\tau \geq 1$,

$$\gamma(1) = a_1\gamma(0) + a_2\gamma(-1),$$

$$\gamma(2) = a_1\gamma(1) + a_2\gamma(0),$$

$$\gamma(3) = a_1\gamma(2) + a_2\gamma(1),$$

$$\vdots$$

In particular, taking also into account that $\gamma(-1) = \gamma(1)$,

$$\gamma(1) = a_1\gamma(0) + a_2\gamma(1), \tag{1.11a}$$

$$\gamma(2) = a_1\gamma(1) + a_2\gamma(0). \tag{1.11b}$$

Equations (1.11) are the Yule–Walker equations for an AR(2). They constitute a linear system relating $\gamma(0)$, $\gamma(1)$, and $\gamma(2)$ to the parameters of the model. Under the assumption that $|\gamma_{yy}(1)| < \gamma_{yy}(0)$, the solution is

$$\hat{a}_1 = \frac{\gamma(0)\gamma(1) - \gamma(1)\gamma(2)}{\gamma(0)^2 - \gamma(1)^2}. \tag{1.12}$$

$$\hat{a}_2 = \frac{\gamma(0)\gamma(2) - \gamma(1)^2}{\gamma(0)^2 - \gamma(1)^2}. \tag{1.13}$$

Finally, for $\tau = 0$, it is easy to see that

$$E[\eta(t)v(t)] = \lambda^2.$$

Therefore,

$$\gamma(0) = a_1\gamma(1) + a_2\gamma(2) + \lambda^2.$$

From such an equation, once a_1 and a_2 have been determined, one can find the variance λ^2 of $\eta(t)$.

For an AR model of order n, the Yule–Walker equations can be generalized as follows:

$$\gamma(\tau) = a_1\gamma(\tau - 1) + a_2\gamma(\tau - 2) + \cdots + a_{n_a}\gamma(\tau - n), \quad \forall \tau \geq 1. \tag{1.14}$$

1.10 ARMA Process

We now merge the two types of models introduced in the previous sections and obtain the so-called *ARMA* model, defined by the difference equation:

$$v(t) = a_1 v(t-1) + a_2 v(t-2) + \cdots + a_{n_a} v(t-n_a) +$$
$$+ c_0 \eta(t) + c_1 \eta(t-1) + c_2 \eta(t-2) + \cdots + c_{n_c} \eta(t-n_c),$$

where $\eta(t) \sim WN(0, \lambda^2)$.

Parameters $a_1, a_2, \ldots, a_{n_a}$ and $c_1, c_2, \ldots, c_{n_c}$ are real.

The first row of this expression corresponds to the (AR) part, the second one to the (MA) part. Correspondingly, integer n_a is named *order of the AR* part and n_c *order of the MA part*. If advisable, one can speak of ARMA model or ARMA(n_a, n_c) model to bring into evidence the orders of the two parts.

Obviously, as a special case, we have:

- MA models:

$$v(t) = c_0 \eta(t) + c_1 \eta(t-1) + c_2 \eta(t-2) + \cdots + c_n \eta(t-n).$$

- AR models:

$$v(t) = a_1 v(t-1) + a_2 v(t-2) + \cdots + a_n v(t-n) + \eta(t).$$

By resorting to the delay operator, the ARMA model can be written as

$$(1 - a_1 z^{-1} - a_2 z^{-2} - \cdots - a_{n_a} z^{-n_a}) v(t)$$
$$= (c_0 + c_1 z^{-1} + c_2 z^{-2} + \cdots + c_{n_c} z^{-n_c}) \eta(t).$$

Then, by introducing the operator polynomials

$$A(z) = (1 - a_1 z^{-1} - a_2 z^{-2} - \cdots - a_{n_a} z^{-n_a}),$$
$$C(z) = c_0 + c_1 z^{-1} + c_2 z^{-2} + \cdots + c_{n_c} z^{-n_c},$$

we can also write

$$A(z)v(t) = C(z)\eta(t).$$

From this expression, the transfer function from η to v is

$$W(z) = \frac{(c_0 + c_1 z^{-1} + c_2 z^{-2} + \cdots + c_{n_c} z^{-n_c})}{(1 - a_1 z^{-1} - a_2 z^{-2} - \cdots - a_{n_a} z^{-n_a})}.$$

Equivalently, $W(z)$ can be written in positive powers of z by multiplying numerator and denominator by z^n, where n is the maximum between integers n_a and n_c, so obtaining

$$W(z) = \frac{(c_0 z^n + c_1 z^{n-1} + c_2 z^{n-2} + \cdots)}{(z^n - a_1 z^{n-1} - a_2 z^{n-2} - \cdots)}.$$

When entering the world of transfer functions, z should be seen as the complex variable. Thus, the numerator and denominator in the above expression are polynomials, whose roots are named zeros (for the numerator) and poles (for the denominator).

An ARMA fed by a white process generates a sequence $v(\cdot)$ of random variables, constituting a stochastic process. In general, such process is not stationary. For example, the AR(1) model $v(t) = 3v(t-1) + \eta(t)$ produces a sequence $v(\cdot)$ with growing variance, as it is easy to check.

A main result is that, if the ARMA is stable (all poles of its transfer function $W(z)$ have absolute values lower than 1), then the generated process is stationary, see Section A.4.3 of Appendix A.

With ARMA models, it is possible to construct a wide family of stationary processes.

1.11 Spectrum of a Stationary Process

Stationary processes can be described in the frequency domain. There are various ways of defining the spectrum. A very simple one is to define it as the Fourier transform of the covariance function, i.e.

$$\Gamma(\omega) = \sum_{\tau=-\infty}^{+\infty} \gamma(\tau)e^{-j\omega\tau}. \tag{1.15}$$

Note that, herein ω is a real variable.

Given $\Gamma(\omega)$, the covariance function $\gamma(\tau)$ can be recovered with the anti-transform formula:

$$\gamma(\tau) = \frac{1}{2\pi} \int_{-\pi}^{+\pi} \Gamma(\omega)e^{-j\omega\tau} \, d\omega. \tag{1.16}$$

1.11.1 Spectrum Properties

The main properties of $\Gamma(\omega)$ are

(a) $\Gamma(\omega)$ is real,
(b) $\Gamma(\omega)$ is periodic of period 2π,
(c) $\Gamma(-\omega) = \Gamma(\omega)$,
(d) $\Gamma(\omega) \geq 0 \quad \forall \omega$,
(e) The area under the curve $\Gamma(\omega)$ between $-\pi$ and π is, up to $1/(2\pi)$, the process variance.

Proof of the Spectrum Properties
The first three properties are easy to prove. Indeed, consider the infinite sum in the definition (1.15), and group the various elements in pairs, by coupling the

term associated to $+\tau$ to that associated to $-\tau$. For example, by coupling the terms relative to $\tau = +1$ and that relative to $\tau = -1$, we have

$$\gamma(1)e^{-j\omega} + \gamma(-1)e^{+j\omega}.$$

Recalling the Euler formula $e^{j\omega} = \cos\omega + j\,\mathrm{sen}\,\omega$ and the fact that $\gamma(\cdot)$ is even, $\gamma(1)e^{-j\omega} + \gamma(-1)e^{-j\omega} = 2\gamma(1)\cos\omega$. This is a real function, even and periodic of period 2π.

The proof of property (d) is postponed to Section 1.13.

Finally, property (e) is a straightforward consequence of the anti-transform formula (1.16). Indeed, if, in such formula, one poses $\tau = 0$, one obtains

$$\gamma(0) = \frac{1}{2\pi} \int_{-\pi}^{+\pi} \Gamma(\omega)d\omega.$$

1.11.2   Spectral Diagram

$\Gamma(\omega)$ being periodic of period 2π, the spectrum can be represented in a graphical way for an angular frequency ω ranging from $\omega = -\pi$ to $\omega = +\pi$. Being the spectrum even, the diagram can be limited between $\omega = 0$ and $\omega = +\pi$. Alternatively, in place of the angular frequency, one can make reference to the frequency $f = \omega/2\pi$ and draw the diagram from $f = -0.5$ and $f = +0.5$ or simply from $f = 0$ to $f = +0.5$.

1.11.3   Maximum Frequency in Discrete Time

As seen above, in discrete time, frequencies may range up to a maximum of $f = 0.5$. This can be easily interpreted. Indeed, the most rapidly varying signal in discrete time is a signal changing its sign passing from one instant to the subsequent one, e.g. signal $v(t) = 1$ for t even and $v(t) = -1$ for t odd. Such function is periodic of period $T = 2$; hence, its frequency is $f = 0.5$ and its pulsation $\omega = 2\pi f = \pi$.

1.11.4   White Noise Spectrum

The spectrum of a white noise is easily computed. Indeed, being $\gamma(\tau) = 0$, $\forall \tau \neq 0$, all terms in (1.15) are null, with the only exception of the term associated with $\tau = 0$. Hence,

$$\Gamma(\omega) = \sum_{\tau=-\infty}^{+\infty} \gamma(\tau)e^{-j\omega\tau} = \gamma(0).$$

Thus, the spectrum is a constant: all frequencies are equally contributing to form the signal. This fact is reminiscent of the constitution of the white light, which, passed through a prism, is decomposed in a variety of different colors associated to different frequencies. Probably, the wording "white noise" comes from this analogy. The presence of all frequencies in the spectrum correspond to the idea of unpredictable signal, in the sense that all frequencies are equally important, without any predominant band.

1.11.5 Complex Spectrum

In analogy with (1.15), we also define the *complex spectrum* as follows:

$$\Phi(z) = \sum_{\tau=-\infty}^{+\infty} \gamma(\tau) z^{-\tau}, \tag{1.17}$$

where z is a complex variable. This definition is useful for computational purposes, as we will see in the sequel.

By comparing this formula with (1.15), it is apparent that

$$\Gamma(\omega) = \Phi(z)|_{z=e^{j\omega}}.$$

1.12 ARMA Model: Stability Test and Variance Computation

We now address the issue of analyzing the stability of an ARMA model and computing the variance of the output process.

A celebrated problem that has vexed many great minds of past centuries is finding the solution to algebraic equations of various degrees. There is of course an explicit formula to solve equations of the second degree. But third-degree equations also admit solutions in a closed form. The history of this latter formula is quite curious. We owe its discovery to two Italian mathematicians, Niccolò Fontana of Brescia, nicknamed Tartaglia (1499–1557), and Scipione del Ferro of Bologna (1465–1526), who reached it independently of one another. Neither Tartaglia nor del Ferro, however, went public with the formula. Tartaglia revealed it in strict confidence to Gerolamo Cardano (1501–1576), while del Ferro confided it to his son-in-law Annibale della Nave, who in turn showed it to Cardano during one of the latter's visits to Bologna. The formula was finally published in Cardano's book *Ars Magna*, a great scientific treatise of the Renaissance. In the book, Cardano admits that he is not the author of the formula and credits it to Tartaglia and del Ferro. Nonetheless, Tartaglia did not take it well, and the episode led to a feud between the two that would last for years. The issue of solving algebraic equations of higher degree has been treated by various outstanding scientists, such as Paolo Ruffini (1765–1822), Niels H. Abel (1802–1829), and Évariste Galois (1811–1832). Thanks to their studies, it is possible to claim that, from the fifth degree on, there is no explicit formula.

Following the above considerations, it would thus seem that the problem of analyzing stability is difficult indeed, especially for complex systems, of a high degree. However, upon further reflection, it turns out the analysis of stability does not require the explicit solution of the characteristic equation; indeed, it requires a great deal less: for discrete-time systems, the issue is establishing that all the solutions have modulus less than 1, while for continuous-time systems, as it is easy to see, establishing that all the solutions have a real negative part.

In continuous time, the problem was solved by Edward J. Routh (1831–1907) in his paper "A Treatise in the Stability of a Given State of Motion," an essay that obtained the Adams prize for the year 1875 in the University of Cambridge, published by MacMillan and Co., Cambridge. Routh had worked out a (necessary and sufficient) condition, thanks to which it was possible to establish whether all of the solutions of an algebraic equation had a negative real part, without the need to find the solutions of the equation itself. The condition only required the construction of a table whose elements could be obtained with elementary calculations based on the coefficients of the initial polynomial. In more recent days, Adolf Hurwitz found another necessary and sufficient condition equivalent to that of Routh. Thus, the famous Routh–Hurwitz criterion was born, which is still taught today in every course on control the world over.

Coming to discrete time, a possibility is to reduce the discrete-time problem into a continuous-time one and then to resort to the Routh–Hurwitz criterion. Such reduction requires some transform that, in the complex plane, maps the discrete-time stability region (namely the inner of the unit disk) into the continuous-time stability region (i.e. the left half-plane). Alternatively, one can use the *Ruzicka criterion*, published in 1962 in Czech in *Strojnicky Casopis* (*Journal of Mechanical Engineering*), Vol. 13, no. 5, pages 395–403. This criterion is based on the analysis of a table, the *Ruzicka stability table*, constructed from the coefficients of the polynomial to be analyzed in a way that is reminiscent of the way the Routh table is constructed for the continuous-time case. By applying this criterion to the polynomial at the denominator of the transfer function of an ARMA model, one can ascertain stability.

The *Ruzicka stability table* can be complemented with another table, the *Ruzicka variance table*, the construction of which is based on the coefficients of both polynomials at the numerator and denominator of the ARMA transfer function. With this second table, it is possible to determine the variance of the stationary process generated by the ARMA system. This method is most important as, with the exception of a few simple cases, the computation of the variance of an ARMA process may be a nontrivial task. We use this algorithm in the case study in Section 11.3.

Note that there are a number of stability criteria in discrete time. The advantage of the *Ruzicka criterion* is that it also enables the computation of the process variance.

Consider then the transfer function of the ARMA model written in positive powers of z, i.e.

$$W(z) = \frac{c_0^{(n)} z^n + c_1^{(n)} z^{n-1} + \cdots + c_n^{(n)}}{a_0^{(n)} z^n - a_1^{(n)} z^{n-1} + \cdots + a_n^{(n)}}.$$

Here, in place of using the usual symbols $c_0, c_1, \ldots, c_n$ for the parameters of the numerator, we have introduced a double indexing: $c_0^{(n)}, c_1^{(n)}, \ldots, c_n^{(n)}$, to put into evidence the order of polynomial. Analogously for the $a_i^{(n)}$'s. For example, for an AR(1) model with

$$W(z) = \frac{1}{1 + 3z^{-1}},$$

we will have

$$W(z) = \frac{z}{z + 3},$$

so that

$$c_0^{(1)} = 1 \quad a_0^{(1)} = 1,$$
$$c_1^{(1)} = 0 \quad a_1^{(1)} = 3.$$

1.12.1 Ruzicka Stability Criterion

Lets first address the stability problem. With the coefficients of the polynomial at the denominator of $W(z)$, construct the so-called *Ruzicka stability table*. This table is constituted by $2n + 1$ rows computed as follows:

$$
\begin{array}{cccc}
a_0^{(n)} & a_1^{(n)} & \cdots & a_{n-1}^{(n)} \quad a_n^{(n)} \\
a_n^{(n)} & a_{n-1}^{(n)} & \cdots & a_1^{(n)} \quad a_0^{(n)} \\
a_0^{(n-1)} & a_1^{(n-1)} & \cdots & a_{n-1}^{(n-1)} \\
a_{n-1}^{(n-1)} & a_{n-2}^{(n-1)} & \cdots & a_0^{(n-1)} \\
\vdots & & & \\
a_0^{(1)} & a_1^{(1)} & & \\
a_1^{(1)} & a_0^{(1)} & & \\
a_0^{(0)} & & &
\end{array}
,
$$

where

$$a_i^{(k-1)} = a_i^{(k)} - \alpha_k a_{k-i}^{(k)},$$
$$\alpha_k = a_k^{(k)} / a_0^{(k)}.$$

This rule admits a simple interpretation.

First row:

The first row is the sequence of polynomial coefficients.

Second row:

This row coincides with the first row with elements in reverse order.

Third row:

Coefficient $a_0^{(n-1)}$, the first element of the third row, is given by

$$a_0^{(n-1)} = \frac{1}{a_0^{(n)}} \det \begin{bmatrix} a_0^{(n)} & a_n^{(n)} \\ a_n^{(n)} & a_0^{(n)} \end{bmatrix}.$$

The 2×2 matrix appearing here is constructed with the following rule: The elements of the first column coincide with the elements of the first column of the previous two rows. The elements of the second column coincide with the elements of the last column of the previous two rows. Then, the determinant of such 2×2 matrix is divided by the first coefficient of the first row.

To determine $a_1^{(n-1)}$, the second element of the third row, the computation to be performed is

$$a_1^{(n-1)} = \frac{1}{a_0^{(n)}} \det \begin{bmatrix} a_0^{(n)} & a_{n-1}^{(n)} \\ a_{n-1}^{(n)} & a_1^{(n)} \end{bmatrix}.$$

This is the determinant of the 2×2 matrix constructed as follows: The first column is constituted by the elements of the first column of the previous two rows. The second column is constituted by the elements of the last but one column of the previous two rows. The determinant of the matrix is then divided by the first coefficient of the first row.

The same rule repeats to compute all the remaining elements of the third row; in any case, one has to construct a 2×2 matrix composed of a first column given by the two elements of the first column of the previous two rows; the second column is given by the elements of the previous two rows in the column located by back stepping from the last column for a suitable number of jumps, 0 for the computation of the first element of the row, 1 for the second element, 2 for the third element, and so on, until the second column of the previous two rows is reached.

Fourth row:

The fourth row is obtained from the third one by reversing the order of parameters.

Fifth row:

Passing to the fifth row, the rule of computation is the same as for the third row, by using the third and fourth rows in place of the first and second one.

The same procedure repeats all over the table.

Example 1.3 (Ruzicka table for a simple polynomial) For the polynomial of order $n = 2$,

$$z^2 - 2$$

the coefficients of the polynomial are $1, 0, -2$. The order being $n = 2$, the Ruzicka table consists of $2n + 1 = 5$ rows.

The first two rows of the table are

$$
\begin{array}{rrr}
1 & 0 & -2 \\
-2 & 0 & 1
\end{array}
$$

The third row is constituted by two elements. The first one is the determinant of the 2×2 matrix obtained with the first and third columns of the first and second rows, divided by the first element of the first column; the second one is obtained as the determinant of the 2×2 matrix given with the first and second columns of the first and second rows, divided by the first element of the first column, i.e.

$$a_0^{(1)} = \frac{1}{1} \det \begin{bmatrix} 1 & -2 \\ -2 & 1 \end{bmatrix} = -3,$$

$$a_1^{(1)} = \frac{1}{1} \det \begin{bmatrix} 1 & 0 \\ -2 & 0 \end{bmatrix} = 0.$$

The fourth row is given by these two elements of the third row in reverse order.

The fifth and last row is constituted of a unique element:

$$a_0^{(0)} = \frac{1}{-3} \det \begin{bmatrix} -3 & 0 \\ 0 & -3 \end{bmatrix} = -3.$$

In conclusion, the table is

$$
\begin{array}{rr}
1 & 0 \quad -2 \\
-2 & 0 \quad 1 \\
-3 & 0 \\
0 & -3 \\
-3 &
\end{array}
$$

Example 1.4 (Ruzicka table for a second-order polynomial) Consider now a generic second-order polynomial

$$z^2 + az + b.$$

The associated table is

$$
\begin{array}{cc}
1 & a \quad b \\
b & a \quad 1 \\
a_0^{(1)} & a_1^{(1)} \\
a_1^{(1)} & a_0^{(1)} \\
a_0^{(0)} &
\end{array}
$$

We leave the computation of the remaining elements of the table to our readers:

$$a_0^{(1)} = 1 - b^2,$$

$$a_1^{(1)} = a - ab,$$

$$a_0^{(0)} = (1 - b)(1 + a + b)(1 + b - a)/(1 + b).$$

We are in a position to state the following.

Proposition 1.1 (Ruzicka stability criterion) *The polynomial* $a_0^{(n)} z^n + a_1^{(n)} z^{n-1} + \cdots + a_n^{(n)} = 0$ *is stable (namely the equation* $a_0^{(n)} z^n + a_1^{(n)} z^{n-1} + \cdots + a_n^{(n)} = 0$ *has all solutions with modulus less than 1) if and only if, for each* $k = n, n - 1, \ldots, 0$, *coefficients* $a_0^{(k)}$ *of the Ruzicka table are all nonzero with the same sign.*

Example 1.5 (Stability of a simple polynomial– Example 1.3 continued) Consider again the polynomial $z^2 - 2$. The first elements of the odd rows of the corresponding Ruzicka table are $1, -3, -3$. These numbers do not have the same sign; according to the Ruzicka stability criterion, this polynomial does not have its two roots inside the unit circle. This is expected as the roots of the polynomial are $\pm\sqrt{2}$.

Example 1.6 (Stability of the second-order polynomial–Example 1.4 continued) Consider again polynomial

$$z^2 + az + b.$$

The first elements of the odd rows Ruzicka table are

$$1,$$

$$a_0^{(1)} = 1 - b^2,$$

$$a_0^{(0)} = (1 - b)(1 + a + b)(1 + b - a)/(1 + b).$$

The Ruzicka stability criterion is satisfied if and only if

$$u_0^{(1)} = 1 - b^2 > 0,$$

$$a_0^{(0)} = (1 - b)(1 + a + b)(1 + b - a)/(1 + b) > 0.$$

Condition $a_0^{(1)} = 1 - b^2 > 0$ is equivalent to $|b| < 1$. Moreover, it is also equivalent to $(1 - b)(1 + b) > 0$, i.e. $(1 - b)$ and $(1 + b)$ must have the same sign. Therefore, the quotient $(1 - b)/(1 + b)$ must be positive, so that the condition

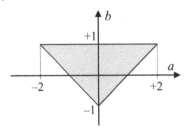

Figure 1.3 Stability region for polynomial $z^2 + az + b$.

$a_0^{(0)} > 0$ is equivalent to $(1 + a + b)(1 + b - a) > 0$, namely $(1 + b)^2 - a^2 > 0$. In conclusion, the stability condition is

$$\begin{cases} |b| < 1, \\ -(1 + b) < a < (1 + b). \end{cases}$$

Summing up, the polynomial is stable when (a, b) belongs to the triangle represented in Figure 1.3.

Notice in passing that the simple polynomial $z^2 - 2$ of Example 1.3 is of the type $z^2 + az + b$ with $a = 0$ and $b = -2$, and the point $(0, -2)$ does not belong to the region depicted in this drawing.

Example 1.7 (Stability of AR(2)) Consider the classical expression of a second-order AR model:

$$v(t) = a_1 v(t - 1) + a_2 v(t - 2) + \eta(t),$$

whose operator description is

$$(z^2 - a_1 z - a_2)v(t - 2) = \eta(t).$$

The stability of this system depends upon polynomial $z^2 - a_1 z - a_2$, which coincides with $z^2 + az + b$ by letting $a = -a_1$ and $b = -a_2$. Therefore, the stability region is now the one of Figure 1.4.

1.12.2 Variance of an ARMA Process

To compute the variance of the process generated by a stable ARMA model, one can proceed by constructing the following array, named *extended*

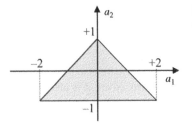

Figure 1.4 Stability region for polynomial $z^2 - a_1 z - a_2$.

Ruzicka table:

$$
\begin{array}{cccccccccc}
a_0^{(n)} & a_1^{(n)} & \cdots & a_{n-1}^{(n)} & a_n^{(n)} & c_0^{(n)} & c_1^{(n)} & \cdots & c_{n-1}^{(n)} & c_n^{(n)} \\
a_n^{(n)} & a_{n-1}^{(n)} & \cdots & a_1^{(n)} & a_0^{(n)} & a_n^{(n)} & a_{n-1}^{(n)} & \cdots & a_1^{(n)} & a_0^{(n)} \\
a_0^{(n-1)} & a_1^{(n-1)} & \cdots & a_{n-1}^{(n-1)} & & c_0^{(n-1)} & c_1^{(n-1)} & \cdots & c_{n-1}^{(n-1)} & \\
a_{n-1}^{(n-1)} & a_{n-2}^{(n-1)} & \cdots & a_0^{(n-1)} & & a_{n-1}^{(n-1)} & a_{n-2}^{(n-1)} & \cdots & a_0^{(n-1)} & \\
\vdots & & & & & \vdots & & & & \\
a_0^{(1)} & a_1^{(1)} & & & & c_0^{(1)} & c_1^{(1)} & & & \\
a_1^{(1)} & a_0^{(1)} & & & & a_1^{(1)} & a_0^{(1)} & & & \\
a_0^{(0)} & & & & & c_0^{(0)} & & & &
\end{array}
$$

The left-hand side of this array is just the previously introduced *Ruzicka stability table*. Turn now to the right part, constituting the *Ruzicka variance table*. Here, we see that the even rows coincide with the even rows of the left-hand side. As for the odd rows, which depend upon both parameters $c_i^{(n)}$ and parameters $a_i^{(n)}$, they can be computed according to the rule below. For the sake of completeness, here we replicate the rule for $a_i^{(k)}$ as well:

$$
a_i^{(k-1)} = a_i^{(k)} - \alpha_k a_{k-i}^{(k)},
$$

$$
c_i^{(k-1)} = c_i^{(k)} - \gamma_k a_{k-i}^{(k)},
$$

where

$$
\alpha_k = a_k^{(k)}/a_0^{(k)},
$$

$$
\gamma_k = c_k^{(k)}/a_0^{(k)}.
$$

Assuming that the stability condition be satisfied, and denoting by λ^2 the variance of the white process at the input of the ARMA model, the variance of the stationary output is given by

$$
\frac{\lambda^2}{a_0^{(n)}} \sum_{i=0}^{n} \frac{[c_i^{(i)}]^2}{a_0^{(i)}}.
$$

Example 1.8 (Variance of an AR(1) process)   Consider the AR(1) process $v(t) = 0.5v(t-2) + \eta(t)$, with $\eta(t) \sim \text{WN}(0,1)$. It is obvious that the system is stable since the characteristic polynomial is $z^2 - 0.5$ (with roots inside the unit circle). We aim at computing the variance of the associated stationary process. First, we perform the computation from the given model by resorting to the definition of variance. Then we use the Ruzicka algorithm.

Using the definition

Considering that $\eta(t)$ and $v(t-2)$ are uncorrelated, it is easy to see that

$$
\text{Var}[v(t)] = 0.25\text{Var}[v(t-2)] + \text{Var}[\eta(t)].
$$

The process being stationary, its variance does not depend upon time. Hence,

$$\text{Var}[v(t)] = 0.25\text{Var}[v(t)] + 1 \quad \text{Var}[v(t)] = 4/3.$$

Using Ruzicka algorithm

The transfer function $W(z)$ from $\eta(t)$ to $v(t)$ is given by

$$W(z) = \frac{1}{1 - 0.5z^{-2}} = \frac{z^2}{z^2 - 0.5}.$$

Hence,

$$c_0^{(2)} = 1 \quad c_1^{(2)} = 0 \quad c_2^{(2)} = 0,$$
$$a_0^{(2)} = 1 \quad a_1^{(2)} = 0 \quad a_2^{(2)} = -0.5.$$

From the definitions of α_k and γ_k, we have

$$\alpha_2 = \frac{a_2^{(2)}}{a_0^{(2)}} = \frac{-1/2}{1} = -\frac{1}{2},$$

$$\gamma_2 = \frac{c_2^{(2)}}{a_0^{(2)}} = \frac{0}{1} = 0,$$

so that

$$a_0^{(1)} = a_0^{(2)} - \alpha_2 a_2^{(2)} = 1 + \frac{1}{2}\left(-\frac{1}{2}\right) = 1 - \frac{1}{4} = \frac{3}{4},$$
$$a_1^{(1)} = a_1^{(2)} - \alpha_2 a_1^{(2)} = 0 + \frac{1}{2}(0) = 0,$$
$$c_0^{(1)} = c_0^{(2)} - \gamma_2 c_2^{(2)} = 1,$$
$$c_1^{(1)} = c_1^{(2)} - \gamma_2 c_1^{(2)} = 0.$$

Proceeding again for the subsequent rows, we have

$$\alpha_1 = \frac{a_1^{(1)}}{a_0^{(1)}} = 0$$

$$\gamma_1 = \frac{c_1^{(1)}}{a_0^{(1)}} = 0$$

from which

$$a_0^{(0)} = a_0^{(1)} - \alpha_1 a_1^{(1)} = \frac{3}{4},$$
$$c_0^{(0)} = c_0^{(1)} - \gamma_1 a_1^{(1)} = 1.$$

The complete Ruzicka table is

$$
\begin{array}{cccccc}
1 & 0 & -\frac{1}{2} & 1 & 0 & 0 \\[4pt]
-\frac{1}{2} & 0 & 1 & -\frac{1}{2} & 0 & 1 \\[4pt]
\frac{3}{4} & 0 & & 1 & 0 & \\[4pt]
0 & \frac{3}{4} & & 0 & \frac{3}{4} & \\[4pt]
\frac{3}{4} & & & 1 & &
\end{array}
$$

The process variance is then given by

$$
\frac{1}{a_0^{(n)}} \sum_{i=0}^{n} \frac{[c_i^{(i)}]^2}{a_0^{(i)}} = \frac{1}{a_0^{(2)}} \left(\frac{[c_2^{(2)}]^2}{a_0^{(2)}} + \frac{[c_1^{(1)}]^2}{a_0^{(1)}} + \frac{[c_0^{(0)}]^2}{a_0^{(0)}} \right) = \frac{1}{1} \left(\frac{1}{\frac{3}{4}} + 0 + 0 \right) = \frac{4}{3}.
$$

Remark 1.6 In such an elementary example, the direct computation based on the definition is simpler. However, when dealing with ARMA processes of high order, the computation may be complex. The Ruzicka procedure offers an effective alternative, easily implementable via software.

The proof of the Ruzicka criterion can be found in the book *Introduction to Stochastic Control Theory*, by Åström (2006), whose original version was published by Academic Press in 1970.

1.13 Fundamental Theorem of Spectral Analysis

According to its definition, the computation of the spectrum of a stationary process requires first the determination of the covariance function and then the application of the transform formula, (1.15) or (1.17), to find $\Gamma(\omega)$ or $\Phi(z)$. This route may be computationally demanding, since, for models of high order, the determination of the covariance function is a nontrivial task. Fortunately, there is a simple alternative based on an important result, the *fundamental theorem of spectral analysis*. This theorem is valid for all stationary processes obtained at the output of a stable linear system fed by any stationary process. We now state and comment on this main result, the proof of which is postponed to Section 1.15. In Section 1.14, we see how the theorem simplifies the spectrum drawing.

Proposition 1.2 (Fundamental theorem of spectral analysis) *Consider a single-input single-output stable system with transfer function $W(z)$ fed by a stationary process $u(\cdot)$ with complex spectrum $\Phi_{uu}(z)$ and real spectrum $\Gamma_{uu}(\omega)$.*

The complex spectrum $\Phi_{yy}(z)$ and the real spectrum $\Gamma_{yy}(\omega)$ of the stationary process $y(\cdot)$ at the output are given by

$$\Phi_{yy}(z) = W(z)W(z^{-1})\Phi_{uu}(z),$$
$$\Gamma_{yy}(\omega) = |W(e^{j\omega})|^2\Gamma_{uu}(\omega).$$

Note that, starting from the expression of the complex spectrum $\Phi_{uu}(z)$, the formula for the real spectrum $\Gamma_{yy}(\omega)$ is an immediate corollary. Indeed, if one assumes $z = e^{j\omega}$, then $z^{-1} = e^{-j\omega}$ is the complex conjugate of z. Therefore, for $z = e^{j\omega}$, $W(z^{-1})$ is the complex conjugate of $W(z)$, so that their product $W(e^{j\omega})W(e^{-j\omega})$ is just the modulus squared.

As a special case, we can consider an ARMA process.

Proposition 1.3 (Fundamental theorem of spectral analysis for ARMA processes) *Consider a single-input single-output stable system with transfer function $W(z)$ fed by a white noise of variance λ^2. The complex spectrum $\Phi_{yy}(z)$ and the real spectrum $\Gamma_{yy}(\omega)$ of the output y are given by*

$$\Phi_{yy}(z) = W(z)W(z^{-1})\lambda^2,$$
$$\Gamma_{yy}(\omega) = |W(e^{j\omega})|^2\lambda^2.$$

A corollary of this last expression is that the spectrum is non-negative at each ω, as anticipated in Section 1.11 (property (d)).

Thanks to this theorem, it is possible to properly understand the meaning of spectrum as a representation of the composition of a signal over various frequencies.

Remark 1.7 (Signal spread over frequencies) To explain this interpretation of the spectrum, we make reference to the *frequency response* of a linear system, see Section A.5 of Appendix A. In particular, we introduce the notion of pass-band filter.

Pass-band filter

Consider the system with transfer function:

$$W(e^{j\omega}) = \begin{cases} 1, & \text{for } \omega_1 < |\omega| < \omega_2, \\ 0, & \text{elsewhere.} \end{cases}$$

In view of the frequency response notion, one can conclude that such system has a peculiar property: the sinusoidal signals with pulsation ω between ω_1 and ω_2 pass through it without any distortion, whereas the sinusoidal signals of pulsation outside this band are fully blocked, in the sense that the corresponding output is null. This is why the system with the above transfer function is called *pass-band filter*.

Suppose now to feed the pass-band filter with a stationary process having a certain spectrum $\Gamma_{uu}(\omega)$. For the fundamental theorem of spectral analysis, the output has a spectrum given by

$$\Gamma_{yy}(\omega) = \begin{cases} \Gamma_{uu}(\omega), & \text{for } \omega_1 < |\omega| < \omega_2, \\ 0, & \text{elsewhere.} \end{cases}$$

In other words, the effect of the pass-band filter is to carve out the input signal spectrum by zeroing the entire diagram outside the band (ω_1, ω_2) and keeping only the strip associated with that band. On the other hand, we know that the area behind the spectrum diagram is proportional to the variance of the process generated by the pass-band filter.

This means that the overall frequency content in the original signal can be partitioned into contributions of various stationary processes whose spectrum is limited to strips of frequency bands.

Example 1.9 (Spectrum of an MA(1) process)   Consider the MA(1) process

$$v(t) = \eta(t) + c\eta(t-1).$$

We already know that its covariance function is

$$\gamma(\tau) = \begin{cases} (1+c^2)\lambda^2, & \tau = 0, \\ c\lambda^2, & \tau = 1, \\ 0, & \tau > 1. \end{cases}$$

From the definition of spectrum, we have

$$\Gamma(\omega) = [1 + c^2 + 2c\,\cos(\omega)]\lambda^2. \tag{1.18}$$

Alternatively, one can resort to the fundamental theorem. Since the transfer function of the system is

$$W(z) = (1 + cz^{-1}),$$

the complex spectrum of the output is

$$\begin{aligned} \Phi(z) &= W(z)W(z^{-1})\lambda^2 \\ &= (1 + cz^{-1})(1 + cz)\lambda^2 \\ &= \{(1 + c^2) + c(z + z^{-1})\}\lambda^2. \end{aligned}$$

Here, replacing z with $e^{j\omega}$, and recalling the Euler formulas, expression (1.18) is re-obtained.

1.14 Spectrum Drawing

The expression of the fundamental theorem is also useful to simplify the graphical drawing of the spectrum. Indeed, write $W(z)$ in its factorized form:

$$W(z) = \beta \frac{(z + \gamma_1)(z + \gamma_2)(\cdots)}{(z + \alpha_1)(z + \alpha_2)(\cdots)}.$$

Then

$$\Gamma(\omega) = \beta^2 \frac{|e^{j\omega} + \gamma_1|^2 |e^{j\omega} + \gamma_2|^2 \cdots}{|e^{j\omega} + \alpha_1|^2 |e^{j\omega} + \alpha_2|^2 \cdots}.$$

Consider one of the factors at numerator and denominator, for instance $e^{j\omega} + \gamma_1$, which is associated to the zero $-\gamma_1$. In the complex plane, $e^{-j\omega}$ is the vector of unit length with phase ω; γ_1 is the vector connecting the point $z = -\gamma_1$ to the origin of the plane (Figure 1.5).

Hence, $e^{j\omega} + \gamma_1$ is the vector connecting the zero in $z = -\gamma_1$ to the point $e^{j\omega}$ of the unit circle. As ω increases from $\omega = 0$ to $\omega = \pi$, such vector moves over the upper half circle counterclockwise from point $+1$ to point -1. Correspondingly, one can have an idea of the behavior over ω of $|e^{j\omega} + \gamma_1|$, the factor appearing in the above expression of $\Gamma(\omega)$. By analyzing the behaviors of all factors of $\Gamma(\omega)$ associated with the various poles and zeros, it is possible to deduce the qualitative shape of the spectrum.

Example 1.10 (Example 1.9 continued)   The transfer function of the MA(1) process of the previous example is

$$W(z) = 1 + cz^{-1} = \frac{z + c}{z},$$

characterized by a zero in $z = -c$ and a pole $z = 0$. We consider first the effect of the pole on the spectrum. Since the pole is located in the origin, the vector

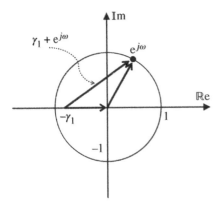

Figure 1.5 The vector $e^{j\omega} + \gamma_1$.

connecting the pole to a generic point on the unit circle has unit modulus for all ω and therefore does not have any influence on the spectrum. Pass now to the zero. If $c > 0$, the zero is located on the negative real axis. Therefore, the vector $e^{j\omega} + c$ has maximum length for $\omega = 0$. The length is decreasing as ω passes from 0 to π. On the contrary, if $c < 0$, vector $e^{j\omega} + c$ has minimum length for $\omega = 0$ and maximum length for $\omega = \pi$. Correspondingly, if $c > 0$, the spectrum will be maximum for $\omega = 0$, whereas, if $c < 0$, it will be minimum at $\omega = 0$. In other words, in the former case, the low frequencies dominate, whereas in the latter case, the high frequencies dominate. Consequently, the realizations of the process will be slowly varying when $c > 0$ and quickly varying when $c < 0$. This is in agreement with the fact that, if $c > 0$, the covariance function $\gamma(\tau) = E[v(t + \tau)v(t)]$ is positive at any τ so that $v(t + \tau)$ tends to have the same sign of $v(\tau)$; if instead $c < 0$, then $\gamma(\tau)$ is negative and $v(t + \tau)$ tends to have the opposite sign of $v(t)$. These considerations are summarized in Figure 1.6.

Remark 1.8 (3D representation of the spectrum as a circus tent) As we have seen, the spectrum $\Gamma(\omega)$ is obtained by evaluating the complex spectrum $\Phi(z)$ over the circle of radius 1 (namely for $z = e^{j\omega}$). Thus, the spectrum can be visualized as follows. Consider the real number $|\Phi(z)|$ as a function of two real variables, the real and the imaginary parts of z. $|\Phi(z)|$ can be represented in a 3D drawing, with the horizontal plane corresponding to the complex plane, while the modulus of the complex spectrum is plotted on the vertical axis. In such a representation, the zeros and poles of $\Phi(z)$ play a prominent role. Indeed, if z is a zero, then $\Phi(z) = 0$, whereas, if z is a pole, $\Phi(z) = \infty$. We can make a parallel with the circus tent, the zeros being the tie-rods and the poles the yards of the big tent. On the other hand, it is obvious that the zeros of $\Phi(z) = W(z)W(z^{-1})\lambda^2$ coincide with the zeros of $W(z)$ and their reciprocals and that the poles of $\Phi(z) = W(z)W(z^{-1})\lambda^2$ coincide with the poles of $W(z)$ and their reciprocals. Summing up, the 3D drawing of the complex spectrum can be compared to a tent anchored to the ground in the zeros of $W(z)$, and their reciprocals, and shoot skyward in the poles, and their reciprocals. The real spectrum $\Gamma(\omega)$ is obtained by intersecting the big tent with a vertical cylinder whose axis coincides with the vertical axis of radius 1 of the 3D drawing.

Example 1.11 (3D spectrum of an ARMA(1,1) process) Given a system with transfer function

$$W(z) = \frac{z - 0.5}{z + 0.5},$$

and input a white noise $WN(0, 1)$, the modulus of the complex spectrum is

$$|\Phi(z)| = |W(z)W(z^{-1})| = \left| \frac{(z - 0.5)\,(z^{-1} - 0.5)}{(z + 0.5)\,(z^{-1} + 0.5)} \right|$$

$$= \left| \frac{(z - 0.5)\,(2 - z)}{(z + 0.5)\,(2 + z)} \right|.$$

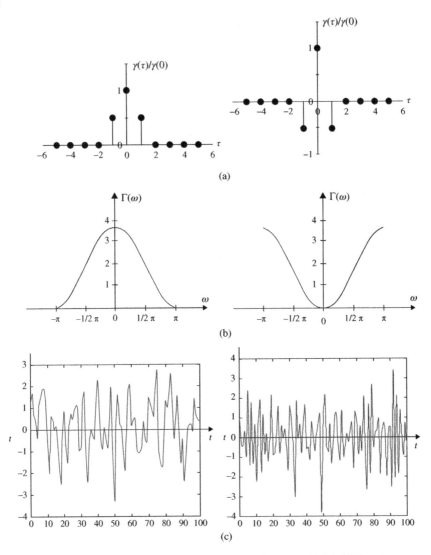

Figure 1.6 MA(1) process features with $c = 0.9$ (left) and $c = -0.9$ (right). (a) Covariance function. (b) Spectrum. (c) Realization.

Such function is null for $z = 0.5$ and $z = 2$ and goes to ∞ for $z = -0.5$ and $z = -2$. The corresponding 3D drawing is depicted in Figure 1.7, in two views with different angles. Note that the real spectrum $\Gamma(\omega)$ is the intersection of the drawing with the radius 1 cylinder encircling the vertical axis (Figure 1.8).

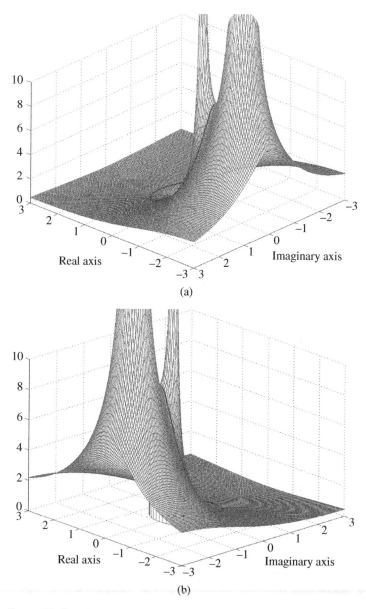

Figure 1.7 Complex spectrum in a 3D representation – Example 1.11.

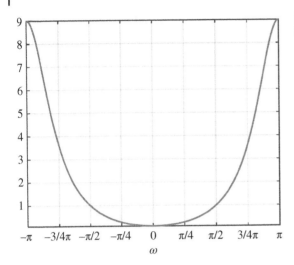

Figure 1.8 The real spectrum $\Gamma(\omega)$ – Example 1.11.

Remark 1.9 (Fundamental theorem of spectral analysis for multivariable systems) The fundamental theorem holds true for multivariable systems too. To be precise, if $u(\cdot)$ is a vector stochastic process with m elements, with zero mean value for simplicity, the covariance function is defined as

$$\gamma_{uu}(\tau) = E[u(t+\tau)u(t)'].$$

This is a square matrix of dimension $m \times m$, function of τ. Analogously, one can define the covariance function of the output, a matrix function of τ of dimension $p \times p$, where p is the number of output signals. The fundamental theorem takes then the expression

$$\Phi_{yy}(z) = W(z)\Phi_{uu}(z)W(z^{-1})'$$

or, in the real case,

$$\Gamma_{yy}(\omega) = W(e^{j\omega})\Gamma_{uu}(\omega)W(e^{-j\omega}).$$

Obviously, for a system with m inputs and p outputs, $W(z)$ is a matrix of dimension $p \times m$.

Remark 1.10 (Adjoint expression of the fundamental theorem) Sometimes, given a transfer matrix $W(z)$, one can use a unique symbol to denote the couple of operations: (i) replacement of z with z^{-1} and (ii) transposition. To be precise, we define as *adjoint matrix* $W(z)^{\sim}$ the matrix

$$W(z)^{\sim} = W(z^{-1})'.$$

The system associated with $W(z)^{\sim}$ is named adjoint system. The complex spectrum can then be written as

$$\Phi_{yy}(z) = W(z)\Phi_{uu}(z)W(z)^{\sim}.$$

1.15   Proof of the Fundamental Theorem of Spectral Analysis

Consider a stable system with transfer function $W(z)$, impulse response $\{w_i\}$, input u, and output y. As is well known, $W(z)$ and $\{w_i\}$ are related to each other as follows:

$$W(z) = w_0 + w_1 z^{-1} + w_2 z^{-2} + \cdots .$$

Correspondingly, the output is given by

$$y(t) = w_0 u(t) + w_1 u(t-1) + w_2 u(t-2) + \cdots$$

$$= \sum_{i=0}^{\infty} w_i u(t-i) .$$

Note that, from such expression, one can conclude that $E[y(t)] = 0$, whenever $u(t)$ is a stationary process with $E[u(t)] = 0$. If $u(\cdot)$ is a white process, this is just the MA(∞) expression of the original ARMA process. In general, when $u(\cdot)$ is a whatever signal, the above formula is known as *convolution product* between the two time functions $u(\cdot)$ e $w_{(\cdot)}$.

To be precise, given the functions $f(\cdot)$ and $g(\cdot)$, the *convolution product* is the function $h(\cdot)$ defined as

$$h(t) = \sum_{i=0}^{\infty} f(i)g(t-i).$$

It can be shown that the *z-transform* $H(z)$ of the function $h(\cdot)$ is the product of the z-transforms of $f(\cdot)$ and $g(\cdot)$, denoted as $F(z)$ and $G(z)$, respectively, i.e.

$$H(z) = F(z)G(z)$$

(see Section A.1.3 of Appendix A for the definition of z-transform).

All that said, suppose that $u(\cdot)$ be a stationary stochastic process, for simplicity with zero mean value. We want to compute the covariance function of the stationary process at the output. To this purpose, we will also compute the input–output cross-covariance.

We start with the convolution formula for $y(t)$ at time t_2

$$y(t) = \sum_{i=0}^{\infty} w_i u(t-i).$$

Multiply both members once by $u(t_1)$ and once by $y(t_1)$, so obtaining

$$u(t_1)y(t_2) = \sum_{i=0}^{\infty} w_i u(t_1)u(t_2-i),$$

$$y(t_1)y(t_2) = \sum_{i=0}^{\infty} w_i y(t_1)u(t_2-i).$$

Applying now the expectation operator, and letting

$$\gamma_{uu}(t_1, t_2) = E[u(t_1)u(t_2)],$$
$$\gamma_{uy}(t_1, t_2) = E[u(t_1)y(t_2)],$$
$$\gamma_{yy}(t_1, t_2) = E[y(t_1)y(t_2)],$$

we have

$$\gamma_{uy}(t_1, t_2) = \sum_{i=0}^{\infty} w_i \gamma_{uu}(t_1, t_2 - i),$$

$$\gamma_{yy}(t_1, t_2) = \sum_{i=0}^{\infty} w_i \gamma_{yu}(t_1, t_2 - i).$$

Since $u(\cdot)$ is stationary, function $\gamma_{uu}(t_1, t_2 - 1)$ depends upon the difference between the two time instants here appearing, namely $t_2 - t_1 - i$. Hence, $\gamma_{uy}(t_1, t_2)$ will depend upon $t_2 - t_1$ only. Consequently, $\gamma_{yy}(t_1, t_2)$ will also depend upon $t_2 - t_1$ only. Letting then $\tau = t_2 - t_1$, one can write

$$\gamma_{uy}(\tau) = \sum_{i=0}^{\infty} w_i \gamma_{uu}(\tau - i), \tag{1.19}$$

$$\gamma_{yy}(\tau) = \sum_{i=0}^{\infty} w_i \gamma_{yu}(\tau - i). \tag{1.20}$$

Interestingly enough, from the first of these expressions, we see that the cross-covariance $\gamma_{uy}(\cdot)$ is given by the convolution product between the impulse response and the covariance function of the input signal. From the second expression, we learn that the convolution product between the impulse response and the input output cross-covariance gives the output covariance function.

Pass now from the time domain to the z domain. For example, we introduce the complex spectra

$$\Phi_{uu}(z) = \sum_{-\infty}^{\infty} \gamma_{uu}(\tau) z^{-\tau},$$

$$\Phi_{yy}(z) = \sum_{-\infty}^{\infty} \gamma_{yy}(\tau) z^{-\tau},$$

$$\Phi_{uy}(z) = \sum_{-\infty}^{\infty} \gamma_{uy}(\tau) z^{-\tau}.$$

The spectra are obtained by evaluating these expressions in $z = e^{j\omega}$. For instance,

$$\Gamma_{uu}(\omega) = \Phi_{uu}(e^{j\omega}).$$

Note that, while $\Gamma_{uu}(\omega)$ and $\Gamma_{yy}(\omega)$ are real, $\Gamma_{uy}(\omega) = \Phi_{uy}(e^{j\omega})$ is in general complex.

Perform now the z-transform of expressions (1.19) and (1.20). On the other side, it can be shown that the z-transform of the function obtained as convolution product of two functions is the product of the z-transforms of the two functions in the product. Hence, one obtains

$$\Phi_{uy}(z) = W(z)\Phi_{uu}(z), \tag{1.21}$$

$$\Phi_{yy}(z) = W(z)\Phi_{yu}(z). \tag{1.22}$$

On the other hand,

$$\gamma_{uy}(\tau) = \gamma_{yu}(-\tau),$$

so that

$$\Phi_{yu}(z) = \Phi_{uy}(z^{-1}). \tag{1.23}$$

From (1.21), (1.22), and (1.23), we have

$$\Phi_{yy}(z) = W(z)W(z^{-1})\Phi_{uu}(z^{-1}).$$

Since

$$\Phi_{uu}(z^{-1}) = \Phi_{uu}(z),$$

we come to the fundamental theorem of spectral analysis:

$$\Phi_{yy}(z) = W(z)W(z^{-1})\Phi_{uu}(z).$$

1.16 Representations of a Stationary Process

Summing up what we have seen in the previous sections, we can conclude that a stationary process with zero mean value can be characterized by resorting to one of these descriptions:

(i) the covariance function $\gamma(\tau)$,
(ii) the spectrum (real or complex, $\Gamma(\omega)$ or $\Phi(z)$),
(iii) the ARMA model with transfer function $W(z)$ and variance λ^2 of the input process.

In the third representation, the process is seen in its time evolution, establishing a bridge between the past and the future. Therefore, it's a *dynamic representation*. By contrast, the first two descriptions can be properly named *static representations*.

The problem of passing from a representation to another takes a variety of forms. Going from (i) to (ii) can be performed via the definition of spectrum.

The passage from (iii) to (ii) is achievable with the fundamental theorem of spectral analysis. The steps from (i) to (iii) require discovering the dynamics hidden in a static representation. This is not an easy topic, giving rise to a research area called *stochastic realization theory* for which we refer to the book by Lindquist and Picci quoted in the bibliography.

To tackle the prediction problem, the appropriate description is the third one. This is not surprising since such representation provides a way to link the future to the past. In Chapter 3, we see how to work out the prediction rule, given representation (iii). Then, in subsequent chapters of this book, we see how to use data to find the dynamic representation.

Before embarking the theory of prediction, however, we are well advised to see how the process characteristics (mean, covariance, and spectrum) can be estimated from the analysis of available data.

2

Estimation of Process Characteristics

2.1 Introduction

In this chapter, we deal with the problem of estimating the main characteristics of a stationary process from data.

We start with a study of the properties of the covariance function. For such function, we have already seen some basic features in Section 1.5. Now, we investigate its properties more in depth (Sections 2.2 and 2.3).

We pass then to the problems of estimating the mean (Section 2.4), the covariance (Section 2.5), and the spectrum (Section 2.6) from observations. The properties of the covariance function seen in Sections 2.2 and 2.3 turn out to be useful to assess the asymptotic features of such estimators.

2.2 General Properties of the Covariance Function

As previously seen in Section 1.5, the covariance function of a stationary process $v(t)$ with zero mean value, defined as

$$\gamma(\tau) = E[v(t)v(t + \tau)],$$

enjoys the following main properties:

(i) $\gamma(0) > 0$,
(ii) $\gamma(-\tau) = \gamma(\tau)$,
(iii) $|\gamma(\tau)| < \gamma(0)$.

We add now an important observation: these properties are not exhaustive, in the sense that there exist functions $\gamma(\cdot)$ satisfying (i) (iii), which cannot be interpreted as covariance functions. The following example, taken from Brockwell and Davis's book, illustrates this fact. The function

$$\gamma(\tau) = \begin{cases} 1, & \tau = 0, \\ \alpha, & |\alpha| < 1, \quad \tau = \pm 1, \\ 0, & |\tau| > 2, \end{cases}$$

Model Identification and Data Analysis, First Edition. Sergio Bittanti.
© 2019 John Wiley & Sons, Inc. Published 2019 by John Wiley & Sons, Inc.

satisfies properties (i)–(iii); however, the associated Fourier transform

$$\frac{1}{2\pi} \sum_{-\infty}^{+\infty} \gamma(\tau)e^{-j\omega\tau} = \frac{1}{2\pi}(1 + 2\alpha \cos \omega)$$

is non-negative for each ω only if $|\alpha| \leq \frac{1}{2}$.

To guarantee that an even function $\gamma(\tau)$ is indeed a covariance function, we need a stronger property than (i)–(iii). To state the appropriate condition, we introduce the *Toeplitz matrix of the covariance function* as

$$\begin{bmatrix} \gamma(0) & \gamma(1) & \gamma(2) & \cdots & \gamma(N-1) \\ \gamma(1) & \gamma(0) & \gamma(1) & \cdots & \gamma(N-2) \\ \gamma(2) & \gamma(1) & \gamma(0) & & \\ \vdots & \vdots & & \ddots & \vdots \\ \gamma(N-1) & \gamma(N-2) & & \cdots & \gamma(0) \end{bmatrix}.$$

An even function $\gamma(\tau)$ is a covariance function of a stationary process if and only if the associated *Toeplitz matrix* is positive semi-definite for each N. In such a case, we also say that $\gamma(\cdot)$ is a *positive semi-definite function*.

It is easy to see that any covariance function satisfies such a condition. Indeed, the matrix above can be written as

$$\begin{bmatrix} \gamma(0) & \gamma(1) & \gamma(2) & \cdots & \gamma(N-1) \\ \gamma(1) & \gamma(0) & \gamma(1) & \cdots & \gamma(N-2) \\ \gamma(2) & \gamma(1) & \gamma(0) & & \\ \vdots & \vdots & & \ddots & \vdots \\ \gamma(N-1) & \gamma(N-2) & & \cdots & \gamma(0) \end{bmatrix}$$

$$= E\left[\begin{bmatrix} v(0) \\ v(1) \\ v(2) \\ \vdots \\ v(N-1) \end{bmatrix} [v(0)v(1)v(2)\cdots v(N-1)] \right].$$

Here, under the brackets, the product of a vector times its transpose appears. Such a product is positive semi-definite, and so it is its expected value.

Vice versa, given a real even function that is positive semi-definite, one can construct a stationary process having that function as covariance function; see the book by Brockwell and Davis quoted in the Bibliography.

Note that properties (i) and (iii) are corollaries of the positive semi-definite condition just stated.

2.3   Covariance Function of ARMA Processes

The covariance function $\gamma(\tau)$ of stationary processes may exhibit a variety of shapes.

Example 2.1 (Constant covariance)   Consider first a random variable $\bar{v}$ with 0 mean value and variance λ^2, and define the process $v(t) = \bar{v}$, $\forall t$. This is a peculiar process, consisting of the replica of the same random variable at all time points. It is straightforward to verify that this is a stationary process. Since its realizations are all constant: $v(1) = v(2) = \cdots = v(N) = \bar{v}$, such process is predictable without any error. This corresponds to the peculiar structure of its covariance function,

$$\gamma(\tau) = \text{constant}.$$

The process introduced in the above example presents such a strong inter-relation among samples at different time points that it is perfectly predictable. Stationary processes of this type are called *deterministic processes* (although it may appear a bit funny to call *deterministic* a *random* process).

On the other hand, we find processes with null-covariance, or low-covariance, among samples at different time points.

Example 2.2 (Vanishing covariance for simple processes)   The covariance function of a moving-average process of order n is null for $\tau > n$.

We also know that, for an autoregressive process of order 1, $v(t) = av(t - 1) + \eta(t)$, $\eta(t) \sim WN(0, \lambda^2)$, $|a| < 1$, we have

$$\gamma(\tau) = a^\tau \gamma(0),$$

so that

$$\lim_{\tau \to \infty} \gamma(\tau) = 0.$$

Moreover, we have seen in Section 1.8.1 that such $|\gamma(\tau)|$ constitutes a geometric series and therefore $\sum_{\tau=-\infty}^{\infty} |\gamma(\tau)|$ is convergent: $\sum_{\tau=-\infty}^{\infty} |\gamma(\tau)| < \infty$.

The natural question is then: What are the general properties of the covariance function of a stationary process? We now prove an important result, valid *for all ARMA processes*.

Proposition 2.1 (Vanishing covariance of any ARMA process)   *The covariance function of any stationary ARMA process is such that*

$$\lim_{\tau \to \infty} \gamma(\tau) = 0. \tag{2.1}$$

Proof: Consider first an AR process of any order, described by the usual difference equation

$$y(t) = a_1 y(t-1) + a_2 y(t-2) + \cdots + a_{n_a} y(t-n) + \eta(t)$$

or, equivalently, by the associated operator representation

$$A(z)y(t) = \eta(t). \tag{2.2}$$

Under the stability condition, this model generates a stationary process, the covariance function of which satisfies, $\forall \tau \geq 1$, the Yule–Walker equation seen in Section 1.9.2:

$$\gamma(\tau) = a_1 \gamma(\tau-1) + a_2 \gamma(\tau-2) + \cdots + a_{n_a} \gamma(\tau-n). \tag{2.3}$$

This equation can be equivalently written in the operator form

$$A(z)\gamma(\tau) = 0. \tag{2.4}$$

Note that, while in Eq. (2.2), unknown $y(\cdot)$ is a sequence of random variables, in Eq. (2.4), unknown $\gamma(\cdot)$ is a deterministic function.

Expression (2.4) tells us that $\gamma(\cdot)$ can be seen as the output of a system with transfer function $1/A(z)$ and null input. Under the stability condition, the output of such system must tend to zero, so that (2.1) must hold for autoregressive processes.

We finally come to ARMA models. For them, formula (2.3) is no more true $\forall \tau \geq 1$. However, as is easy to verify, it is still valid for large values of τ, to be precise for $\tau > n_c$, n_c being the order of the MA part. Hence, the previous analysis for AR models can be straightforwardly extended.

According to the study in Section A.4.3 of Appendix A, Proposition 2.1 can be strengthened by claiming that the covariance function of an ARMA process tends to zero *exponentially*, i.e. there exists an α with $|\alpha| \leq 1$ such that, for some ρ, $|\gamma(\tau)| \leq \rho\alpha^\tau$. In turn, this implies that

$$\sum_{-\infty}^{\infty} |\gamma(\tau)| < \infty. \tag{2.5}$$

Remark 2.1 (Wold decomposition) It can be shown that any stationary process can be seen as the sum of two uncorrelated processes, a deterministic process and an $MA\,(\infty)$ process. This is known as *Wold decomposition*. ARMA processes have a null deterministic part.

2.4 Estimation of the Mean

Given a stationary process $v(\cdot)$, assume to have N samples of one of its realizations. A main question is whether it is possible to estimate the process mean with such data and with which accuracy. To avoid possible confusions between random variables and numerical samples, in this section, we denote by $x(1)$, $x(2), \ldots, x(N)$ the samples and by $v(1), v(2), \ldots, v(N)$ the associated random variables. In other words, $x(t)$ is the value taken by random variable $v(t)$ for a given realization of the process.

To estimate the mean, the typical formula is

$$\hat{m}(x) = \frac{1}{N} \sum_{1}^{N} x(t),$$

where x denotes the set $x = \{x(1), x(2), \ldots, x(N)\}$. If we replace $x(t)$ with $v(t)$, then we have the estimator

$$\hat{m}(v) = \frac{1}{N} \sum_{1}^{N} v(t).$$

While the estimate $\hat{m}(x)$ is a real number, estimator $\hat{m}(v)$ is a random variable. It may be useful to bring into evidence the number of samples; in that case, we write

$$\hat{m}_N(x) = \frac{1}{N} \sum_{1}^{N} x(t)$$

and

$$\hat{m}_N(v) = \frac{1}{N} \sum_{1}^{N} v(t). \tag{2.6}$$

To probe the validity of an estimator, statisticians usually resort to the concepts of *unbiasedness* and *consistency*. Denoting by m the mean value of the process, the estimator is said to be

- *unbiased* if

$$E[\hat{m}(v)] = m,$$

- *asymptotically unbiased* if

$$E[\hat{m}_N(v)] \xrightarrow{N \to \infty} m,$$

- *consistent* if

$$E[(\hat{m}_N(v) - m)^2] \xrightarrow{N \to \infty} 0.$$

It is easy to verify that the sample mean estimator is unbiased for each N:

$$E[\hat{m}_N(v)] = \frac{1}{N} \sum_{1}^{N} E[v(t)] = m.$$

As for *consistency*, we have

$$E[(\hat{m}_N(v) - m)^2] = \frac{1}{N^2} \sum_{i,j}^{N} E[(v(i) - m)(v(j) - m)]. \tag{2.7}$$

Note that the quantities $E[(v(i) - m)(v(j) - m)]$ appearing here are just the covariances $\gamma(\tau)$ with $\tau = j - i$.

If process $v(\cdot)$ is a white noise, then the only non-null elements in such sum are those associated with condition $i = j$, and they all coincide with $\gamma(0)$. Therefore, for a white noise,

$$E[(\hat{m}_N(v) - m)^2] = \frac{1}{N^2} N\gamma(0) = \frac{1}{N}\gamma(0).$$

This quantity tends to 0 when $N \to \infty$, so that the sample estimator of the mean is consistent when $v(\cdot) \sim WN(m, \lambda^2)$.

Passing now to any stationary process, observe that, in the double sum

$$\sum_{i,j}^{N} E[(v(i) - m)(v(j) - m)]$$

appearing in Eq. (2.7), there are N terms with $j = i$, $N - 1$ terms with $|j - i| = 1$, $N - 2$ terms with $|j - i| = 2$, and so on. Hence,

$$E[(\hat{m}_N(v) - m)^2] = \frac{1}{N} \sum_{|\tau| < N} \left(1 - \frac{|\tau|}{N}\right) \gamma(\tau).$$

This quantity tends to 0 when $\lim_{\tau \to \infty} \gamma(\tau) = 0$.

Furthermore, it can be shown that, if $\sum_{\tau=-\infty}^{+\infty} \gamma(\tau)$ is finite, then

$$\lim_{N \to \infty} N\, E[(\hat{m}_N(v) - m)^2] = \sum_{-\infty}^{+\infty} \gamma(\tau).$$

Thus, for a large number N of samples, its variance is approximately given by

$$\frac{1}{N} \sum_{-\infty}^{+\infty} \gamma(\tau).$$

Recalling the definition of spectrum of a stationary process,

$$\Gamma(\omega) = \sum_{\tau=-\infty}^{+\infty} \gamma(\tau)e^{-j\omega\tau},$$

we observe that

$$\sum_{-\infty}^{+\infty} \gamma(\tau) = \Gamma(0).$$

Summing up, given a stationary ARMA process, the mean estimator (2.6) is unbiased and consistent. Moreover, for a large number of data, the variance of such estimator can be approximately assessed as

$$E[(\hat{m}_N(v) - m)^2] \cong \frac{1}{N} \sum_{-\infty}^{+\infty} \gamma(\tau) = \frac{1}{N}\Gamma(0).$$

Remark 2.2 (Law of large numbers) The convergence of the sample mean to the expected value has been studied since time immemorial, an early statement going back to the Renaissance book *Liber de Ludo Aleae*, by the polymath Gerolamo Cardano (1501–1576). In probability and statistics, this result is the celebrated *law of large numbers*. For those readers who are not familiar with the Latin language, we can say that *Liber de Ludo Aleae* can be roughly translated as *On Games of Chance*. To know more about this book, we refer to *Decoding Cardano's Liber de Ludo Aleae*, by Bellhouse (2005).

Remark 2.3 (Convergence in mean square) In general, given a sequence of random variables $v(i)$, we say that the sequence tends to a random variable v in mean square if, for i tending to ∞,

$$E[(v(i) - v)^2] \longrightarrow 0.$$

Thus, the above definition of consistency is equivalent to saying that, as N tends to ∞, the random variable *sample mean* $\hat{m}_N(v)$ defined by formula (2.6) tends to the expected value m in mean square. Sometimes, one speaks of *consistency in mean square* or *convergence in mean square* to avoid possible confusions with other types of random convergences.

2.5   Estimation of the Covariance Function

Again we assume to have N samples of a realization of a stationary process $v(\cdot)$. As before, we denote by $x(1), x(2), \dots, x(N)$ the samples and by $v(1), v(2), \dots, v(N)$ the associated random variables.

If the process mean m is known, the classical formulas to estimate $\gamma(\tau)$ are

$$\hat{\gamma}_N(\tau, x) = \frac{1}{N} \sum_{1}^{N-\tau} {}_t (x(t) - m)(x(t + \tau) - m), \tag{2.8}$$

$$\hat{\gamma}_N(\tau, x) = \frac{1}{N - \tau} \sum_{1}^{N-\tau} {}_t (x(t) - m)(x(t + \tau) - m), \tag{2.9}$$

where $\tau \geq 0$ (and, obviously, $\tau < N$). For $\tau < 0$, the estimate is simply obtained as $\hat{\gamma}_N(-\tau, x) = \hat{\gamma}_N(\tau, x)$.

When the mean has to be estimated as well, then the typical estimation formulas are

$$\hat{\gamma}_N(\tau, x) = \frac{1}{N} \sum_{1}^{N-\tau} (x(t) - \hat{m}(x))(x(t + \tau) - \hat{m}(x)), \qquad (2.10)$$

$$\hat{\gamma}_N(\tau, x) = \frac{1}{N - \tau} \sum_{1}^{N-\tau} (x(t) - \hat{m}(x))(x(t + \tau) - \hat{m}(x)), \qquad (2.11)$$

where $\hat{m}(x)$ is the estimate of the mean.

The number of product elements of the type $(x(t) - m)(x(t + \tau) - m)$ or $(x(t) - \hat{m}(x))(x(t + \tau) - \hat{m}(x))$ appearing in the sums (2.8)–(2.11) is decreasing with τ. This is why we expect that the reliability of such estimates decreases with τ. A rule of thumb proposed by Box and Jenkins in their book is that N should be at least 50 and $\tau < N/4$.

It is straightforward to verify that only estimator (2.9) is unbiased for each N. However, under weak assumptions, all estimators (2.8)–(2.11) are asymptotically unbiased.

We finally point out an important feature of the above covariance estimators, concerning the positive semi-definite property introduced in Section 2.2. It is easy to verify that, for the estimator (2.10),

$$\begin{bmatrix} \hat{\gamma}_N(0, x) & \hat{\gamma}_N(1, x) & \hat{\gamma}_N(2, x) & \cdots & \hat{\gamma}_N(N-1, x) \\ \hat{\gamma}_N(1, x) & \hat{\gamma}_N(0, x) & \hat{\gamma}_N(1, x) & \cdots & \hat{\gamma}_N(N-2, x) \\ \hat{\gamma}_N(2, x) & \hat{\gamma}_N(1, x) & \hat{\gamma}_N(0, x) & \cdots & \hat{\gamma}_N(N-3, x) \\ \vdots & & & & \\ \hat{\gamma}_N(N-1, x) & \hat{\gamma}_N(N-2, x) & \hat{\gamma}_N(N-3, x) & \cdots & \hat{\gamma}_N(0, x) \end{bmatrix} = \frac{1}{N} T\, T',$$

where T is the $N \times 2N$ matrix:

$$T = \begin{bmatrix} 0 & \cdots & & 0 & \delta x(1) & \delta x(2) & \cdots & \delta x(N) \\ \vdots & & & & & & & \vdots \\ 0 & \cdots & & 0 & \delta x(1) & \delta x(2) & \cdots & \delta x(N) & 0 \\ \vdots & & & & & & & & \vdots \\ 0 & \cdots & 0 & \delta x(1) & \delta x(2) & \cdots & \delta x(N) & 0 & 0 \\ \vdots & & & & & & & & \vdots \\ 0 & & \delta x(1) & \delta x(2) & \cdots & \delta x(N) & 0 & \cdots & 0 \end{bmatrix},$$

where $\delta x(i) = x(i) - \hat{m}(x)$. Thus, if one resorts to the estimator (2.10), the semi-definiteness property is guaranteed. Therefore, we mainly focus on (2.10) as covariance estimator.

2.6 Estimation of the Spectrum

We now come to the problem of determining the spectrum of a stationary process. For simplicity, we assume from the beginning that the mean value of the process is null.

There are two alternatives:

- The first possibility is to use the available data to identify a suitable ARMA model via black box techniques. Then the spectrum is obtained by means of the fundamental theorem of spectral analysis. This is named *parametric estimation procedure* or *indirect estimation procedure,* as the vector of ARMA parameters has to be preliminarily identified. The obtained spectrum is called *maximum entropy spectrum.*
- Alternatively, one can derive the spectrum directly from data, with the so-called *nonparametric estimation procedure* (also known as *direct estimation procedure*). We concisely introduce here this second option, as the first one is an outcome of the black box identification approach we will see in subsequent chapters.

For direct estimation, one starts by elaborating data according to (2.10) to compute

$$\hat{\gamma}_N(\tau, x) = \frac{1}{N} \sum_{t}^{N-\tau} x(t)x(t + \tau)$$

for $\tau = 0, 1, \ldots, N - 1$. Then, after setting $\hat{\gamma}_N(-\tau, x) = \hat{\gamma}_N(+\tau, x)$, one can introduce the sampled truncated version of the general formula of the spectrum

$$\Gamma(\omega) = \sum_{-\infty}^{+\infty} \gamma(\tau)e^{-j\omega\tau}, \tag{2.12}$$

as

$$\hat{\Gamma}_N(\omega, x) = \sum_{-(N-1)}^{(N-1)} \hat{\gamma}_N(\tau, x)e^{-j\omega\tau}. \tag{2.13}$$

Function $\hat{\Gamma}_N(\cdot, x)$ is named *periodogram.* The periodogram is a rough assessment of the spectrum. It suffers from two types of approximations. First, the horizon described by index τ in (2.13) ranges from $-(N - 1)$ to $+(N - 1)$ only. Second, the covariance function $\gamma(\tau)$ is replaced by its sampled surrogate $\hat{\gamma}_N(\tau, x)$.

Is the periodogram a fair estimate of the spectrum? In this regard, one can prove the following results:

$$E[\hat{\Gamma}_N(\omega, x)] \to \Gamma(\omega),$$

$$E[(\hat{\Gamma}_N(\omega, x) - \Gamma(\omega))^2] \to \Gamma(\omega)^2, \quad \omega \neq 0,$$

$$E[(\hat{\Gamma}_N(\omega, x) - \Gamma(\omega))(\hat{\Gamma}_N(\tilde{\omega}, x) - \Gamma(\tilde{\omega}))] \to 0, \quad \tilde{\omega} \neq \omega.$$

From these statements, we see that

- For $N \to \infty$, the expected value of $\hat{\Gamma}_N(\omega, x)$ tends to $\Gamma(\omega)$. In other words, the estimator is asymptotically unbiased.
- However, $\hat{\Gamma}_N(\omega, x)$ is not consistent, namely the variance of $\hat{\Gamma}_N(\omega, x) - \Gamma(\omega)$ does not vanish as $N \to \infty$. Together with the previous statement, this implies that, even for a very large number of data, the periodogram exhibits ups and downs about the spectrum.
- Two samples of the periodogram at different frequencies are uncorrelated, even for frequencies very close to each other, so reinforcing the zigzag fluctuating character of the periodogram line.

Example 2.3 (Periodogram of an AR(3) process)   Consider the process

$$v(t) = W(z)\eta(t), \quad \eta(t) \sim \text{WN}(0, 1),$$

where

$$W(z) = \frac{z^3}{z^3 - a_1 z^2 - a_2 z - a_3} = \frac{z^3}{(z - p_1)(z - p_2)(z - p_3)},$$

with

$$p_1 = 0.9,$$

$$p_2, p_3 = 0.8 \quad e^{\pm j \frac{3}{4}\pi}.$$

The spectrum $\Gamma(\omega)$, derived as

$$\Gamma(\omega) = |W(e^{j\omega})|^2,$$

and the periodogram, computed from 1024 snapshots $\{x(1), x(2), \dots, x(1024)\}$ generated by simulation, are compared in Figure 2.1.

To obtain a smoother nonparametric estimate, there are various remedies. One of them, the *Bartlett method*, consists in partitioning sequence $\{x(1), x(2), \dots, x(N)\}$ into a number of subsequences. For example, if $N = r\bar{N}$, it is possible to consider r subsequences of $\bar{N}$ snapshots, for each of which one can compute the associated periodogram:

$$1) \ x^{(1)} = \{x(1), x(2), \dots, x(\bar{N})\} \qquad \Rightarrow \hat{\Gamma}_{\bar{N}}(\omega, x^{(1)}),$$

$$2) \ x^{(2)} = \{x(\bar{N} + 1), x(\bar{N} + 2), \dots, x(2\bar{N})\} \Rightarrow \hat{\Gamma}_{\bar{N}}(\omega, x^{(2)}),$$

$$\vdots$$

$$r) \ x^{(r)} = \{x((r - 1)\bar{N} + 1), \dots, x(N)\} \qquad \Rightarrow \hat{\Gamma}_{\bar{N}}(\omega, x^{(r)}).$$

Figure 2.1 Spectrum and periodogram of the AR(3) process of Example 2.3.

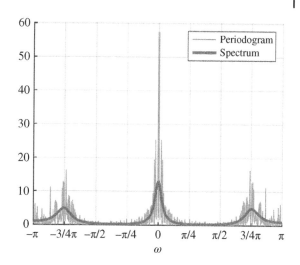

Then the spectrum is computed by averaging

$$\hat{\Gamma}(\omega, x) = \frac{1}{r} \sum_{1}^{r} {}_{i}\, \hat{\Gamma}_{\bar{N}}(\omega, x^{(i)}),$$

so obtaining a smoothed periodogram.

Example 2.4 (Spectrum estimation with the Bartlett method – Example 2.3 continued) Consider again the AR(3) process of Example 2.3. In Figure 2.2, one can see the periodogram obtained by the Bartlett method, with 4 subsequences with 256 snapshots each, 8 subsequences with 126 snapshots, and 16 subsequences with 64 snapshots ((b)–(d), respectively).

In Figure 2.2a, one can compare the result with the true spectrum (obtained via the fundamental theorem of spectrum analysis from the generating mechanism equation).

2.7 Whiteness Test

We end this chapter with the presentation of a statistical method to test if a given sequence of data can be interpreted as a white noise. Such a test is important in model checking, since if a model is appropriate, then the corresponding prediction error should be white, as discussed at the beginning of the book in Section 1.2.

The distinctive feature of a white noise is that its covariance function $\gamma(\tau)$ is null for any $\tau \neq 0$. The whiteness tests are all based on the assessment of the validity of such feature. We focus here on the so-called *Anderson test*.

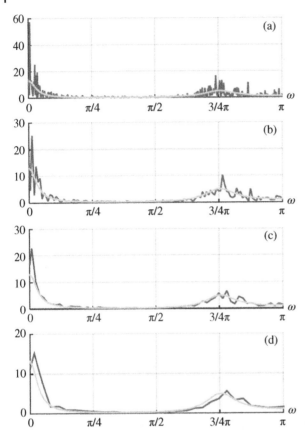

Figure 2.2 Spectrum estimate for Example 2.3: (a) the periodogram, (b)–(d) the smoothed periodogram obtained with the Bartlett method with subsequences constituted by 256, 128, and 64 snapshots, respectively.

The *Anderson test* is based on the analysis of the *sample covariance function* deduced from data according to (2.10). To be precise, denoting by $\varepsilon(\cdot)$ the signal to be tested, assumed to be zero mean, compute

$$\hat{\gamma}(\tau) = \frac{1}{N} \sum_{t}^{N-\tau} \varepsilon(t)\varepsilon(t+\tau),$$

where N is the number of available snapshots. $\hat{\gamma}(0)$ is an estimate of the variance, whereas $\hat{\gamma}(\tau)$ is an estimate of the cross-variance between $\varepsilon(t)$ and $\varepsilon(t+\tau)$. The function

$$\hat{\rho}(\tau) = \frac{\hat{\gamma}(\tau)}{\hat{\gamma}(0)}$$

is the *normalized sample covariance function*.

It can be shown that, if $\varepsilon(\cdot)$ is white, then for $\tau > 0$, $\hat{\rho}(\tau)$ has a probability distribution that, for high values of N, tends to a Gaussian with zero mean and

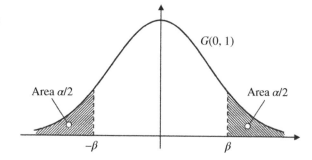

Figure 2.3 Whiteness test – determining β from α in the standard unit variance Gaussian.

variance $1/N$. In order to make this statement precise, we introduce the symbol As $\sim G(0, \sigma^2)$ to denote a random variable having an asymptotic distribution given by a Gaussian with zero mean value and variance σ^2. Then we can write the previous statement as follows:

$$\sqrt{N}\hat{\rho}(\tau) \sim \text{As } G(0, 1), \quad \tau > 0.$$

Moreover, it can be proved that $\hat{\rho}(i)$ and $\hat{\rho}(j)$ are asymptotically uncorrelated for $i \neq j$.

Correspondingly, the Anderson test operates as follows. Choose a confidence level α in the interval $(0, 1)$, for example, $\alpha = 0.05$. With reference to the standard Gaussian with zero mean and unit variance, find the value β for which the tails of the distribution outside interval $(-\beta, +\beta)$ cover an area α (such an area is the dash zone in Figure 2.3). Then, one counts the number of points of $\hat{\rho}(\tau)$ falling outside the interval

$$\left(\frac{-\beta}{\sqrt{N}}, \frac{+\beta}{\sqrt{N}}\right).$$

For example, if one takes into consideration $\hat{\rho}(1), \hat{\rho}(2), \dots, \hat{\rho}(30)$, namely 30 points of $\hat{\rho}(\tau)$, one counts the number of such samples outside this interval and computes the fraction with respect to the total of 30 points. If such a fraction is lower than α, the whiteness assumption can be accepted.

We will use the Anderson test in the case study on the Kobe earthquake data analysis dealt with in Section 11.2.

3

Prediction

3.1   Introduction

We can now tackle the problem posed at the beginning of the book, namely the problem of estimating the future value of a signal $v(\cdot)$ from the observation of its past. We shall indicate with the symbol $\hat{v}(t + r|t)$ or, simply, as $\hat{v}(t + r)$, the optimal predictor at time $t + r$, given the observations up to time t. Integer r is the *prediction horizon*.

The derivation is based on the fundamental assumption that the process is stationary, so that we can rely on the theory seen in Chapter 1. To be precise, we assume to know the dynamical representation of the process as the output of a system with a given transfer function fed by white noise, as discussed in Section 1.16. The derivation of such representation from experimental data is postponed to the subsequent chapter on identification.

The prediction method we present is mainly owed to the studies of two masters of the past century, Andrey Kolmogorov (1903–1987) and Norbert Wiener (1894–1964), the former working at the University of Moscow and the latter at MIT in Cambridge – Boston. Therefore, the theory is also named *K-W theory*.

Remark 3.1 (Kolmogorov and Wiener: two scholars independently working at prediction theory thousands of miles removed from one another) Among the many outstanding contributions of Andrey Kolmogorov, a masterpiece is his monograph *Theory of Probability*, about 60 pages, published by Springer in 1933. This monograph, which was written in German and translated into English much later, in 1950, had the merit of posing probability theory on solid mathematical grounds. In this way, Kolmogorov was able to tackle challenging problems with transparent mathematical tools. In 2006, the American Mathematical Society published the book *Kolmogorov in Perspective*, with a number of articles written by previous students and colleagues on his scientific contributions and personal life.

Thousands of miles away, Norbert Wiener was also studying stochastic processes. One of the early motivations that raised his interest goes back

Model Identification and Data Analysis, First Edition. Sergio Bittanti.
© 2019 John Wiley & Sons, Inc. Published 2019 by John Wiley & Sons, Inc.

to 1942, during the war, when, upon a request from the National Defense Research Committee of Unites States, he worked at a data filtering problem and wrote the report *Extrapolation, Interpolation, and Smoothing of Stationary Time Series: With Engineering Applications*, where he provided the main tools for prediction via data filtering. This report, originally classified, was rather difficult to read, so much so that it was nicknamed *the yellow peril*, the yellow referring to the color of its cover and peril to the general difficulty in reading it. What a coincidence: In the same years, Kolmogorov was independently working on analogous issues. Wiener made many contributions to the nascent field of information and control science. In particular, in the book *Cybernetics* (MIT Press, 1948), he explored the analogies between control and communications systems acting in animals and machines, as the books subtitle clearly states. For references on Wiener's work and character, the paper "Recollections of Norbert Wiener and the First IFAC World Congress" by Bernard Widrow, published in 2001 in the *IEEE Control Systems Magazine* is worth reading. The acronym IFAC means International Federation of Automatic Control. This scientific association, organized as federation of nations, was created in 1957, and it is now constituted by some 50 states. One of the top event of IFAC is the World Congress, held once every three years in various cities. The title of the paper refers to the first IFAC World Congress, which was held in 1960 in Moscow.

We will solve the prediction problem in two steps. First, in Section 3.2, we deal with a "fake problem," where the predictor is worked out under the assumption that the available observations are the samples of the white process feeding the dynamic representation of the process. Then, in Section 3.5, we pass to the real problem, where the available observations are the past data of the process itself. Between these two topics, there is an intermediate discussion to clarify fundamental aspects on the dynamical representation of a stationary process, in order to pose the prediction problem from data on proper grounds (Sections 3.3 and 3.4).

3.2 Fake Predictor

We assume to know the model of a stationary process $v(t)$ as the output of a dynamical system fed by a white noise $\eta(t)$ of null mean value and known variance:

$$v(t) = W(z)\eta(t), \quad \eta(t) \sim WN(0, \lambda^2).$$

The expansion of $W(z)$ in negative powers of z will be useful in our next computations:

$$W(z) = w_0 + w_1 z^{-1} + w_2 z^{-2} + \cdots.$$

Correspondingly, we can write

$$v(t) = w_0\eta(t) + w_1\eta(t-1) + w_2\eta(t-2) + \cdots .$$

We face the following problem: Suppose that $\eta(t), \eta(t-1), \eta(t-2), \ldots$, namely the "past of η," is measured; use such information to estimate $v(t+r)$ for $r \geq 1$.

To solve it, we write the unknown $v(t+r)$ as

$$v(t+r) = w_0\eta(t+r) + w_1\eta(t+r-1) + \cdots + w_{r-1}\eta(t+1)$$
$$+ w_r\eta(t) + w_{r+1}\eta(t-1) + \cdots .$$

In this infinite sum, we can isolate two terms:

$$\alpha(t) = w_0\eta(t+r) + w_1\eta(t+r-1) + \cdots + w_{r-1}\eta(t+1),$$
$$\beta(t) = w_r\eta(t) + w_{r+1}\eta(t-1) + \cdots .$$

$\alpha(t)$ depends upon η from $t+1$ to $t+r$, whereas $\beta(t)$ depends upon η up to t only. η being a white noise, $\alpha(t)$ and $\beta(t)$ are uncorrelated random variables. While $\beta(t)$ can be computed once the past of η is known, $\alpha(t)$ is determined by future values of η. Hence, $\alpha(t)$ is unpredictable from that past; we can only estimate its mean value:

$$E[\alpha(t)] = w_0 E[\eta(t+r)] + w_1 E[\eta(t+r-1)] + \cdots + w_{r-1}E[\eta(t+1)] = 0.$$

Hence, the optimal predictor is

$$\hat{v}(t+r|t) = \beta(t) = w_r\eta(t) + w_{r+1}\eta(t-1) + \cdots . \tag{3.1}$$

Remark 3.2 (Prediction error)   The prediction error is

$$v(t+r) - \hat{v}(t+r|t) = \alpha(t).$$

This is just an MA process; its variance is given by

$$\text{Var}[v(t+r) - \hat{v}(t+r|t)] = (w_0^2 + w_1^2 + \cdots + w_{r-1}^2)\lambda^2.$$

This quantity is monotonically increasing with r. In other words, as expected, the uncertainty in prediction increases with the prediction horizon.

When $r = 1$, the variance is simply given by $w_0^2\lambda^2$. If $w_0 = 1$, then the variance coincides with the variance of noise $\eta(t)$.

When the prediction horizon tends to ∞, such variance becomes $(w_0^2 + w_1^2 + \cdots)\lambda^2$. As seen in Remark 1.2, this infinite sum is just the variance of stationary process $v(t)$.

Summing up, if $w_0 = 1$:

- $r = 1$:   Var [prediction error] $= \text{Var}[\eta(t)]$,
- $r \to \infty$:   Var [prediction error] $\to \text{Var}[v(t)]$.

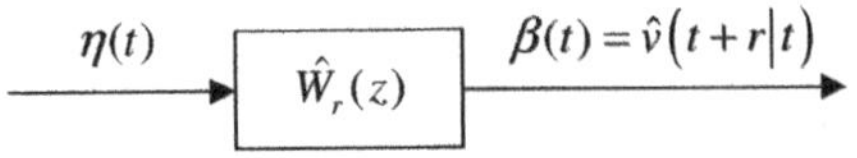

$$\eta(t) \quad \boxed{\hat{W}_r(z)} \quad \beta(t) = \hat{v}(t+r|t)$$

Figure 3.1 Optimal predictor from $\eta(t)$.

3.2.1 Practical Determination of the Fake Predictor

The practical determination of the predictor can be carried out by constructing its transfer function (from $\eta(t)$ to $\beta(t)$) as follows. Observe that (3.1) can be written as

$$\hat{v}(t+r|t) = [w_r + w_{r+1}z^{-1} + w_{r+2}z^{-2} + \cdots] \, \eta(t). \tag{3.2}$$

Then, by defining

$$\hat{W}_r(z) = w_r + w_{r+1}z^{-1} + w_{r+2}z^{-2} + \cdots , \tag{3.3}$$

the optimal predictor can be represented as the dynamical system of Figure 3.1.

Transfer function $\hat{W}_r(z)$ can be constructed from $W(z)$. Indeed, considering again the expansion of $W(z)$ in negative powers of z in the form

$$W(z) = w_0 + w_1 z^{-1} + \cdots + w_{r-1} z^{-(r-1)} + w_r z^{-r} + w_{r+1} z^{-r-1} + \cdots \tag{3.4}$$

and comparing (3.3) with (3.4), it is apparent that

$$W(z) = w_0 + w_1 z^{-1} + \cdots + w_{r-1} z^{-(r-1)} + z^{-r} \hat{W}_r(z).$$

This formula tells us that the transfer function of the optimal r-steps-ahead predictor can be determined via the long division algorithm performed for r steps, by dividing the numerator by the denominator of $W(z)$.

Example 3.1 (Fake predictor for an AR(1) process) Consider an AR(1) process, with transfer function

$$W(z) = \frac{1}{1 - az^{-1}}.$$

In Remark 1.5, we have already seen that the long division algorithm, performed for one, two, and three steps, leads to the following results:

One step:

$$W(z) = 1 + \frac{az^{-1}}{1 - az^{-1}}$$
$$= 1 + z^{-1} \left\{ \frac{a}{1 - az^{-1}} \right\}.$$

Two steps:

$$W(z) = 1 + az^{-1} + \frac{a^2 z^{-2}}{1 - az^{-1}}$$
$$= 1 + az^{-1} + z^{-2} \left\{ \frac{a^2}{1 - az^{-1}} \right\}.$$

Three steps:

$$W(z) = 1 + az^{-1} + a^2 z^{-2} + \frac{a^3 z^{-3}}{1 - az^{-1}}$$

$$= 1 + az^{-1} + a^2 z^{-2} + z^{-3} \left\{ \frac{a^3}{1 - az^{-1}} \right\}.$$

Hence, the fake optimal predictor is

One step:

$$\hat{W}_1(z) = \frac{a}{1 - az^{-1}}.$$

Two steps:

$$\hat{W}_2(z) = \frac{a^2}{1 - az^{-1}}.$$

Three steps:

$$\hat{W}_3(z) = \frac{a^3}{1 - az^{-1}}.$$

For a generic prediction horizon r, the predictor is

$$\hat{W}_r(z) = \frac{a^r}{1 - az^{-1}}. \tag{3.5}$$

Note that the denominators of all predictors coincide with the denominator of $W(z)$.

The variance of the prediction error is given by

One step:

$$(1^2)\lambda^2.$$

Two steps:

$$(1^2 + a^2)\lambda^2.$$

Three steps:

$$(1^2 + a^2 + a^4)\lambda^2.$$

And, for a generic prediction horizon r,

$$(1^2 + a^2 + \cdots + a^{2(r-1)})\lambda^2. \tag{3.6}$$

From expression (3.5), we see that $W_r(z) \to 0$, when $r \to \infty$. In other words, $\hat{v}(t + r|t) \to 0$, when $r \to \infty$. Moreover, from (3.6), we see that

$$\lim_{r \to \infty} \mathrm{Var}[v(t + r) \hat{-} v(t + r|t)] = \frac{1}{1 - a^2}\lambda^2 = \mathrm{Var}\,[v(t)].$$

These conclusions have a simple intuitive explanation: When the prediction horizon becomes large, the prediction task is more difficult since the variable to be estimated refers to a time point at large distance ahead than data. Hence, in the long run ($r \to \infty$), the information brought by data is so weak that the only reasonable estimate is the trivial one, namely the mean of the unknown, i.e. $E[v(t)] = 0$. Correspondingly, the variance of the prediction error tends to that of the process $v(t)$.

3.3 Spectral Factorization

We have seen how the optimal r-steps-ahead predictor can be derived under the assumption that the past of white process $\eta(t)$ is measurable. To tackle the problem of finding the predictor from data, we must find a way to link the past of $\eta(\cdot)$ to the past of $v(\cdot)$. To this purpose, we have to preliminarily discuss the so-called *spectral factorization problem*, which is the subject of this section.

Focus again on a stationary process with zero mean value. In view of all what we have seen so far, the process can be described by specifying one of the following three representations: (i) the covariance function $\gamma(\tau)$, (ii) the spectrum, or (iii) the dynamic (ARMA) model.

As already discussed, to solve the prediction problem, the appropriate representation is the third one as it brings into light the dynamical evolution of the process. When using the dynamic representation, however, we encounter a problem: *There exist many different ARMA models associated with the same stationary process.* Stated another way, there exist many different ARMA models giving rise to the same spectrum (or, equivalently, the same covariance function).

Indeed, consider the formula of the fundamental theorem of spectral analysis:

$$\Phi(z) = W(z)W(z^{-1})\lambda^2.$$

The reasons behind the multiplicity of ARMA representations are the following ones:

First reason of multiplicity:

If we divide transfer function $W(z)$ by a constant, say α, and multiply the variance of the input noise by α^2, the spectrum does not change. Indeed, if we let

$$\tilde{W}(z) = \left(\frac{1}{\alpha}\right) W(z),$$
$$\tilde{\lambda}^2 = \alpha^2 \lambda^2,$$

we have

$$\tilde{\Phi}(z) = \tilde{W}(z)\tilde{W}(z^{-1})\tilde{\lambda}^2 = \left(\frac{1}{\alpha^2}\right) W(z)W(z^{-1})\alpha^2 \lambda^2 = \Phi(z).$$

This result is no surprise: We may change the gain of the transfer function without modifying the spectrum by resizing the variance of the input.

Second reason of multiplicity:

Letting

$$\tilde{W}(z) = z^{-k} W(z),$$
$$\tilde{\lambda}^2 = \lambda^2,$$

we obtain

$$\tilde{\Phi}(z) = z^{-k} W(z) z^k W(z^{-1}) \lambda^2 = \Phi(z).$$

So, the two representations $\tilde{W}(z)$ and $W(z)$ are different but describe the same process. This fact can be easily interpreted. Multiplying a given transfer function by z^{-k} is the same as translating the process in the time domain by k steps; in other words, if $v(t)$ is the process with dynamical representation $(W(z), \lambda^2)$, then $(\tilde{W}(z), \tilde{\lambda}^2)$ is the dynamical representation of $v(t - k)$. On the other hand, it is apparent that a temporal translation does not alter the statistics of a stationary process.

Third reason of multiplicity:

It is obvious that, if we multiply numerator and denominator of $W(z)$ by the same polynomial (so increasing the number of poles and zeros), the spectrum does not change. This is a trivial modification without any effect on the spectrum.

Fourth reason of multiplicity:

Consider a system with transfer function

$$T(z) = \rho \frac{(z + \alpha)}{\left(z + \frac{1}{\alpha}\right)},$$

which is characterized by a zero reciprocal of the pole. An easy computation leads to

$$T(z)T(z^{-1}) = \rho^2 \alpha^2.$$

Hence, by taking $\rho^2 = 1/\alpha^2$, the output has the same spectrum of the input: No distortion occurs in the frequency distribution of the harmonic components of the process. This is why the considered system is named *all-pass filter*.

With this notion, consider now the stationary process $\tilde{v}(\cdot)$ obtained by passing a given process $v(\cdot)$ through an all-pass filter $T(z)$ as indicated in Figure 3.2. The overall transfer function is obtained by multiplying the transfer functions of the two subsystems:

$$\tilde{W}(z) = W(z)T(z).$$

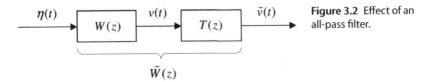

Figure 3.2 Effect of an all-pass filter.

Hence, being $T(z)T(z^{-1}) = 1$, the spectrum of $\tilde{v}(\cdot)$ is given by

$$\tilde{\Phi}(z) = W(z)T(z)T(z^{-1})W(z^{-1})\lambda^2 = \Phi(z).$$

This means that, although described by different models, $v(t)$ and $\tilde{v}(t)$ are the same stochastic processes.

The above considerations show that a stationary ARMA process admits infinitely many dynamical descriptions. To properly address the prediction problem, we introduce suitable conditions in order to isolate a *unique* dynamic representation, enabling the solution of the prediction problem.

Proposition 3.1 (Spectral factorization theorem) *Consider a stationary ARMA process. There exists a unique representation satisfying these conditions:*

- *numerator and denominator*
  - *are monic[1]*
  - *have the same degree*
  - *are coprime[2]*
- *poles and zeros are inside the unit disk of the complex plane.*

The polynomial monicity inhibits the first cause of multiplicity; the coincidence between the degrees of numerator and denominator inhibits the second one; finally, the condition on the location of poles and zeros in the complex plane avoids the presence of an all-pass filter in the transfer function, and the coprimeness prevents the existence of common factors.

The unique transfer function defined by this theorem is denoted with the symbol

$$\hat{W}(z)$$

and is named *canonical spectral factor*. The corresponding dynamic model is the called *canonical representation* of the given process.

---

1 A polynomial is said to be monic if the leading coefficient, namely the coefficient of the highest power, is equal to 1.
2 Two polynomials are said coprime if they do not have common factors, so that no simplification is possible.

Remark 3.3 (Why a spectral factor is named spectral factor?) The basic formula we are dealing with is

$$\Phi(z) = W(z)W(z^{-1})\lambda^2.$$

At the second member, there is the product of two factors, $W(z)$ and $W(z^{-1})$. Finding a dynamic representation from $\Phi(z)$ is therefore the same as extracting the "factor" $W(z)$ from such expression.

The white process at the input of the canonical ARMA representation is a special white process, sometimes characterized with an *ad hoc* symbol, such as $\xi(t)$. For example, we write an ARMA(1,1) model in generic representation as

$$v(t) = a_1 v(t-1) + c_0 \eta(t) + c_1 \eta(t-1),$$

whereas the same model in canonical form is written as

$$v(t) = a_1 v(t-1) + c_0 \xi(t) + c_1 \xi(t-1).$$

Example 3.2 (Canonical representation of an MA process) Consider the process

$$v(t) = 2\eta(t-1) + 4\eta(t-2) \quad \eta \sim WN(0, \lambda^2).$$

The associated transfer function is

$$W(z) = 2z^{-1} + 4z^{-2} = \frac{2z+4}{z^2}.$$

This is not a canonical representation. Indeed, first of all, the denominator and the numerator do not have the same degree; second, the numerator in not monic; finally, there is a zero in $z = -2$, outside the unit disk. To derive the canonical model, first, we multiply the denominator by z^{-1}; second, we divide the numerator by 2; third, we replace the zero in $z = -2$ with its reciprocal $z = -1/2$. Thus, we have

$$\hat{W}(z) = \frac{z+0.5}{z}.$$

Returning to the time domain, we have

$$v(t) = \xi(t) + 0.5\xi(t-1),$$

where $\xi(\cdot)$ is the new white process.

In the new description, the variance of the input noise must be adjusted in order to have a full equivalence with the original process, in particular to ensure that the output be a process with unchanged variance. To this purpose, let's compare the two MA(1) representations:

$$v(t) = 2\eta(t-1) + 4\eta(t-2), \quad \eta \sim WN(0, \lambda^2),$$
$$v(t) = \xi(t) + 0.5\xi(t-1), \quad \xi \sim WN(0, \mu^2).$$

From the first expression, we can straightforwardly compute the variance of $v(t)$:

$$\text{Var}[v(t)] = (2^2 + 4^2)\text{Var}[\eta(t)] = 20\lambda^2.$$

Analogously, from the second expression, we have

$$\text{Var}[v(t)] = (1^2 + 0.5^2)\text{Var}[\xi(t)] = 1.25\mu^2.$$

To obtain identical variances, we must take

$$\mu^2 = 16\lambda^2.$$

We leave to our readers the (easy) task to verify that, with such a choice, not only the variance but also the whole spectra (and therefore the covariance functions) associated to the original and to the new representation coincide.

3.4 Whitening Filter

Consider now a stationary process in its canonical representation, Figure 3.3a.

Since numerator and denominator of $\hat{W}(z)$ have the same degree, the inverse $\hat{W}(z)^{-1}$ can be also interpreted as a transfer function of a dynamical system. The possibility would not be valid if the denominator had a greater degree.

Then, define

$$\check{W}(z) = \hat{W}(z)^{-1}.$$

Obviously, the zeros of $\hat{W}(z)$ are the poles of $\check{W}(z)$ and the poles of $\hat{W}(z)$ are the zeros of $\check{W}(z)$. Since the zeros of $\hat{W}(z)$ lie inside the unit disk, $\check{W}(z)$ is stable.

With such preliminary considerations in mind, let's focus on a system with transfer function $\check{W}(z)$, the inverse of the transfer function of the canonical representation. If we feed this system with process $v(t)$, the output will be the signal at the input of the canonical representation, namely the white process $\xi(t)$ (Figure 3.3b). Since $\xi(t)$ is a white process, the system with transfer function $\check{W}(z)$ is called *whitening filter*. Indeed, it generates a white process from the given stationary process.

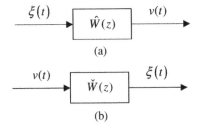

(a)

(b)

Figure 3.3 Canonical representation (a) and whitening filter (b).

3.5 Optimal Predictor from Data

We are now in a position to solve the prediction problem of a stationary ARMA process. The starting assumption is that its canonical representation is available. The determination of such representation from data is left to a subsequent chapter.

As seen in Section 3.2, the optimal r-steps-ahead predictor from $\xi(t)$ can be found by making reference to the following expression of the canonical representation, obtained via the long division algorithm:

$$\hat{W}(z) = w_0 + w_1 z^{-1} + \cdots + w_{r-1} z^{-(r-1)} + z^{-r} \hat{W}_r(z).$$

Here, $\hat{W}_r(z)$ is the transfer function of the optimal r-steps-ahead fake optimal predictor, namely the predictor we could form if $\xi(t)$ could be measured. Note that the denominator of $\hat{W}_r(z)$ coincides with that of $\hat{W}(z)$. In other words, if

$$\hat{W}(z) = \frac{C(z)}{A(z)},$$

then $\hat{W}_r(z)$ is of the following type:

$$\hat{W}_r(z) = \frac{C_r(z)}{A(z)}.$$

Now, the remote white process $\xi(t)$ can be extracted from data by means of the whitening filter. This leads to the following idea:

- Filter data by the whitening filter to recover the remote white process.
- Filter the recovered remote white process with the fake optimal predictor.

The combination of these two steps will provide the optimal predictor from data. This is illustrated in Figure 3.4. Here we see three drawings:

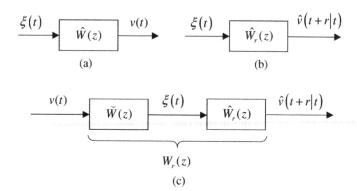

Figure 3.4 (a) Canonical representation. (b) Fake optimal predictor. (c) Optimal predictor from data.

Figure 3.4a – the canonical factor, (b) – the fake optimal predictor, and (c) – the optimal predictor from data. As indicated in this last drawing, the optimal predictor from data is the cascade of the whitening filter and the fake predictor; therefore, the overall transfer function is

$$W_r(z) = \check{W}_r(z)\hat{W}_r(z) = (\hat{W}_r(z))^{-1}\hat{W}_r(z).$$

In conclusion, the optimal predictor from data is given by

$$W_r(z) = \frac{A(z)}{C(z)}\frac{C_r(z)}{A(z)} = \frac{C_r(z)}{C(z)}.$$

Thus, we come to the following conclusion:

The optimal predictor from $v(t)$ coincides with that from $\xi(t)$ provided that its denominator is replaced with the numerator of $\hat{W}(z)$.

Example 3.3 (Optimal predictor of an AR(1) process) Let's again consider an AR(1) process with transfer function

$$\hat{W}(z) = \frac{1}{1 - az^{-1}}.$$

Under the assumption that $a < 1$, the model is in canonical form.

As already seen, the long division leads to

One step:

$$W(z) = 1 + z^{-1}\frac{a}{1 - az^{-1}} \Rightarrow C_1(z) = a.$$

Two steps:

$$W(z) = 1 + az^{-1} + z^{-2}\frac{a^2}{1 - az^{-1}} \Rightarrow C_2(z) = a^2.$$

Hence, considering that $A(z) = 1 - az^{-1}$ and $C(z) = 1$, the predictor is given by

	From ξ	From v
One step	$\hat{W}_1(z) = \dfrac{a}{1 - az^{-1}}$	$W_1(z) = a$
Two steps	$\hat{W}_2(z) = \dfrac{a^2}{1 - az^{-1}}$	$W_2(z) = a^2$

In the time domain, the optimal prediction rule from data is therefore

One step:

$$\hat{v}(t + 1|t) = av(t).$$

Two steps:

$$\hat{v}(t + 2|t) = a^2v(t).$$

By squaring the coefficients of the long division result and summing them, we can compute the prediction error variance:

One step:

$$Var[v(t+1) - \hat{v}(t+1|t)] = 1^2 \lambda^2.$$

Two steps:

$$Var[v(t+2) - \hat{v}(t+2|t)] = (1^2 + a^2)\lambda^2,$$

where, obviously, λ^2 is the variance of ξ. Not surprisingly, these variances coincide with those previously computed for the fake predictor.

In general, the r-steps-ahead optimal predictor is

$$\hat{v}(t+r|t) = a^r v(t), \tag{3.7}$$

with an error variance

$$Var[v(t+r) - \hat{v}(t+r|t)] = (1 + a^2 + \cdots + a^{2(r-1)})\lambda^2.$$

As already discussed in Section 3.2.1 for the optimal fake predictor, from (3.7), we see that the $\hat{v}(t+r|t)$ tends to 0 when the prediction horizon tends to ∞ and, from (3.3), we see that the variance of the error tends to that of the process.

Remark 3.4 (Optimal predictor for an MA(1) process) Consider the MA(1) process

$$v(t) = \xi(t) + c\xi(t-1), \quad |c| < 1, \quad \xi(\cdot) \sim WN(0, \lambda^2).$$

Being $|c| < 1$, the model is in canonical form.

The transfer function from $\xi(t)$ to $v(t)$ is given by

$$W(z) = 1 + cz^{-1}.$$

Such expression is by itself the expansion of $W(z)$ in negative powers of z. Therefore, no long division is required; we see immediately that the optimal fake predictor, i.e. the predictor from the noise ξ is

$$\hat{W}_1(z) = c,$$

$$\hat{W}_k(z) = 0, \quad \forall k > 1,$$

and the optimal predictor from data is

$$W_1(z) = \frac{c}{1 + cz^{-1}},$$

$$W_k(z) = 0, \quad \forall k > 1.$$

In the time domain,

$$\hat{v}(t+1|t) = -c\hat{v}(t|t-1) + v(t),$$

$$\hat{v}(t+k|t) = 0, \quad \forall k > 1.$$

The fact that the prediction is null for $k > 1$ is not surprising. Indeed, an MA(1) process has a covariance $\gamma(\tau)$ such that $\gamma(\tau) = 0$, $\forall |\tau| > 1$. Hence, the knowledge of past data up to time t does not bring information to estimate the process two or more steps ahead. The only possible estimate is the mean value. Correspondingly, the variance of the prediction error coincides with the variance of the process for $k > 1$.

Remark 3.5 (Importance of the canonical representation) We now stress the importance of resorting to the canonical representation to find the proper predictor, by focusing on the MA(1) process:

$$y(t) = \eta(t) + \frac{6}{5}\eta(t-1), \quad \eta(\cdot) \sim WN(0,1).$$

The corresponding polynomial $C(z)$ is given by

$$C(z) = 1 + \frac{6}{5}z^{-1}.$$

There is a zero in $z = -\frac{6}{5}$, outside the unit disk, so that the representation is not canonical. Nevertheless, suppose to use the standard formulas for the predictor. We would obtain

$$\hat{y}(t) = -\frac{6}{5}\hat{y}(t-1) + \frac{6}{5}y(t-1).$$

In Figure 3.5, the simulation of the process $y(t)$ is compared with that of the corresponding predictor $\hat{y}(t)$. As it can be seen, the error is growing and growing.

This is not surprising since the predictor we have (erroneously) designed is unstable. Thus, the predictor generates diverging signals, whereas the given process is stationary: we have come to a nonsense as we predict a stationary process with a nonstationary one!

If instead we replace the initial representation with the canonical one, namely

$$y(t) = \xi(t) + \frac{5}{6}\xi(t-1), \quad \xi(\cdot) \sim WN\left(0, \frac{36}{25}\right),$$

then the predictor is

$$\hat{y}(t) = -\frac{5}{6}\hat{y}(t-1) + \frac{5}{6}y(t-1),$$

with associated diagrams as in Figure 3.6.

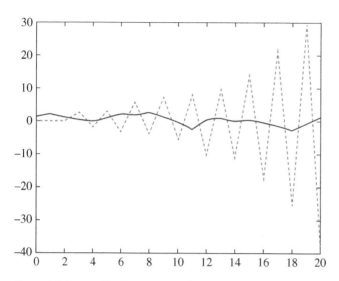

Figure 3.5 Process MA(1) (continuous line) and its prediction (dotted line) with the incorrect predictor.

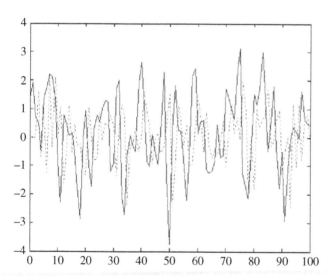

Figure 3.6 Process MA(1) (continuous line) and its prediction (dotted line) obtained with the correct optimal rule.

3.6 Prediction of an ARMA Process

We now derive a general formula for the one-step-ahead predictor of a generic ARMA process:

$$A(z)v(t) = C(z)\xi(t),$$

where $A(z)$ and $C(z)$ are the usual polynomials in negative powers of z:

$$A(z) = 1 - a_1 z^{-1} - a_2 z^{-2} - \cdots - a_{n_a} z^{-n_a}, \tag{3.8}$$

$$C(z) = 1 + c_1 z^{-1} + c_2 z^{-2} - \cdots + c_{n_c} z^{-n_c}. \tag{3.9}$$

Denoting by n the maximum between n_a and n_c, the transfer function, in positive powers of z, is

$$\frac{z^n C(z)}{z^n A(z)} = \frac{z^n + c_1 z^{n-1} + \cdots}{z^n - a_1 z^{n-1} - \cdots}.$$

Here, both numerator and denominator are monic. Moreover, they have the same degree. By further assuming that the zeros and the poles are located inside the unit disk, and that no zero coincide with a pole, we can conclude that the representation is canonical.

We then apply the theory seen in previous sections, according to which the one-step-ahead optimal predictor is obtained by resorting to the long division algorithm. Since the polynomials are monic, the first outcome of this division will be 1 and we write

$$\frac{C(z)}{A(z)} = 1 + \frac{C(z) - A(z)}{A(z)} = 1 + z^{-1} \frac{z(C(z) - A(z))}{A(z)}.$$

The optimal one-step-ahead predictor from $v(\cdot)$ is

$$W_1(z) = \frac{z(C(z) - A(z))}{C(z)},$$

with associated difference equation

$$C(z)\hat{v}(t + 1|t) = z(C(z) - A(z))v(t)$$
$$= (C(z) - A(z))v(t + 1). \tag{3.10}$$

Note that, both $A(z)$ and $C(z)$ being monic, the polynomial $C(z) - A(z)$ has null known term:

$$C(z) - A(z) = (a_1 + c_1)z^{-1} + (a_2 + c_2)z^{-2} + \cdots. \tag{3.11}$$

This is why the second member of the predictor formula (3.10) does not depend upon $v(t + 1)$, as it could appear at a first sight; it depends upon $v(\cdot)$ up to time t at most.

In the time domain, the predictor equation is

$$\hat{v}(t+1|t) = -c_1 \hat{v}(t|t-1) - c_2 \hat{v}(t|t-2) - \cdots - c_{n_c} \hat{v}(t-n_c+1|t-n_c)$$
$$+ (a_1 + c_1)v(t) + (a_2 + c_2)v(t-1) + \cdots . \tag{3.12}$$

Finally, observe that the stability of the predictor is related to polynomial $C(z)$ only.

Remark 3.6 (Simplified derivation of the predictor) The predictor formula can be easily derived by a simplified procedure–a kind of shortcut. By adding and subtracting $C(z)v(t)$ at one of the sides of $A(z)v(t) = C(z)\xi(t)$, and re-organizing the various terms, one obtains

$$C(z)v(t) = (C(z) - A(z))v(t) + C(z)\xi(t).$$

Dividing by $C(z)$ the two sides of this expression, we have

$$v(t) = \frac{C(z) - A(z)}{C(z)} v(t) + \xi(t). \tag{3.13}$$

Let's now analyze the right-hand side of this formula. Polynomial $C(z) - A(z)$ is given by expression (3.11), a polynomial without known term. Therefore, dividing it by $C(z)$, we obtain a transfer function of the type

$$\frac{C(z) - A(z)}{C(z)} = w_1 z^{-1} + w_2 z^{-2} + \cdots$$

This implies that the first term at the right-hand side of (3.13) does not depend upon $v(t)$; it is a function of the past of $v(\cdot)$, i.e. a function of

$$v(t-1), v(t-2), v(t-3), \ldots .$$

The second term of (3.13) is instead a sample of the white process at time t; as such, it is unpredictable from the past of $v(\cdot)$.

From these two observations, it follows that the optimal predictor is simply obtained by canceling $\xi(t)$ in (3.13); in this way, prediction formula (3.12) is re-obtained.

This type of shortcut is frequently used for its simplicity. Its weak side is that the importance of starting from the canonical factor is not evident.

3.7 ARMAX Process

Thanks to a generalization of the ARMA model, it is possible to describe a phenomena subject to the influence of external variables, named *exogenous variables*. This generalization enables one to apply such type of models, which originated in the realm of statistics and time series analysis, to the world of systems and control, so obtaining a remarkable enlargement of their applicability.

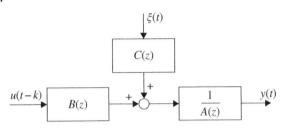

Figure 3.7 Block diagram of the ARMAX model (3.16).

To emphasize such passage to the systems and control realm, we use the symbol $y(t)$ to denote the signal under study, in place of $v(t)$.

To be precise, we pass from the classical ARMA model seen so far,

$$A(z)v(t) = C(z)\xi(t),$$

with $A(z)$ and $C(z)$ given by (3.8) and (3.9), to the model

$$A(z)y(t) = C(z)\xi(t) + f(t),$$

where $f(t)$ is a given function of an exogenous variable. As a typical case, we may think of a system subject to a control variable $u(t)$, with

$$f(t) = b_1 u(t-1) + b_2 u(t-2) + \cdots + b_{n_b} u(t - n_b).$$

Then, by letting

$$B(z) = b_1 z^{-1} + b_2 z^{-2} + \cdots + b_{n_b} z^{-n_b}, \tag{3.14}$$

we can write

$$A(z)y(t) = B(z)u(t) + C(z)\xi(t), \tag{3.15}$$

where $A(z)$ and $C(z)$ are again given by (3.8) and (3.9).

More in general, the delay with which the input acts on the output may be any integer k, $k \geq 1$. Correspondingly, the model becomes

$$A(z)y(t) = B(z)u(t-k) + C(z)\xi(t), \tag{3.16}$$

the block scheme of which is represented in Figure 3.7.

3.8 Prediction of an ARMAX Process

To determine the optimal predictor of the ARMAX model, rewrite expression (3.16) at time $t + k$:

$$A(z)y(t+k) = B(z)u(t) + C(z)\xi(t+k). \tag{3.17}$$

Perform now the long division of $C(z)$ by $A(z)$ for k steps. Denoting by $E(z)$ the result of such division, $C(z)$ can be written as

$$C(z) = A(z)E(z) + z^{-k}\tilde{F}(z), \tag{3.18}$$

with

$$E(z) = e_0 + e_1 z^{-1} + \cdots + e_{k-1} z^{-k+1}. \tag{3.19}$$

Note that, $A(z)$ and $C(z)$ being monic, the leading coefficient of $E(z)$ is unitary, $e_0 = 1$. Equation (3.18) is called *Diophantine equation*.

Multiply now both members of Eq. (3.17) by $E(z)$. We have

$$A(z)E(z)y(t + k) = B(z)E(z)u(t) + C(z)E(z)\xi(t + k).$$

Adding at both members $C(z)y(t + k)$, it follows that

$$C(z)y(t + k) = B(z)E(z)u(t) + (C(z) - A(z)E(z))y(t + k) + C(z)E(z)\xi(t + k).$$

Then, by dividing both members by $C(z)$, and taking into account the Diophantine equation (3.18),

$$y(t + k) = \frac{\tilde{F}(z)}{C(z)}y(t) + \frac{B(z)E(z)}{C(z)}u(t) + E(z)\xi(t + k). \tag{3.20}$$

This is the key expression of $y(t + k)$ for prediction. Indeed, at the right-hand side of (3.20), there is the sum of three terms:

(i) The first term is the output of a system with transfer function $\tilde{F}(z)/C(z)$ fed by input $y(t)$. Therefore, it depends upon $y(\cdot)$ up to time t at most.
(ii) The second term depends upon $u(\cdot)$ up to time t at most.
(iii) Recalling expression (3.19) of polynomial $E(z)$, the third term is given by

$$E(z)\xi(t + k) = e_0\xi(t + k) + e_1\xi(t + k - 1) + \cdots + e_{k-1}\xi(t + 1).$$

This is a linear combination of the snapshots of $\xi(\cdot)$ associated with the time points $t + 1, t + 2, \ldots, t + k$.

Hence, in (3.20), we see a split between the past (on which the first two terms depend upon) and the future (the third term, unpredictable from the past). Consequently, the optimal predictor is

$$C(z)\hat{y}(t + k|t) = \tilde{F}(z)y(t) + B(z)E(z)u(t). \tag{3.21}$$

In particular, for $k = 1$, $E(z) = e_0 = 1$ and $\tilde{F}(z) = z(C(z) - A(z))$, so that Eq. (3.18) simply becomes

$$C(z) = A(z) + z^{-1}z(C(z) - A(z)).$$

Consequently,

$$C(z)\hat{y}(t + 1|t) = (C(z) - A(z))y(t + 1) + B(z)u(t),$$

which is an obvious extension of predictor (3.10).

4

Model Identification

4.1 Introduction

Capturing in mathematical form phenomena of the real world has been a main objective of human beings since time immemorial. An example is the description of the motion of planets of the solar system, which captured the attention of many scholars over several centuries. With the growth of the industrial world, systems modeling has become essential for proper design in all fields. In this chapter, we present the main methods enabling the construction of mathematical models from data.

Often models are obtained by describing every constitutive element with an appropriate *mathematical law*, found in the history of this or that discipline. For instance, in the mathematical model of a ship, there will be variables corresponding to the driving force of the propeller, to the position of the rudder, to the external disturbances (wind and waves), and to the movement of the ship. The relationships that link these variables will be arrived at by drawing on the wealth of laws of mechanics and hydraulics. Though rather basic, this approach to constructing models can come up against a series of difficulties. The link between certain variables can be uncertain, either in its analytic characterization or due to the unknown value of a relevant parameter; in other words, one does not always have at hand a *law* to describe a given phenomenon with the necessary accuracy. Another difficulty is related to the notable complexity of some systems. The mathematical model that one would obtain from the analysis of their constitutive parts and from the related mathematical descriptions might be excessively complex (too many equations) to be of any practical use.

There is a valid alternative for cases such as these, which is to set up the model starting from elaborations direct from experimental data, measured over a determined period of observation of the variables of interest in the system under study. In this way, the data themselves will provide, through appropriate elaboration, a suitable mathematical model to describe the input–output of the system. The description arrived at in this way in large part leaves out

Model Identification and Data Analysis, First Edition. Sergio Bittanti.
© 2019 John Wiley & Sons, Inc. Published 2019 by John Wiley & Sons, Inc.

the physical context of the phenomenon, limiting itself to providing a plausible interpretation of the existing link between the input and output variables, according to what appears from the observations carried out. In the extreme event in which a model diverges completely from the physical context, we speak of a *black box model*. On the other hand, classical models – in which the mathematical description of the system is arrived at by constitutive equations (laws) of the parts that make it up – are called *white box models*.

Between the two approaches briefly sketched out here, however, there are many intermediary possibilities. In particular, the analysis of experimental data can be very useful for physical models, in order to deal with any residual uncertainties. For example, it often happens that a given model contains some parameters the values of which are uncertain, or even completely unknown; in such cases, the parameter can be estimated by analyzing through experimentation how the system functions under determined working conditions.

Another major problem regarding identification is estimating one or more unmeasurable variables starting from data relating to one or more observable variables. This is known as the problem of *signal identification*. It is one of the problems of this kind that has challenged humanity since time immemorial: calculating the point in which, moment by moment, a navigator finds himself. Up until just a few decades ago, before the advent of satellite systems, this calculation was made by integrating two pieces of information, the distance traveled since the last berthing and the observation of heavenly bodies by means of a sextant. Both of these pieces of information, gathered in this manner, are subject to imprecision; moreover, the use of the sextant (a tool that works based on the reflection of rays of light) is limited when the sky is cloudy. The determination of the unknown thus calls for the analysis and elaboration of measures affected by uncertainty.

The collection of ideas and procedures for the complete determination of a model or a signal on the basis of experimentation constitutes the *science of identification*.

In this chapter, we deal with identification problems in general. Chapters 5 and 6 deal with the black box identification of input–output models, while black box identification of state space models is the subject of Chapter 7. In Chapter 9, we treat signal identification problems. Finally, parameter estimation in a given model is discussed in Chapter 10.

4.2 Setting the Identification Problem

4.2.1 Learning from Maxwell

While still a student at Cambridge University, James Clerk Maxwell (born in Edinburgh in 1831, whose portrait is provided in Figure 4.1) became interested

Figure 4.1 James Clerk Maxwell.

in Saturn's rings and set out to discover the nature of their composition. His research was not based on astronomical observations but rather on pure mathematical modeling.

Following the law of universal gravitation, Maxwell constructed various models that explained the rings' nature, each deduced according to a given working hypothesis. For instance, one model was constructed on the hypothesis that the ring was a single solid crown; another was based on the hypothesis that they were formed by a gaseous crown, and so forth. Once Maxwell had built the models, he passed to the model *validation* phase, namely he set about to determine which one was correct by comparing the hypotheses with the established facts of life. The models with characteristics that contradicted reality were discarded, and with them the hypotheses upon which they were based. As his basic tool for validation, Maxwell used the notion of stability, starting from the assumption that, since the rings had existed in that configuration for millennia, the model obtained had to be stable necessarily. Unstable models were to be rejected, along with their underlying hypotheses. This approach, which anticipated today's techniques of model identification, is magisterially summarized by the author as follows: *By rejecting every hypothesis which leads to conclusions at variance with the facts, we may learn more on the nature of these distant bodies than the telescope can yet ascertain.*

Interestingly, one of the family of models considered by Maxwell in his study was based on the idea that the ring was composed by a set of multiple elements that are not rigidly connected, as though the ring were made up of various independent satellites. Naturally, in this case, the modeling required the consideration, in addition to the force of attraction exercised by Saturn and the centrifugal force, also the mutual effect of one satellite on the others. The study was conducted based on several simplifying conditions: (i) there are a number of satellites, each satellite describing a circular trajectory; (ii) their speed of rotation is uniform; (iii) the various satellites of the ring are all identical to one another; and (iv) the transversal dimension of the ring is negligible. In his analysis, Maxwell considered the number of satellites (indicated by the Greek letter μ), as a parameter. He showed that, for a certain range of values of μ, the system was stable, so that the starting hypothesis was not contradicted. On the contrary, the hypothesis that the ring was a single solid crown led to an unstable model. Thus, Maxwell came to the conclusion that the rings had to be composed of many solid elements, unattached to one another, in rotation around the planet, a deduction that, over a century later, has been confirmed by modern space missions (*Pioneer 11, Voyager 1, Voyager 2, Cassini-Huyghens*, which passed near the "Lord of the Rings" in the years 1979, 1980, 1981, and 1997, respectively).

The essay on the rings of Saturn was the scientific debut of Maxwell, who was then 24 years old. If we consider his later fundamental contributions to electromagnetism and to other fields, it is understandable why many hail Maxwell as the greatest scientist of nineteenth century. At the commemoration for centenary of his birth, Albert Einstein said of Maxwell that he had been "the most profound and the most fruitful scientist that physics has experienced since the time of Newton." I myself, a more humble admirer, have often wondered what other contributions this illustrious scientist might have left us if a tumor had not killed him in 1879, at just 48 years of age.

For more on the work of Maxwell in identification and control, we refer to the paper "James Clerk Maxwell, A Precursor of System Identification and Control Science" by the author of this book, published in the *International Journal of Control* in 2015.

4.2.2 A General Identification Problem

To tackle the identification problem in a general framework, we start by formulating a candidate family of models. One of the simplest situations occurs when the various models within the family admit a common analytical description, a single model being simply characterized by a parameter. For example, in Maxwell's study on Saturn rings, under the working assumption that the rings were composed by a set of multiple elements that are not rigidly connected, a main parameter of the model was the number of pieces composing the ring. In general, we write a family of models as follows:

$$\mathfrak{M} = \{\mathcal{M}(\theta) | \theta \in \Theta\},$$

where θ denotes the parameter vector and Θ is the set of admissible parameters. More concisely, we can alternatively write

$$\mathfrak{M} = \{\mathcal{M}(\theta)\}.$$

The identification problem consists then in the determination of the appropriate θ. The search is based on the analysis of data, normally constituted by the measurements of one or more variables of the system under consideration. We denote such system (also referred to as the *data generation mechanism*) with the symbol $\mathcal{S}$.

4.3 Static Modeling

A most frequent identification problem is to find the static relationship between two (or more) observed variables. One can then start by making a conjecture on the type of relation (linear, quadratic, etc.), and find the best model by a fitting technique, where the data provided by the model are compared with observations. One of the most effective "comparison techniques" is the least squares method, conceived by Carl Friedrich Gauss (1777–1855) in 1801.

4.3.1 Learning from Gauss

The evening of 1 January 1801, astronomer Giuseppe Piazzi, while working in his observatory in Palermo (Sicily, Italy), detected a "new" heavenly body rotating around the Sun, between Mars and Jupiter. He called it Ceres, the Roman goddess who was the protectress of Sicily. Ceres has diameter of less than 1000 km (for comparison, the diameter of the Earth is 12 700 km). Today, it has been classified as an asteroid since its features do not meet the requirements for planets established by the *International Astronomical Union* in 2006. Anyhow, at the opening of the nineteenth century, the mass of the new body was still unknown (and the International Astronomical Union did not exist yet). So, the solar system enriched with a "new" planet. Enthusiastic about his discovery, Piazzi followed the Ceres trajectory for some time, until it disappeared in a conjunction with the Sun. How to predict the position of the new planet when it came out of the conjunction became a topic of great debate, capturing both the attention of the scientific world and the popular imagination.

The problem caught the interest of Carl F. Gauss (1777–1855), one of the greatest minds in the history of mathematics. Gauss had the idea of extrapolating the trajectory of Ceres as observed by Piazzi in order to predict the subsequent path of the asteroid. It was to this end that Gauss conceived of the *least squares* method, the underlying idea of which was to assume that the measurements made were affected by some imprecisions due to errors in astronomical observation, and therefore to proceed to estimate Ceres' trajectory by minimizing, on average, the interpolation errors made. The calculations made by Gauss did indeed make it possible to estimate Ceres' position when it emerged from the Sun (December 1801) with admirable precision. The event brought great fame to Gauss, even outside the academic world.

4.3.2 Least Squares Made Simple

We introduce the least squares method by referring to a very common problem, that of finding the trend and/or the seasonal component in a sequence. We shall denote by $y(t)$ a scalar real sequence, with t discrete time.

4.3.2.1 Trend Search

One may postulate that there is a hidden linear trend such as

$$\hat{y}(t) = a_0 + a_1 t$$

or, more in general, a hidden polynomial trend:

$$\hat{y}(t) = a_0 + a_1 t + a_2 t^2 + \cdots + a_n t^n,$$

where the a_i's are parameters to be determined. Such parameters can be organized in the vector θ defined as

$$\theta = [a_0 \ a_1]'$$

for a linear trend or

$$\theta = [a_0 \ a_1 \cdots a_n]'$$

for the general polynomial trend. By posing, respectively,

$$\phi(t) = [1 \ t]',$$

$$\phi(t) = [1 \ t \ \cdots \ t^n]',$$

we can rewrite the trend expression as

$$\hat{y}(t) = \phi(t)'\theta.$$

4.3.2.2 Seasonality Search

Certain time series are clearly seasonal, as they exhibit a sort of intrinsic periodicity. If the periodicity reduces to a sinusoidal component of period T with a possibly non-null mean value, then we can write

$$\hat{y}(t) = a_0 + a_{1c} \cos(\omega t) + a_{1s} \sin(\omega t),$$

where

$$\omega = \frac{2\pi}{T}.$$

If there are various sinusoids, then the underlying seasonal signal can be written as

$$\begin{aligned}
\hat{y}(t) = {} & a_0 + a_{1c} \cos(\omega t) + a_{1s} \sin(\omega t) \\
& + a_{2c} \cos(2\omega t) + a_{2s} \sin(2\omega t) \\
& + \cdots \\
& + a_{nc} \cos(n\omega t) + a_{ns} \sin(n\omega t).
\end{aligned}$$

The parameters can be organized in a unique vector

$$\theta = [a_0 \; a_{1c} \; a_{1s}]'$$

for a single sinusoid or

$$\theta = [a_0 \; a_{1c} \; a_{1s} \; \cdots \; a_{nc} \; a_{ns}]'$$

for a set of sinusoids. Then by posing, respectively,

$$\phi(t) = [1 \; \cos(\omega t) \; \sin(\omega t)]',$$

$$\phi(t) = [1 \; \cos(\omega t) \; \sin(\omega t) \; \cdots \; \cos(n\omega t) \; \sin(n\omega t)]',$$

we can rewrite the seasonal expression as

$$\hat{y}(t) = \phi(t)'\theta.$$

We let our readers describe the situation when there is both a trend and a seasonal bottom line in data.

4.3.2.3 Linear Regression

As seen above, the search for both a trend and a seasonality lead to the same type of model. Indeed, in both cases, the interpolated signal $\hat{y}(t)$ can be written as

$$\hat{y}(t) = \phi(t)'\theta,$$

although the meaning of $\phi(t)$ and θ is different. Actually, this model applies to many different situations and is therefore of large and primary interest in identification. It is called *linear regression* model since $\hat{y}(t)$ is linear in θ.

We now discuss how it can be estimated from data, taking into consideration that the snapshots of the observed sequence are usually affected by imprecisions due to many different reasons, such as measurement errors, distortions, approximations, and so on. Thus, when dealing with real data, finding a parameter vector θ leading to perfect fitting is Utopian. Rather, we expect that, whatever the choice of the parameters may be,

$$\epsilon(t) = y(t) - \hat{y}(t) \tag{4.1}$$

will be non-null. In general, $\epsilon(t)$ will be positive for some t and negative for other, with small intensity (absolute value) in some cases and large in other. Gauss' idea was to choose the parameters so as to minimize the average squared errors over the observation horizon. If the data are available over the interval from 1 to N, this means minimizing

$$J(\theta) = \frac{1}{N} \sum_{t}^{N} \epsilon(t)^2,$$

with respect to θ.

Thus, we have to optimize a function (J) of a vector (θ). Here, we use the notion of first and second derivative of a scalar with respect to a vector as outlined in Section B.9 of Appendix B.

The determination of the minimum of performance index $J(\theta)$ is simple since, $\epsilon(t)$ being linear in θ, $J(\theta)$ is quadratic in θ:

$$J(\theta) = \frac{1}{N} \sum_{t}^{N} (y(t) - \phi(t)'\theta)^2.$$

Hence, the derivative of $J(\theta)$ with respect to θ is given by

$$-\frac{2}{N} \sum_{t}^{N} (y(t) - \phi(t)'\theta)\phi(t)'.$$

Setting such derivative to 0, one easily comes to the following equation, known as *normal equation*:

$$\left[\sum_{t}^{N} \phi(t)\phi(t)' \right] \theta = \sum_{t}^{N} \phi(t)y(t). \tag{4.2}$$

All and only the solutions of this equation are the points where the derivative of the performance index is null. Actually, these are all and only *points of minimum* of $J(\theta)$. Indeed, denote by $\bar{\theta}$ any solution and let's develop $J(\theta)$ around $\bar{\theta}$:

$$J(\theta) = J(\bar{\theta}) + \left.\frac{dJ(\theta)}{d\theta}\right|_{\theta=\bar{\theta}} (\theta - \bar{\theta}) + \frac{1}{2}(\theta - \bar{\theta})' \left.\frac{d^2J(\theta)}{d\theta^2}\right|_{\theta=\bar{\theta}} (\theta - \bar{\theta}) + h(\theta),$$

where $h(\theta)$ is the remainder term containing the higher order (third order, fourth order, ...) derivatives of $J(\theta)$ with respect to θ. Now, two main observations are in order. First of all, $J(\theta)$ is quadratic; therefore, all higher derivatives are null, so that $h(\theta) = 0$. Second, $\bar{\theta}$ is a solution of the normal equation; this means that the first derivative of $J(\theta)$ evaluated in $\bar{\theta}$ must be null too. Therefore, the previous expression simplifies as follows:

$$J(\theta) = J(\bar{\theta}) + \frac{1}{2}(\theta - \bar{\theta})' \left.\frac{d^2J(\theta)}{d\theta^2}\right|_{\theta=\bar{\theta}} (\theta - \bar{\theta}).$$

Here, the second derivative of $J(\theta)$ with respect to θ appears. This square matrix, whose dimension is determined by the number of parameters in vector θ, is also called *Hessian matrix*. It is easy to see that

$$\frac{d^2J(\theta)}{d\theta^2} = \frac{2}{N} \left[\sum_{t}^{N} \phi(t)\phi(t)' \right].$$

Interestingly enough, this expression shows that the Hessian matrix does not depend upon the particular point θ dealt with.

In conclusion, we can write

$$J(\theta) = J(\bar{\theta}) + (\theta - \bar{\theta})'R(N)(\theta - \bar{\theta}),$$

where

$$R(N) = \frac{1}{N}\left[\sum_{t}^{N}\phi(t)\phi(t)'\right].$$

Observe the peculiar structure of matrix $R(N)$, a sum of matrices of the form $\phi\phi'$. Each of these $\phi\phi'$ matrices is positive semi-definite, so that the whole sum $R(N)$ is positive semi-definite. We distinguish two cases, depending on the fact that

$$\sum_{t}^{N}\phi(t)\phi(t)'$$

is non-singular or singular. In the former case, $R(N)$ is positive definite; in the second case, it is positive semi-definite but not positive definite (see Appendix B).

Summing up, if $R(N)$ is non-singular, $J(\theta)$ is a paraboloid as shown in Figure 4.2a. There is a unique point of minimum, given by the unique solution of the normal equation:

$$\hat{\theta} = \left[\sum_{t}^{N}\phi(t)\phi(t)'\right]^{-1}\sum_{t}^{N}\phi(t)y(t).$$

Such $\hat{\theta}$ is the *least squares estimate* of θ.

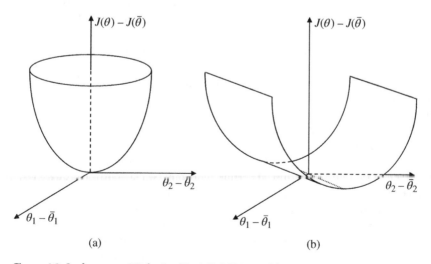

Figure 4.2 Performance $J(\theta)$ for $\theta = [\theta_1 \ \theta_2]'$. $R(N)$ invertible (a); $R(N)$ singular (b).

If instead matrix $R(N)$ is singular, then $J(\theta)$ degenerates into a valley with horizontal bottom (Figure 4.2b). At such bottom, there are the infinitely many solutions of the normal equation. We have then many *least squares estimates* of θ. These estimates are all equivalent to each other since they all have identical performance index $J(\theta)$. In other words, the multiplicity of *least squares estimates* provides the same "degree of fitting" of the given data.

4.3.3 Estimating the Expansion of the Universe

The study of galaxies brings information on the evolution of the Universe. A main question is whether these heavenly systems are moving or not, in particular whether the Universe is stationary, contracting, or expanding. In the first case, a natural question would be establishing the *size of the Universe*. In the second option, by estimating the speed of contraction, it would be possible to find the *final collapsing day*, the day when the Universe will end its existence. Finally, in the third case, the Universe would be smaller and smaller as the arrow of time is reversed. Hence, by solving a back prediction problem, it would be possible to determine the moment of the so-called Big Bang and the *age of the Universe*.

The issue is then: How can we evaluate the distances of galaxies and establish if they are constant or time varying? Since there is no ruler for such distances, we have to make an indirect estimation by exploiting some auxiliary data. A main source of information for such questions is the light emitted from galaxies. This light is analyzed with an instrument called spectroscope, which provides the spectrum of the galactic signal, by showing its energy distribution over frequencies. In particular, the spectroscope highlights the characteristic sharp peaks (emission lines) produced at various wavelengths by the intrinsic chemical-physical phenomena taking place in the source of light.

The first studies on the emission lines of galaxies go back to astronomers George Lemaitre (1894–1966) and Edwin Powell Hubble (1889–1953), which culminated in some publications around the period 1927–1929. These studies led to a striking discovery: The galaxies are subject to a recession movement, implying that the Universe is expanding! Moreover, the recessional velocity of a galaxy increases with its distance from the Earth.

This fact is the natural conclusion of the following observation. If a far galaxy would be located at a constant distance from us, its emission lines would stay in fixed positions of its spectrum, at certain (constant) wavelengths (frequencies). If instead a galaxy is moving, then the lines are subject to a shift: The wavelength becomes shorter if the distance is decreasing (approaching source) and longer if it is increasing (receding source). These two situations are known as *blueshift* and *redshift*, respectively. An analogous effect is well known for the propagation of sound waves: If an ambulance is approaching, the sound of its siren has an higher pitch than a receding one (*Doppler effect*). The same occurs for light sources, leading to a frequency shift in a direction (toward the blue) or in the opposite direction (toward the red) depending on the fact that the galaxy is approaching or recessing.

A main contribution on this matter can be found in a Hubble's paper published in 1929 in the *Proceedings of the National Academy of Sciences of the United States of America*. By means of the redshift data collected during his activity at Mount Wilson Observatory, in the surroundings of Los Angeles, Hubble estimated the velocity of some galaxies, and made a correlation with their distance from the Earth, so constructing a diagram that is replicated in Figure 4.3.

Note that the velocity is measured in kilometers per second and the distance in PARSEC. The PARSEC, often indicate with the symbol pc, is a unit of distance used in astronomy; 1 pc is approximately equal to 3.26 light-years (namely 340 560 billion kilometers). The symbol Mpc denotes MegaPARSEC.

As can be seen, the diagram is rather scattered, but it exhibits a clear linear trend, with positive slope. Hence, the larger the distance, the larger the regression velocity. In other words, denoting by v the velocity and by d the distance, their relationship is

$$v = Hd.$$

This is known as *Hubble's law* and the proportionality coefficient H is named *Hubble constant*. The slope of the interpolating linear trend, determined via the least squares method, led Hubble to find an estimate of the value of H, about 500 km s^{-1} Mpc^{-1}.

The value of H has been subject to a long controversy, with a recurrent series of estimates and corrections. One could even say that the Hubble constant is

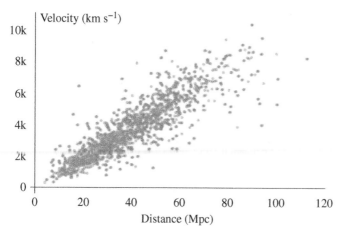

Figure 4.3 Hubble law: The recession velocity of galaxies is proportional to their distances.

the least constant of constants. Today, it is believed that H is about 65 km s^{-1} Mpc^{-1}.

Having ascertained that the Universe is currently expanding, a challenging question is under scrutiny in the last decades: What is the dynamic evolution of such expansion? Correspondingly, what is the ultimate fate of the Universe? There are three possibilities. In the cosmological jargon, they are called *flat Universe*, *open Universe*, and *closed Universe*. The Universe is said to be flat if its rate of expansion becomes smaller and smaller as time increases so that, asymptotically, when time tends to infinite, the rate of expansion slows down to zero and the velocity of expansion becomes a constant. The Universe is said to be open if the rate of expansion does not tend to zero. It is said to be closed if the expansion would eventually stop and then a reverse phenomenon starts leading to a recollapse, with the end of the Universe, possibly followed by a new Big Bang. These topics are dealt with in astronomy research today.

4.4 Dynamic Modeling

We now overview the main discrete time dynamic models we deal with for identification purposes. They are can be given in *external representation* or in *internal representation*.

4.5 External Representation Models

4.5.1 Box and Jenkins Model

By *external representation* models we mean those models in the equations of which only the input and output variables appear, without any auxiliary signals. Here, we restrict to linear models. In the case of time series, a typical model is a difference equation or a transfer function relating the output variable to a white noise input, as in Figure 4.4a, where the white noise is denoted by $\xi(\cdot)$, the signal to be modeled by $y(\cdot)$, and the transfer function by $W(z)$.

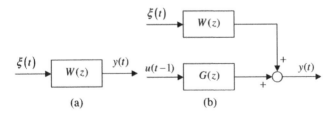

Figure 4.4 Modeling a time series (a) or a cause–effect system (b).

If we pass to cause–effect systems, the basic scheme is that of Figure 4.4b, where $G(z)$ is the input–output transfer function and $v(\cdot)$ is the residual signal (disturbance) to capture all types of errors. In turn, $v(\cdot)$ is seen as a process generated by a white noise $\xi(t)$, feeding a system with transfer function $W(z)$.

We can also write

$$y(t) = W(z)\xi(t) \tag{4.3}$$

in the case of time series and

$$y(t) = G(z)u(t) + W(z)\xi(t) \tag{4.4}$$

for cause–effect systems. Often, the output at time t depends upon the input at time $t - 1$ at most. If so, the model is written as

$$y(t) = G(z)u(t - 1) + W(z)\xi(t). \tag{4.5}$$

In this general framework, many of the models introduced in previous chapters can be encompassed, as we will see now on.

Model 4.5 is known as *Box–Jenkins model*. Such denomination comes from the names of statisticians George E. P. Box (1919–2013) and Gwilym M. Jenkins (1932–1982) who published in 1970 a mainstream book entitled *Time Series Analysis: Forecasting and Control*. This reminds us what George Box used to say: *All models are wrong, but some are useful.*

4.5.2 ARX and AR Models

As already seen in Chapter 3, ARX models are defined by the difference equation

$$\begin{aligned}
y(t) = {}& a_1 y(t - 1) + a_2 y(t - 2) + \cdots + a_{n_a} y(t - n_a) \\
& + b_1 u(t - 1) + b_2 u(t - 2) + \cdots + b_{n_b} u(t - n_b) + \xi(t),
\end{aligned} \tag{4.6}$$

where $\xi(\cdot) \sim \mathrm{WN}(0, \lambda^2)$ is a white noise with zero mean, $E[\xi(t)] = 0$, $\forall t$, and variance λ^2.

Here, one recognizes the autoregressive (AR) part, constituted by the sequence of past $y(\cdot)$ terms, while the eXogenous part is formed by the sequence of inputs $u(\cdot)$. Integers n_a and n_b are called orders of the AR and the X part, respectively. we also use the symbol $\mathrm{ARX}(n_a, n_b)$.

By means of polynomials (3.8) and (3.15), we can write

$$A(z) \quad y(t) = B(z) \quad u(t - 1) + \xi(t). \tag{4.7}$$

From (4.7), it appears that this model is of the Box–Jenkins type (4.5), with

$$G(z) = \frac{B(z)}{A(z)}, \quad W(z) = \frac{1}{A(z)}.$$

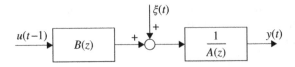

Figure 4.5 Block scheme of the ARX model (4.7).

In these expressions, at the numerator and denominator, we find polynomials in powers of z^{-1}. It is possible to pass to polynomials in positive powers of z by multiplying numerator and denominator by z^n, where $n = \max(n_a, n_b)$.

The corresponding block scheme is depicted in Figure 4.5.

When there is no exogenous signal, the above model becomes

$$y(t) = a_1 y(t-1) + a_2 y(t-2) + \cdots + a_{n_a} y(t - n_a) + \xi(t) \tag{4.8}$$

or, in operator form,

$$A(z)y(t) = \xi(t).$$

This is the AR model, already discussed in the previous chapter.

The solution of the difference equation (4.8) is a stochastic process. If all roots of polynomial

$$z^{n_a} A(z) = z^{n_a} - a_1 z^{n_a - 1} - a_2 z^{n_a - 2} - \cdots - a_{n_a} \tag{4.9}$$

belong to the open unit disk in the complex plane, then the solution of (4.8) asymptotically tends to a unique *stationary* stochastic process whatever the initial condition be.

As for the ARMA model (4.7), assuming that the stability condition concerning polynomial (4.9) is satisfied, if $u(\cdot)$ is a stationary process, then $y(\cdot)$ asymptotically tends to a unique stationary process. In case $u(\cdot)$ has zero mean value (and the white noise has null mean), then such process will have zero mean value too.

4.5.3 ARMAX and ARMA Models

Consider now the difference equation

$$\begin{aligned} y(t) = {} & a_1 y(t-1) + a_2 y(t-2) + \cdots + a_{n_a} y(t - n_a) \\ & + b_1 u(t-1) + b_2 u(t-2) + \cdots + b_{n_b} u(t - n_b) + v(t), \end{aligned} \tag{4.10}$$

where $v(t)$ is the *moving average* process:

$$v(t) = \xi(t) + c_1 \xi(t-1) + \cdots + c_{n_c} \xi(t - n_c). \tag{4.11}$$

Expression (4.10) is analogous to (4.6), the only difference being the fact that $v(t)$ (the *equation residual*) is no more a white noise but instead an MA process.

The model defined by Eqs. (4.10) and (4.11) is known as *ARMAX* model, as anticipated in Section 4.5.3.

Recalling definitions (3.8), (3.9), and (3.15), formulas (4.10) and (4.11) lead to

$$A(z)y(t) = B(z)u(t-1) + C(z)\,\xi(t). \tag{4.12}$$

In case there is no exogenous variable, the model becomes the ARMA model

$$y(t) = a_1 y(t-1) + a_2 y(t-2) + \cdots + a_{n_a} t(t - n_a)$$
$$+ \xi(t) + c_1 \xi(t-1) + c_2 \xi(t-2) + \cdots + c_{n_c} \xi(t - n_c), \tag{4.13}$$

namely

$$A(z)y(t) = C(z)\xi(t). \tag{4.14}$$

For both ARMAX and ARMA models, the stability depends upon polynomial $A(z)$: if all roots of (4.9) have modulus less than 1, the solution of (4.14) is, at steady state, a stationary process. The same holds true for model (4.12) if $u(\cdot)$ is stationary.

As for polynomial $C(z)$, note that $v(t) = C(z)\xi(t)$ is a linear combination of white noise samples; hence, it is a stationary process, whatever the value of parameters c_i. In view of the *spectral factorization theorem* (Proposition 3.1), without loss of generality, we can assume that all roots of

$$z^{n_c} C(z)$$

have modulus lower than 1.

Remark 4.1 (Further input–output models)   In the literature, one can find many other types of external representation models. A couple of them are now mentioned. Again, they can be seen as special cases of the *ARMAX* or Box–Jenkins models.

Output error models:
   In the model

$$A(z)\tilde{y}(t) = B(z)u(t-1),$$
$$y(t) = \tilde{y}(t) + \xi(t),$$

the output $y(t)$ is given by the output of a deterministic system (with transfer function $B(z)/A(z)$) corrupted by additive noise ($\xi(t)$). It is therefore referred to as *output error model*.

A simple substitution shows that the above equations can be given the ARMAX form:

$$A(z)\,y(t) = B(z)\,u(t-1) + C(z)\,\xi(t),$$

with

$$C(z) = A(z).$$

ARXAR models:

In the ARMAX model, the residual is seen as an MA process. An alternative is to see it as an AR process thus obtaining the so-called *ARXAR* model:

$$\mathcal{M}(\theta) : A(z)y(t) = B(z)u(t-1) + (1/D(z))\xi(t).$$

We can rewrite it as an ARMAX model as follows:

$$\mathcal{M}(\theta) : A(z)D(z)y(t) = B(z)D(z)u(t-1) + \xi(t).$$

4.5.4 Multivariable Models

The previous models can be generalized to multivariable systems. In particular, for a time series composed by p signals, $y(\cdot) = [y_1(\cdot)\, y_2(\cdot) \cdots y_p(\cdot)]'$, a multivariable ARMA model can be defined as

$$y(t) = A_1 y(t-1) + A_2 y(t-2) + \cdots + A_{n_a} y(t-n_a)$$
$$+ C_0 \xi(t) + C_1 \xi(t-1) + C_2 \xi(t-2) + \cdots + C_{n_c} \xi(t-n_c),$$

where $A_1, A_2, \ldots, A_{n_a}$ are square matrices of dimension $p \times p$; by assuming that $\xi(\cdot)$ is a vector white noise with p components, matrices $C_0, C_1, C_2, \ldots, C_{n_c}$ are also $p \times p$ matrices. Usually, one takes $C_0 = I$. By vector white noise, we mean a stochastic process such that

$$E[\xi(t)] = \bar{\xi},$$

$$E[(\xi(t_1) - \bar{\xi})(\xi(t_2) - \bar{\xi})'] = \begin{cases} 0, & t_1 \neq t_2, \\ \Lambda, & t_1 = t_2. \end{cases}$$

Hence, by defining the $p \times p$ identity matrix I, and introducing the polynomial matrices

$$A(z) = I - A_1 z^{-1} - A_2 z^{-2} - \cdots - A_{n_a} z^{-n_a},$$
$$C(z) = C_0 + C_1 z^{-1} + C_2 z^{-2} + \cdots + C_{n_c} z^{-n_c},$$

the corresponding transfer matrix is

$$W(z) = A(z)^{-1} C(z).$$

This is a $p \times p$ matrix, the (i,j) element of which, $W_{ij}(z)$, is a rational function representing the z-transfer function from the jth component of $\xi(t)$ to the ith component of $y(\cdot)$.

4.6 Internal Representation Models

A (scalar or vector) time series can be seen as the output of a linear *state space model* fed by a white noise $\xi(t)$ as follows:

$$x(t+1) = Fx(t) + K\xi(t), \tag{4.15a}$$

$$y(t) = Hx(t) + \xi(t). \tag{4.15b}$$

Here, $x(t)$ è is the so-called *state* of the system. From (4.15b), it follows that the dimension of $\xi(t)$ coincides with that of $y(t)$. In particular, if $y(t)$ is scalar, $\xi(t)$ is scalar.

More in general, the state description may be based on two white noise signals $v_1(t)$ and $v_2(t)$, leading to

$$x(t+1) = Fx(t) + v_1(t),$$
$$y(t) = Hx(t) + v_2(t).$$

This last description is the *Markovian representation* of $y(\cdot)$. Equation (4.15) is a special case named *innovation representation*. Such representation plays a main role in Kalman filtering, as we see in Section 9.8.

If there is an exogenous variable $u(t)$, then the typical models are

$$x(t+1) = Fx(t) + Gu(t) + K\xi(t)$$
$$y(t) = Hx(t) + \xi(t)$$

or

$$x(t+1) = Fx(t) + Gu(t) + v_1(t)$$
$$y(t) = Hx(t) + v_2(t).$$

Vector $u(t)$ has a dimension determined by the number of external variables influencing the system. Signals $v_1(t)$ and $v_2(t)$ are the state disturbance and output disturbance, respectively. Matrix F is the *dynamic matrix*, H is the *output transformation*, and G is the *input transformation*.

Remark 4.2 (Internal vs external representation)   From (4.15), we have

$$zx(t) = Fx(t) + K\xi(t) \quad \Rightarrow \quad x(t) = (zI - F)^{-1}K\xi(t),$$
$$y(t) = Hx(t) + \xi(t) \quad \Rightarrow \quad y(t) = [H(zI - F)^{-1}K + I]\xi(t).$$

Hence, the transfer matrix from ξ to y is

$$W(z) = [H(zI - F)^{-1}K + I]. \tag{4.16}$$

Given the state representation, this formula enables the determination of the external representation. The inverse problem of passing from an input–output representation to a state model is called *realization problem*. It is equivalent to solving (4.16) with respect to the triple (F, H, K), given $W(z)$. Each (F, H, K) such that the associated state space model has transfer matrix equal to $W(z)$ is named *realization*.

Example 4.1 (State space realization of an AR model)   For the AR(2) difference equation

$$y(t) = a_1 y(t-1) + a_2 y(t-2) + \xi(t),$$

we can pose

$$x_1(t) = y(t-1),$$
$$x_2(t) = y(t-2).$$

Then

$$x_1(t+1) = a_1 x_1(t) + a_2 x_2(t) + \xi(t),$$
$$x_2(t+1) = x_1(t),$$
$$y(t) = a_1 x_1(t) + a_2 x_2(t) + \xi(t),$$

which can be written in the state space form (4.15) by letting

$$F = \begin{bmatrix} a_1 & a_2 \\ 1 & 0 \end{bmatrix},$$

$$K = \begin{bmatrix} 1 \\ 0 \end{bmatrix},$$

$$H = [a_1 \quad a_2].$$

Example 4.2 (State space realization of an ARMA model)   Consider the ARMA(2, 2) model

$$y(t) = a_1 y(t-1) + a_2 y(t-2) + \xi(t) + c_1 \xi(t-1) + c_2 \xi(t-2),$$

the transfer function of which is

$$W(z) = \frac{C(z)}{A(z)},$$

with

$$A(z) = 1 - a_1 z^{-1} - a_2 z^{-2}, \quad C(z) = 1 + c_1 z^{-1} + c_2 z^{-2}.$$

The state space realization can be obtained through the following steps. First, perform the long division and write the transfer function as

$$W(z) = 1 + \frac{N(z)}{A(z)},$$

with

$$N(z) = (a_1 + c_1)z^{-1} + (a_2 + c_2)z^{-2}.$$

Introduce then signal $y'(t)$ as the output of the system with transfer function $N(z)/A(z)$ fed by $\xi(t)$:

$$y'(t) = \frac{N(z)}{A(z)} \xi(t).$$

In this way, we have

$$y(t) = y'(t) + \xi(t).$$

We can also write

$$y'(t) = N(z)y''(t),$$

where

$$y''(t) = \frac{1}{A(z)}\xi(t).$$

This last signal can be described in the time domain with the standard AR difference equation

$$y''(t) = a_1 y''(t-1) + a_2 y''(t-2) + \xi(t).$$

Then, by the same rationale used in Example 4.1, define

$$x_1(t) = y''(t-1),$$
$$x_2(t) = y''(t-2),$$

so that

$$x_1(t+1) = a_1 x_1(t) + a_2 x_2(t) + \xi(t),$$
$$x_2(t+1) = x_1(t).$$

Now, coming back to signal $y'(t)$:

$$y'(t) = N(z)y''(t) = (a_1 + c_1)y''(t-1) + (a_2 + c_2)y''(t-2).$$

Taking into account the previous definitions of state variables $x_1(t)$ and $x_2(t)$, we can rewrite this expression as

$$y'(t) = N(z)y''(t) = (a_1 + c_1)x_1(t) + (a_2 + c_2)x_2(t).$$

Finally, we come to the output $y(t)$:

$$y(t) = y'(t) + \xi(t) = (a_1 + c_1)x_1(t) + (a_2 + c_2)x_2(t) + \xi(t).$$

In this way, the realization of the ARMA(2,2) model is

$$x_1(t+1) = a_1 x_1(t) + a_2 x_2(t) + \xi(t),$$
$$x_2(t+1) = x_1(t),$$
$$y(t) = (a_1 + c_1)x_1(t) + (a_2 + c_2)x_2(t) + \xi(t),$$

which can be written in the form (4.15) with matrices

$$F = \begin{bmatrix} a_1 & a_2 \\ 1 & 0 \end{bmatrix},$$

$$K = \begin{bmatrix} 1 \\ 0 \end{bmatrix},$$

$$H = [a_1 + c_1 \quad a_2 + c_2].$$

4.7   The Model Identification Process

Setting up a model from data is a process composed of various steps, which can be outlined as follows:

(a) *data gathering and analysis*
(b) *choice between a physical model and a synthetic model*
(c) *selection of the family of models*
(d) *determination of the appropriate complexity*
(e) *parameter estimation*
(f) *critical analysis of the achieved results (validation).*

As for the data, the situation may be quite different depending on the specific problem at hand. In the field of industrial processes, the only available data are normally measurements taken during the functioning of the plant; sometimes, some small variations to this or that signal are allowed to probe the modes of the system. Such variations may include temporary steps in the opening of a valve, small variations in the composition of a gas, and so on. In the biological area, short duration perturbations may be allowed. For example, the Conard curve for the glucose tolerance test can be seen as the impulse response of a human being to a glucose injection. Often data contain outliers, namely snapshots that do not meet the nature of the phenomenon under study, and must be discharged or replaced by a likely value. In other cases, there are *missing data*, in the sequence of measurements, due to malfunctioning of sensors during a certain interval of time.

As for the choice between a physical model and a synthetic one, the intended use of the model is to be taken into account. In particular, if the system is very complex, an accurate physical model might require a remarkable effort to be constructed from first principles, eventually in many equations, so that the final description might be not easy to manage and simulate.

Coming to the choice of the family of models, in time series analysis, a basic decision concerns whether to adopt an AR, an MA, or an ARMA model. In this regard, in Section 6.6, we see an algorithm enabling to advise about this issue.

Another important problem to deal with is the choice of the complexity, discussed in Chapter 6. In ARMAX models, this calls for the determination of orders n_a, n_b, and n_c of the AR, X, and MA parts. Often, this is accomplished via the estimation of models of different complexities and then, by comparing their performances, the selection of the appropriate orders.

The described procedure ends in a (provisional) choice of the *optimal* model in the selected class. The last question is then: Is such *optimal* model a fair interpretation of the given data? This leads to the phase of *validation*. Often this phase amounts to testing the performance of the obtained model in experimental conditions different than those associated to the experiment providing the considered data set. This idea is related to the belief that a model is good to the extent to which it is of general validity, namely can explain data in all possible working conditions. If, for some reasons, the final analysis shows some unsatisfactory behavior in certain operative conditions, then one can reconsider the whole procedure to pinpoint the critical passage(s) and revise it(them).

The whole procedure is illustrated in detail in the case study discussed in Section 11.2.

4.8 The Predictive Approach

Consider a family of models

$$\mathfrak{M} = \{\mathcal{M}(\theta) | \theta \in \Theta\}.$$

Each model of the family (with a fixed complexity) is characterized by a vector of parameters θ. We assume that the data are the snapshots of measurable variables, taken over an interval, say from time 1 to time N. To be precise, the data are the sequence of output $y(\cdot)$ for time series and the two sequences of the input $u(\cdot)$ and the output $y(\cdot)$ for scalar systems with an exogenous variable.

The problem we address is the choice of the *best model* within the given family of candidates. Since each model is associated to a vector of parameters, this amounts to choosing the *best parameter* θ in the feasible set Θ.

In this regard, a basic problem is how to compare the data with the signals populating a model. Such comparison requires relating numerical sequences (the data) with stochastic processes (the variables of the model).

A possibility is to adopt a predictive approach, according to which a model is appropriate to the extent to which the associated prediction error is – in some sense to be specified – *small*. To be more concrete, take a time series $y(\cdot)$, measured over a certain interval of time. To describe the series, one can use a model such as an ARMA model described by a certain vector of parameters θ. As we know, in this model, $y(\cdot)$ is seen as a stationary process generated via a white noise input. The corresponding predictor supplies $\hat{y}(t + 1/t)$ as a function of past data. If we adopt the symbol y^t to denote the sequence $\{y(t), y(t - 1), y(t - 2), \dots\}$, we can write

$$\mathcal{M}(\theta) : \quad \hat{y}(t + 1/t) = f(y^t, \theta).$$

Here, the input is the past sequence of $y(\cdot)$. Analogously, in case of systems, we write

$$\mathcal{M}(\theta) : \quad \hat{y}(t + 1/t) = f(u^t, y^t, \theta),$$

where u^t denotes the sequence $\{u(t), u(t - 1), u(t - 2), \dots\}$.

In both cases of time series and systems, we see that the predictor is fed by past data. Such feeding inputs being numerical (past data are measured), the prediction is also numerical. As such a comparison with current data is now possible. For each time point, the prediction error of a given model can

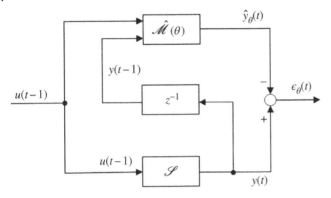

Figure 4.6 The prediction error identification rationale.

therefore be evaluated. The *best* model is the one for which the prediction error is – in some sense to be specified – *small*.

To quantify the amount of the identification error, one can adopt various criteria. One of the most widely considered is

$$J(\theta) = \frac{1}{N} \sum_{t}^{N} \epsilon(t)^2.$$

According to such a criterion, the best model in the family is the one corresponding to parameter θ within the admissible set Θ minimizing $J = J(\theta)$.

The rationale of prediction error methods is summarized in Figure 4.6.

Remark 4.3 (Reflection on the assessment of the model quality)   The introduction of performance index $J(\theta)$ enables to pose the identification problem as an optimization problem. In this way, one can proceed to estimation by devising *ad hoc* algorithms tailored for this or that family of models, as we do in the next points. However, the assessment of model performance should take into consideration a number of further items. In particular, the statistical properties of the prediction error of the estimated model should be taken into account. In this regard, our readers will remember that in the Introduction of this book, we claimed that a model is good when the prediction error is white. Following this basic concept, at the end of the optimization procedure, the whiteness of the prediction error is often tested to have a final confirmation of the quality of identification. This can be done with the test of Section 2.7.

4.9   Models in Predictive Form

As seen above, in the framework of predictive identification, what really matters is the model in predictive form. We now exploit the methods of

Chapter 3 to work out the main predictive models for input–output models. Most of such predictors were already derived; however, here we put them in a common framework for the benefit of future developments of identification techniques. Determining the predictor for a state space model is the subject of Chapter 9.

4.9.1 Box and Jenkins Model

For the Box and Jenkins model

$$\mathcal{M}(\theta) : y(t) = G(z)u(t-1) + W(z)\xi(t), \tag{4.17}$$

where, as usual, $\xi(t)$ is a zero mean white noise, the predictor can be worked out by dividing the two members of above expression (4.17) by $W(z)$ and then adding and subtracting $y(t)$, so obtaining

$$y(t) = [1 - (1/W(z))]y(t) + G(z)/W(z)u(t-1) + \xi(t). \tag{4.18}$$

Since $W(z)$ is given by the quotient of two monic polynomial of the same degree, the polynomial at the numerator and denominator of $1/W(z)$ are also monic with the same degree. Therefore, by performing the long division, we have

$$1/W(z) = 1 + \alpha_1 z^{-1} + \alpha_2 z^{-2} + \cdots$$

Consequently, $[1 - (1/W(z))]y(t)$ does not depend upon $y(t)$; it depends upon $y(t-1), y(t-2), \dots$ only. Analogously, $G(z)/W(z)u(t-1)$ depends on $u(t-1), u(t-2), \dots$ We can therefore rewrite (4.18) in the form

$$y(t) = f(y^{t-1}) + g(u^{t-1}) + \xi(t),$$

where we have set $[1 - (1/W(z))]y(t) = f(y^{t-1})$ and $G(z)/W(z)u(t-1) = g(u^{t-1})$. $\xi(t)$ being a white noise, the knowledge of the past values of $u(\cdot)$ and $y(\cdot)$ does not bring any useful information to evaluate $\xi(t)$. Hence, the optimal predictor is

$$\mathcal{M}(\theta) : \hat{y}(t) = [1 - (1/W(z))]y(t) + G(z)/W(z)u(t-1). \tag{4.19}$$

Here, symbol $\hat{y}(t)$ is a shorthand for $\hat{y}(t/t-1)$.

As a corollary of expression (4.19), we can easily work out the predictors for the following families of models.

4.9.2 ARX and AR Models

For an ARX model,

$$\mathcal{M}(\theta) : A(z)y(t) = B(z)u(t-1) + \xi(t),$$

we have $G(z) = B(z)/A(z)$ and $W(z) = 1/A(z)$, so that, substituting these expressions into (4.19),

$$\mathcal{M}(\theta) : \hat{y}(t) = [1 - A(z)]y(t) + B(z)u(t - 1).$$

Of course, from this result, we can deduce the predictor for a time series described with an AR model, in which case we have

$$\mathcal{M}(\theta) : \hat{y}(t) = [1 - A(z)]y(t).$$

4.9.3 ARMAX and ARMA Models

In an ARMAX model

$$\mathcal{M}(\theta) : A(z)y(t) = B(z)u(t - 1) + C(z)\xi(t),$$

$G(z) = B(z)/A(z)$ and $W(z) = C(z)/A(z)$. From (4.19), we have

$$\mathcal{M}(\theta) : C(z)\hat{y}(t) = [C(z) - A(z)]y(t) + B(z)u(t - 1). \qquad (4.20)$$

By setting $B(z) = 0$, the predictor for an $ARMA$ model is

$$\mathcal{M}(\theta) : C(z)\hat{y}(t) = [C(z) - A(z)]y(t). \qquad (4.21)$$

Remark 4.4 (Linearity in the parameters) As seen above, the predictor of an ARX model is

$$\hat{y}(t) = [1 - A(z)]y(t) + B(z)u(t - 1)$$
$$= a_1 y(t - 1) + a_2 y(t - 2) + \cdots + b_1 u(t - 1) + b_2 u(t - 2) + \cdots .$$

Therefore, $\hat{y}(t)$ a linear combination of $y(t - 1), y(t - 2), \ldots$ and $u(t - 1), u(t - 2), \ldots$, with coefficients of the combination given by the parameters of the polynomials $A(z)$ and $B(z)$. Stated another way, the predictor is linear in the model parameters.

On the contrary, the predictors for models with a nontrivial moving average part, such as an ARMAX model with $C(z) \neq 1$, are nonlinear in the parameters. For instance, in an ARMAX(1, 1, 1):

$$A(z) = 1 - az^{-1}, \quad B(z) = b, \quad C(z) = 1 + cz^{-1},$$

by the long division algorithm, we have

$$(C(z) - A(z))/C(z) = (a + c)/(1 + cz^{-1}) = (a + c) - c(a + c)z^{-1} + \cdots ,$$
$$B(z)/C(z) = b/(1 + cz^{-1}) = b - cbz^{-1} + \cdots .$$

Hence, the predictor is

$$\hat{y}(t) = (a + c)y(t - 1) - c(a + c)y(t - 2) + \cdots$$
$$+ bu(t - 1) - cbu(t - 2) + \cdots .$$

In this expression, the coefficient multiplying $y(t-2)$ contains the product ca and the square c^2. Similarly, the coefficient multiplying $u(t-2)$ is given the product cb. Hence, $\hat{y}(t)$ is nonlinear in the parameters.

In an ARXAR model,

$$\mathcal{M}(\theta) : A(z)y(t) = B(z)y(t-1) + (1/D(z))\xi(t),$$

the predictor is given by

$$\hat{\mathcal{M}}(0) : \hat{y}(t) - [1 - A(z)D(z)]y(t) + B(z)D(z)u(t-1).$$

Here $A(z)D(z)$ and $B(z)D(z)$ are the polynomial product; hence, the predictor is nonlinear in the parameters. However, if the parameters of polynomial $D(z)$ are fixed, then the linearity with respect to the parameters of $A(z)$ and $B(z)$ is recovered; analogously, if $A(z)$ and $B(z)$ are fixed, the predictor is linear in the parameters of $D(z)$.

In the next chapter, we present two main identification methods, one for the estimation of AR and ARX models and one for the estimation of ARMA and ARMAX models.

5

Identification of Input–Output Models

5.1   Introduction

In this chapter, we present two of the most widely used methods for the estimation of dynamical models in input–output form, the *least squares* and the *maximum likelihood* (ML) methods. The former is conceived for the estimation of AR or ARX models and the latter for the estimation of ARMA or an ARMAX models.

5.2   Estimating AR and ARX Models: The Least Squares Method

Since the AR model is a special case of the ARX model, we focus on the latter class of models,

$$\mathcal{M}(\theta) : y(t) = a_1 y(t-1) + a_2 y(t-2) + \cdots + a_{n_a} y(t-n_a)$$
$$+ b_1 u(t-1) + b_2 u(t-2) + \cdots + b_{n_b} u(t-n_b) + \xi(t),$$

where, as usual, $u(\cdot)$ is the system input, $y(\cdot)$ is the system output, and $\xi(t) \sim WN(0, \lambda^2)$. By introducing the vector of parameters

$$\theta = [a_1 \; a_2 \; \cdots \; a_{n_a} \; \vdots \; b_1 \; b_2 \; \cdots \; b_{n_b}]'$$

and the vector of the observations

$$\phi(t) = [y(t-1) \; y(t-2) \; \cdots \; y(t-n_a) \; \vdots \; u(t-1) \; u(t-2) \; \cdots \; u(t-n_b)]',$$
$$(5.1)$$

the model can be written as

$$\mathcal{M}(\theta) : y(t) = \phi(t)'\theta + \xi(t).$$

The associated predictor is

$$\hat{\mathcal{M}}(\theta) : \hat{y}(t) = \phi(t)'\theta,$$

Model Identification and Data Analysis, First Edition. Sergio Bittanti.
© 2019 John Wiley & Sons, Inc. Published 2019 by John Wiley & Sons, Inc.

so that the prediction error is given by

$$\epsilon(t) = y(t) - \phi(t)'\theta. \tag{5.2}$$

Assuming as estimation criterion

$$J(\theta) = \frac{1}{N} \sum_{t}^{N} \epsilon(t)^2, \tag{5.3}$$

we are in the same framework already discussed for static modeling in Section 4.3, the difference being the meaning of the observation vector. We can therefore refer to the *normal equation* (4.2) to conclude that all and only the parameters leading to the minimum of $J(\theta)$ are the solutions of

$$\left[\sum_{t}^{N} \phi(t)\phi(t)' \right] \theta = \sum_{t}^{N} \phi(t)y(t) \tag{5.4}$$

or, equivalently,

$$\left[\frac{1}{N} \sum_{t}^{N} \phi(t)\phi(t)' \right] \theta = \frac{1}{N} \sum_{t}^{N} \phi(t)y(t). \tag{5.5}$$

Letting

$$S(N) = \left[\sum_{t}^{N} \phi(t)\phi(t)' \right] \tag{5.6}$$

and

$$R(N) = \frac{1}{N} \left[\sum_{t}^{N} \phi(t)\phi(t)' \right], \tag{5.7}$$

if $S(N)$ (or equivalently $R(N)$) is invertible, then there is a unique solution of the normal equations, the least squares estimator

$$\hat{\theta}_N = S(N)^{-1} \left[\sum_{t}^{N} \phi(t)y(t) \right], \tag{5.8}$$

$$\hat{\theta}_N = R(N)^{-1} \left[\frac{1}{N} \sum_{t}^{N} \phi(t)y(t) \right]. \tag{5.9}$$

If $S(N)$ is singular, then, as already seen in the static modeling case, the system of normal equations admits infinitely many solutions, all equivalent to each other in the sense that the value of the associated performance index is the same.

Remark 5.1 (Using least squares for different models)   The least squares method is not limited to the AR or ARX models. It is useful in all cases when the estimation problem can be reduced to the linear-in-the-parameters case, as illustrated in the following examples.

Example 5.1 Consider the problem of estimating parameter d in model

$$y(t) = 0.5u(t-1) + \frac{1}{1 + dz^{-1}}\xi(t), \quad \xi(\cdot) \sim \text{WN}(0, \lambda^2).$$

This model can be written as

$$(1 + dz^{-1})y(t) = 0.5(1 + dz^{-1})u(t-1) + \xi(t),$$

namely

$$y(t) = -dy(t-1) + 0.5u(t-1) + 0.5du(t-2) + \xi(t).$$

This is an ARX(1,2) model of the type

$$y(t) = a_1 y(t-1) + b_1 u(t-1) + b_2 u(t-2) + \xi(t).$$

Then, one can define

$$\phi(t) = [y(t-1) \; u(t-1) \; u(t-2)] \quad \text{and} \quad \theta = [a_1 \; b_1 \; b_2]$$

and estimate the three unknown parameters of the ARX(1,2) model via least squares and deduce d from the estimates $\hat{a}_1$ and $\hat{b}_2$ of a_1 and b_2 by recalling that $a_1 = -d$ and $b_2 = 0.5d$.

Alternatively, one can observe that the original model can be written as

$$y(t) = 0.5u(t-1) + d[-y(t-1) + 0.5u(t-2)] + \xi(t).$$

Here, by defining

$$\tilde{\phi}(t) = -y(t-1) + 0.5u(t-2),$$
$$\tilde{y}(t) = y(t) - 0.5u(t-1),$$
$$\tilde{\theta} = d,$$

the linear regression model

$$\tilde{y}(t) = \tilde{\theta}\tilde{\phi}(t) + \xi(t)$$

is obtained, where the only unknown is the scalar $\tilde{\theta} = d$.

Example 5.2 In the nonlinear model

$$y(t) = ay(t-1)^2 + b_1 u(t-3) + b_2 u(t-5)^3 + \xi(t), \quad \xi(\cdot) \sim \text{WN}(0, \lambda^2),$$

the parameters can be estimated via least squares provided that we adopt as observation vector $\phi(t) = [y(t-1)^2 \; u(t-3) \; u(t-5)^3]$.

5.3 Identifiability

A main question is whether a system can be uniquely identified from data. Identifiability is determined by two circumstances: the structure of the adopted model (structural identifiability) and the characteristics of the experiment during which the data have been collected (experimental identifiability), as illustrated in the subsequent examples.

Example 5.3 (A system described by many models)   Assume that the data are generated by a deterministic system with transfer function

$$\mathcal{S} : G^0(z) = \frac{z}{(z + 0.5)(z + 0.8)}. \tag{5.10}$$

For identification, the adopted model is

$$\mathcal{M}(\theta) : y(t) = a_1 y(t - 1) + a_2 y(t - 2) + a_3 y(t - 3) + b_1 u(t - 1) + b_2 u(t - 2). \tag{5.11}$$

The parameter vector of such model is

$$\theta = [a_1 \ a_2 \ a_3 \ b_1 \ b_2]'$$

and the corresponding transfer function

$$\mathcal{M}(\theta) : G(z) = \frac{b_1 z^2 + b_2 z}{z^3 - a_1 z^2 - a_2 z - a_3}.$$

$G(z)$ can also be written as

$$\mathcal{M}(\theta) : G(z) = \frac{z(b_1 z + b_2)}{(z - \gamma_1)(z - \gamma_2)(z - \gamma_3)},$$

where γ_1, γ_2, and γ_3 are the roots of $z^3 - a_1 z^2 - a_2 z - a_3$.

System $\mathcal{S}$ is included in the class of models. Indeed, $G^0(z)$ can be obtained from $G(z)$ via a simplification of a common term at numerator and denominator. This can be achieved by posing $(b_1 z + b_2) = (z - \gamma_3)$ so that the model becomes

$$\mathcal{M}(\theta) : G(z) = \frac{z}{(z - \gamma_1)(z - \gamma_2)}.$$

However, the condition $(b_1 z + b_2) = (z - \gamma_3)$ is satisfied by infinitely many triplets (b_1, b_2, γ_3) so that we expect that infinitely many models can describe the same system. The reason is that the model structure is oversized with respect to what would suffice.

Example 5.4 (Insufficient information in data)   Consider again system (5.10), assumed to be stable, and model (5.11). If the input of the system is a constant signal, $u(t) = \bar{u}$, and the output is measured at steady state, the input–output relationship is $\bar{y} = \mu^0 \bar{u}$, where μ^0 is the system gain

$$\mu^0 = G^0(z)|_{z=1}$$

(see Appendix A). On the other hand, the model gain is

$$\mu = G(z)|_{z=1} = \frac{b_1 + b_2}{1 - a_1 - a_2 - a_3}.$$

Hence, the value of the individual parameters of the model cannot be ascertained from the only knowledge of the gain. In other words, the considered experimental conditions are not suitable for the identification of all parameters, only the gain can be found.

We now elaborate the above considerations with the objective of finding a general condition of experimental identifiability. To this purpose, we study the matrix $R(N)$ previously defined in (5.7): Should this matrix be invertible, then the estimate would be unique. We investigate this point by focusing on the asymptotic expression

$$\lim_{N \to \infty} R(N) = \bar{R} = E[\phi(t)\phi(t)'].\tag{5.12}$$

When $\bar{R}$ is invertible, then, for N sufficiently large, $R(N)$ is invertible too, and the least squares estimate is unique. Thus, we focus on matrix $\bar{R}$ and examine its structure starting from the ARX(1, 1) case.

5.3.1 The $\bar{R}$ Matrix for the ARX(1, 1) Model

If

$$\phi(t) = [y(t-1)\ u(t-1)],$$

then

$$\phi(t)\phi(t)' = \begin{bmatrix} y(t-1)^2 & y(t-1)u(t-1) \\ u(t-1)y(t-1) & u(t-1)^2 \end{bmatrix},$$

$$S(N) = \begin{bmatrix} \Sigma y(t-1)^2 & \Sigma y(t-1)u(t-1) \\ \Sigma u(t-1)y(t-1) & \Sigma u(t-1)^2 \end{bmatrix},$$

$$R(N) = \begin{bmatrix} (1/N)\Sigma y(t-1)^2 & (1/N)\Sigma y(t-1)u(t-1) \\ (1/N)\Sigma u(t-1)y(t-1) & (1/N)\Sigma u(t-1)^2 \end{bmatrix},$$

where all summations extend from 1 to N.

The first element on the diagonal of $R(N)$

$$(1/N)\Sigma y(t-1)^2$$

is the sample mean of $y(\cdot)$. If the output is stationary with null mean, then, under mild assumptions, this quantity will converge to the variance of $y(\cdot)$, namely to $\gamma_{yy}(0)$, the covariance function of $y(\cdot)$ evaluated at $\tau = 0$.

Analogously, the second element along the diagonal

$$(1/N)\Sigma u(t-1)^2$$

will converge to $\gamma_{uu}(0)$.

As for the off diagonal $(1, 2)$ and $(2, 1)$ elements, they will tend to $\gamma_{uy}(0)$. In conclusion, matrix $R(N)$ will converge to

$$\bar{R} = \begin{bmatrix} \gamma_{yy}(0) & \gamma_{uy}(0) \\ \gamma_{yu}(0) & \gamma_{uu}(0) \end{bmatrix}.$$

5.3.2 The $\bar{R}$ Matrix for a General ARX Model

We now pass to the general case of an ARX model of any complexity. In view of the structure of the observation vector $\phi(t)$, see Eq. (5.1), matrix $\bar{R}$ defined in (5.12) can be partitioned into four blocks. To be precise, let

$$\bar{R}_{uu}^{(n_b)} = E\left[\begin{bmatrix} u(t-1) \\ u(t-2) \\ \vdots \\ u(t-n_b) \end{bmatrix} [u(t-1)u(t-2)\cdots u(t-n_b)]\right], \qquad (5.13)$$

$$\bar{R}_{yy}^{(n_a)} = E\left[\begin{bmatrix} y(t-1) \\ y(t-2) \\ \vdots \\ y(t-n_a) \end{bmatrix} [y(t-1)y(t-2)\cdots y(t-n_a)]\right]. \qquad (5.14)$$

Then

$$\bar{R} = \begin{bmatrix} \bar{R}_{yy}^{(n_a)} & \bar{R}_{yu} \\ \bar{R}_{uy} & \bar{R}_{uu}^{(n_b)} \end{bmatrix},$$

where the obvious definition of the cross-matrices $\bar{R}_{uy}$ and $\bar{R}_{yu}$ is left to the readers.

Focus now on the diagonal blocks. An easy computation shows that $R_{uu}^{(n_a)}$ is given by the following matrix evaluated for $n = n_b$:

$$\bar{R}_{uu}^{(n)} = \begin{bmatrix} \gamma_{uu}(0) & \gamma_{uu}(1) & \cdots & \gamma_{uu}(n-1) \\ \gamma_{uu}(1) & \gamma_{uu}(0) & \cdots & \gamma_{uu}(n-2) \\ \cdots & \cdots & \cdots & \cdots \\ \gamma_{uu}(n-1) & \gamma_{uu}(n-2) & \cdots & \gamma_{uu}(0) \end{bmatrix}.$$

Note the peculiar structure of such array: along the main diagonal there is $\gamma_{uu}(0)$, along the first super-diagonal $\gamma_{uu}(1)$, and so on. An analogous expression holds for $R_{yy}^{(n_b)}$. Matrices of this type are called Toeplitz matrices.

We discuss now the non-singularity of $\bar{R}$ as main identifiability condition. From (5.12), it is apparent that $\bar{R}$ is positive semi-definite by construction. Thus, $\bar{R}$ is non-singular if and only if it is positive definite, i.e. $x'\bar{R}x > 0, \forall x \neq 0$ (see Appendix B). By taking vector x with the first n_a entries null, i.e. as $x' = [0' \ \tilde{x}']$, where 0 is a the null vector with dimension n_a, we have $x'\bar{R}x - \tilde{x}'R_{uu}^{(n)}\tilde{x}$. Therefore, in order for $\bar{R}$ to be positive definite, it is necessary that $\bar{R}_{uu}^{(n_a)}$ be positive definite as well, or equivalently, $\bar{R}_{uu}^{(n_a)}$ be invertible.

From this observation, we conclude that a necessary condition for the invertibility of $\bar{R}$ is that block $\bar{R}_{uu}^{(n_a)}$ should be invertible.

Interestingly enough, this last condition involves the input signal $u(\cdot)$ only. This motivates the following definition.

Definition 5.1 (Persistent excitation)   Input $u(\cdot)$ is said to be persistently exciting of order n if matrix $\bar{R}_{uu}^{(n)}$ is invertible.

Thus, in order to uniquely identify an ARX(n_a, n_b) model from data, it is necessary that $u(\cdot)$ be persistently exiting of order n_b.

Example 5.5 (Excitation induced by the white noise)   Consider a white noise input, $u(\cdot) \sim \text{WN}(0, \lambda^2)$. The corresponding matrix $\bar{R}_{uu}^{(n)}$ is simply given by

$$\bar{R}_{uu}^{(n)} = \lambda^2 I^{(n)},$$

where $I^{(n)}$ is the $n \times n$ identity matrix. Hence, a white noise is persistently exiting of any order.

Remark 5.2 (Order of persistence of excitation) $\bar{R}_{uu}^{(n-1)}$ is a submatrix of $\bar{R}_{uu}^{(n)}$. As a consequence, if a signal is persistently exciting of order n, then it is also persistently exciting of order $n - 1$.

Remark 5.3 (Persistence of excitation in the frequency domain)   The persistent excitation condition can be given a useful frequency domain interpretation. Consider any vector of dimension n, say

$$\alpha = [\alpha_1 \ \alpha_2 \ \cdots \ \alpha_n]'$$

and the signal $\tilde{u}(\cdot)$ generated by filtering process $u(\cdot)$ as follows:

$$\tilde{u}(t) = \alpha_1 u(t-1) + \cdots + \alpha_n u(t-n).$$

Since the transfer function of such filter is

$$H_\alpha(z) = \alpha_1 z^{-1} + \cdots + \alpha_n z^{-n} = \frac{\alpha_1 z^{n-1} + \cdots + \alpha_n}{z^n},$$

we can write

$$\tilde{u}(t) = H_\alpha(z)u(t). \tag{5.15}$$

Recall now the concept of *generalized moving average (GMA)* process introduced in Remark 1.4. If we assume that $u(\cdot)$ is a stationary process, then $\tilde{u}(\cdot)$ is a GMA stationary process, the variance of which can be expressed in the quadratic form

$$\text{Var}[\tilde{u}(t)] = \alpha' \bar{R}_{uu}^{(n)} \alpha.$$

On the other hand, the variance of $\tilde{u}(t)$ can be computed from its spectrum $\Gamma_{\tilde{u}\tilde{u}}$ as

$$\text{Var}[\tilde{u}(t)] = \frac{1}{2\pi} \int\limits_{-\pi}^{+\pi} \Gamma_{\tilde{u}\tilde{u}}(\omega)d\omega.$$

In turn, the *fundamental theorem of spectral analysis* applied to (5.15) leads to

$$\Gamma_{\tilde{u}\tilde{u}}(\omega) = |H_\alpha(e^{j\omega})|^2 \Gamma_{uu}(\omega).$$

In conclusion,

$$\alpha' \bar{R}_{uu}^{(n)} \alpha = \frac{1}{2\pi} \int\limits_{-\pi}^{+\pi} |H_\alpha(e^{j\omega})|^2 \Gamma_{uu}(\omega)d\omega.$$

This is the key formula for the spectral interpretation of the persistent excitation notion, thanks to which we can say that $u(\cdot)$ is not persistently exciting of order n (and therefore $n + 1, n + 2, \ldots$) when the condition

$$\int\limits_{-\pi}^{+\pi} |H_\alpha(e^{j\omega})|^2 \Gamma_{uu}(\omega)d\omega = 0 \tag{5.16}$$

is met with for some $\alpha \neq 0$. Note that filter $H(z)$ has $n - 1$ zeros. If these zeros are located on the unit disk, at $e(j\bar{\omega}_i)$, $i = 1, 2, \ldots, n - 1$, and Γ_{uu} has $n - 1$ spectral lines located exactly at ω_i, $i = 1, 2, \ldots, n - 1$, then (5.16) is satisfied and input $u(\cdot)$ is not persistently exiting of order n. If the spectrum of $u(\cdot)$ is non-null for at least n distinct frequencies, then $u(\cdot)$ is persistently exciting of order n. In particular, one can conclude that an ARMA process is persistently exciting of any order.

5.4   Estimating ARMA and ARMAX Models

Consider again the ARMAX model

$$\mathcal{M}(\theta) : A(z) \quad y(t) = B(z)u(t-1) + C(z)\xi(t), \tag{5.17}$$

where, as usual,

$$A(z) = 1 - a_1 z^{-1} - a_2 z^{-2} - \cdots - a_{n_a} z^{-n_a},$$

$$B(z) = 1 + b_1 z^{-1} + b_2 z^{-2} + \cdots + b_{n_b} z^{-n_b},$$

$$C(z) = 1 + c_1 z^{-1} + c_2 z^{-2} + \cdots + c_{n_c} z^{-n_c}.$$

As seen in Section 4.9, its predictive form is

$$\hat{\mathcal{M}}(\theta) : C(z)\hat{y}(t) = [C(z) - A(z)]y(t) + B(z)u(t-1). \tag{5.18}$$

To estimate its parameters, we adopt the standard prediction error criterion:

$$J(\theta) = \frac{1}{N} \sum_{1}^{N} \epsilon(t)^2.$$

The expression of $\epsilon(t) = y(t) - \hat{y}(t)$ can be easily deduced from (5.17) and (5.18), by a subtraction member by member, so obtaining

$$C(z)\epsilon(t) = A(z)y(t) - B(z)u(t-1). \tag{5.19}$$

As seen in Remark 4.4, the predictor is no more linear in the parameters. Consequently, the identification criterion, although quadratic in the prediction error, is not quadratic in the unknown parameter vector. Hence, the minimization problem cannot be reduced to a simple set of linear equations such as the normal equations we encountered in least squares.

We adopt an iterative minimization procedure, the so-called *Gauss–Newton iterative method*. First, we describe the general rationale of this method, working out its basic updating formula. Then we go into detail to tailor the general procedure to the identification problem; in particular, we see how the available data have to be elaborated to compute the entries of the updating formula. As we will see, this requires the processing of the available data through a set of filters, whose parameters must be updated at each iteration.

In general, the proposed algorithm does not guarantee achieving the global minimum of the criterion. This is why the procedure is often repeated by starting from different initializations; the optimal estimate is selected by comparing the values taken by the criterion in the various convergence points.

This rationale can be put into practice as follows. According to Newton's ideas, to find the minimum of a complex function $J(\theta)$, we replace $J(\theta)$ with a simpler function, $V(\theta)$, approximating $J(\theta)$ about a certain point. As proxy, a quadratic function is considered. The advantage of such a choice is that it is possible to provide its minimum in closed form, as in least squares.

To be precise, denoting by $\theta^{(r)}$ the parameter vector determined at iteration r, we define the proxy $V(\theta)$ by imposing the following conditions:

1) Function $V(\theta)$ evaluated in $\theta^{(r)}$ coincides with $J(\theta^{(r)})$.
2) The gradient of $V(\theta)$ evaluated in $\theta^{(r)}$ coincides with the gradient of $J(\theta)$ in $\theta^{(r)}$.
3) The Hessian of $V(\theta)$ in $\theta^{(r)}$, namely the matrix of the second derivatives of $V(\theta)$ in $\theta^{(r)}$, coincides with the Hessian of $J(\theta)$ in $\theta^{(i)}$.

The approximating function around $\theta^{(r)}$ is therefore given by

$$V(\theta) = J(\theta^{(r)}) + \frac{dJ(\theta^{(r)})}{d\theta}(\theta - \theta^{(r)}) + \frac{1}{2}(\theta - \theta^{(r)})'\frac{d^2J(\theta^{(r)})}{d\theta^2}(\theta - \theta^{(r)}).$$

As depicted in Figure 5.1, we take the point of minimum of $V(\theta)$, $\theta^{(r+1)}$, as new candidate point of minimum of $J(\theta)$. Assuming that the Hessian is invertible, such point is given by

$$\theta^{(r+1)} = \theta^{(r)} - \left(\frac{d^2J(\theta^{(r)})}{d\theta^2}\right)^{-1}\left(\frac{dJ(\theta^{(r)})}{d\theta}\right)'. \tag{5.20}$$

This formula is known as Newton formula.

Note that the variation $\Delta\theta = \theta^{(r+1)} - \theta^{(r)}$ is computed from the vector of *steepest descent* (namely $-dJ/d\theta$) by premultiplying it by the inverse of the Hessian.

The line connecting $V(\theta^{(r)})$ to $V(\theta^{(r+1)})$ is named *Newton direction* (see Figure 5.2).

After the determination of point $\theta^{(r+1)}$, a further iteration is performed with a new proxy of $J(\theta)$ evaluated about it.

The iterative procedure may exhibit convergence, for instance, the variation

$$\Delta\theta = \theta^{(r+1)} - \theta^{(r)}$$

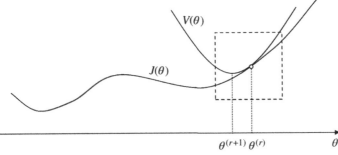

Figure 5.1 Newton method.

Figure 5.2 Newton
method – zoom.

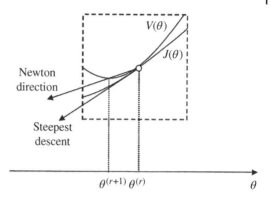

or/and the variation

$$\Delta J = J(\theta^{(r+1)}) - J(\theta^{(r)})$$

may tend to zero for increasing r. The point to which sequence $\theta^{(\cdot)}$ converges can be seen as a candidate for the global minimum of $J(\theta)$.

Note that, for a generic function $J(\theta)$, the Newton method does not guarantee the convergence to the global minimum. In particular, proxy $V(\theta)$ may lead toward an ascent direction when the Hessian is negative semi-definite, see Figure 5.3.

In practice, various initializations of the algorithms are considered. Once converge is reached starting from each of them, the values taken by $J(\theta)$ in those points are compared, and the best parametrization is selected as the one corresponding to the minimum of the various minima.

5.4.1 Computing the Gradient and the Hessian from Data

To put into practice the Newton procedure for ARMAX models, we have to specify how to compute the various ingredients appearing in the basic iteration formula (5.20). To this end, we preliminarily introduce the *gradient vector*

$$\psi(t) = -\frac{d\epsilon(t)'}{d\theta}. \tag{5.21}$$

Then, we have

$$J(\theta) = (1/N) \sum_{1}^{N} \epsilon(t)^2,$$

$$\frac{dJ(\theta^{(r)})}{d\theta} = -(2/N) \sum_{t}^{N} \epsilon(t)\psi(t)',$$

$$\frac{d^2 J(\theta^{(r)})}{d\theta^2} = (2/N) \left[\sum_{1}^{N} \psi(t)\psi(t)' - \sum_{1}^{N} \epsilon(t)\frac{d^2\epsilon(t)}{d\theta^2} \right].$$

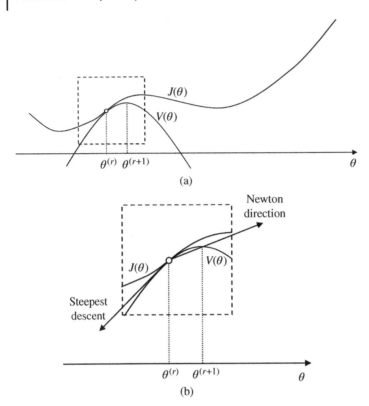

Figure 5.3 Newton method in an another point of curve $J(\theta)$ (a) and the corresponding zoom (b).

In this last expression, we see that the Hessian is given by the sum of two matrices, namely $(2/N) \sum_{t1}^{N} \psi(t)\psi(t)'$ and $-(2/N) \sum_{t1}^{N} \epsilon(t)\frac{d^2\epsilon(t)}{d\theta^2}$. Often the second term is neglected and one takes

$$\frac{d^2 J(\theta^{(r)})}{d\theta^2} = (2/N) \sum_{1}^{N} \psi(t)\psi(t)'.$$

Such cancellation guarantees that the second derivative of $J(\theta^{(r)})$ is positive semi-definite. Correspondingly, variation $\Delta\theta$ takes place in a descending direction, namely the angle between $\Delta\theta$ and the steepest descent is acute.

By simplifying the Hessian in this way, the iterative formula then takes the expression

$$\theta^{(r+1)} = \theta^{(r)} + \left(\sum_{1}^{N} \psi(t)\psi(t)' \right)^{-1} \left(\sum_{1}^{N} \epsilon(t)\psi(t) \right),$$

sometimes called *Gauss–Newton iterative formula*.

We now discuss how to compute $\epsilon(t)$ and $\psi(t)$ from data.

To determine $\epsilon(\cdot)$, we start from expression (5.19). According to this equation, the prediction error $\epsilon(\cdot)$ is obtained by processing $y(\cdot)$ and $u(\cdot)$ via filters $A(z)/C(z)$ and $B(z)/A(z)$, respectively, as shown in Figure 5.4.

Passing to $\psi(\cdot)$, observe that the entries of vector θ are parameters a_i of polynomial $A(z)$, parameters b_i of $B(z)$, and parameters c_i of $C(z)$. Correspondingly, $\psi(\cdot)$ can be written as

$$\psi(t) = [\alpha_1(t)\alpha_2(t)\cdots\alpha_{na}(t) \;\vdots\; \beta_1(t)\beta_2(t)\cdots\beta_{nb}(t) \;\vdots\; \gamma_1(t)\gamma_2(t)\cdots\gamma_{nc}(t)]',$$

where $\alpha_i(t)$ is the opposite of the derivative of the prediction error $\epsilon(t)$ with respect to a_i, $\beta_j(t)$ is the opposite of the derivative of the prediction error $\epsilon(t)$ with respect to b_j, and $\gamma_k(t)$ is the opposite of the derivative of the prediction error $\epsilon(t)$ with respect to c_k.

We now work out the equations for the computation of $\alpha_i(t)$. To this purpose, we perform the derivative of the two sides of Eq. (5.19) with respect to parameter a_i. The derivative of the left-hand side of that equation with respect to a_i is $-C(z)\alpha_i(t)$. As for the derivative of the right-hand side, recall that

$$A(z)y(t) = y(t) - a_1 y(t-1) - \cdots - a_{na}y(t-n_a),$$

so that the derivative is $-y(t-i)$. Hence,

$$C(z)\alpha_i(t) = y(t-i).$$

In particular, this implies that there is a chain linking the various derivatives in the sense that

$$\alpha_2(t) = \alpha_1(t-1); \quad \alpha_3(t) = \alpha_1(t-2); \dots.$$

Analogously, if we derive the prediction error equation with respect to b_j, and then with respect to c_k, we obtain

$$C(z)\beta_j(t) = u(t-j),$$
$$C(z)\gamma_k(t) = \epsilon(t-k).$$

Again we see a chain relating the various $\beta_i(t)$ to each other:

$$\beta_2(t) = \beta_1(t-1); \quad \beta_3(t) = \beta_1(t-2); \dots$$

and the various $\gamma_i(t)$ to each other:

$$\gamma_2(t) = \gamma_1(t-1); \quad \gamma_3(t) = \gamma_1(t-2); \dots.$$

These observations suggest to introduction of signals $\alpha(t)$, $\beta(t)$, and $\gamma(t)$ as the solutions of difference equations

$$C(z)\alpha(t) = y(t),$$
$$C(z)\beta(t) = u(t),$$
$$C(z)\gamma(t) = \epsilon(t).$$

In this way, we can write

$$\alpha_i(t) = \alpha(t - i),$$
$$\beta_j(t) = \beta(t - j),$$
$$\gamma_k(t) = \gamma(t - k),$$

and compose the gradient vector $\psi(t)$ from filtered data as

$$\psi(t) = [\alpha(t - 1) \cdots \alpha(t - n_a) \; \vdots \; \beta(t - 1) \cdots \beta(t - n_b) \; \vdots \; \gamma(t - 1) \cdots \gamma(tn_c)]'.$$
(5.22)

We sum up the ARMAX model estimation algorithm by detailing the passage from step r to step $r + 1$.

a) At step r, the estimate $\theta^{(r)}$ of vector θ is available. With the entries of $\theta^{(r)}$, the estimates $A(z)^{(r)}, B(z)^{(r)}$, and $C(z)^{(r)}$ of polynomials $A(z), B(z)$, and $C(z)$ can be constructed.

b) Filter the available data $y(\cdot)$ and $u(\cdot)$ to work out signals $\alpha(\cdot)$ and $\beta(\cdot)$ as

$$C(z)^{(r)}\alpha(t) = y(t),$$
$$C(z)^{(r)}\beta(t) = u(t).$$

We obtain $\alpha(\cdot)^{(r)}, \beta(\cdot)^{(r)}$.

c) Find signal $\epsilon(t)$ by data filtering as

$$C(z)^{(r)}\epsilon(t) = A(z)^{(r)}y(t) - B(z)^{(r)}u(t - 1).$$

We obtain $\gamma(\cdot)^{(r)}$.

d) With $\alpha(\cdot)^{(r)}, \beta(\cdot)^{(r)}$, and $\gamma(\cdot)^{(r)}$, fill the entries of the vector $\psi(\cdot)^{(r)}$, as

$$\psi(t) = [\alpha(t - 1)^{(r)} \cdots \alpha(t - n_a)^{(r)} \; \vdots \; \beta(t - 1)^{(r)} \cdots \beta(t - n_b)^{(r)} \; \vdots$$
$$\gamma(t - 1)^{(r)} \cdots \gamma(t - n_c)^{(r)}]'.$$

e) By means of $\epsilon(\cdot)^{(r)}$ (point (c)) and $\psi(\cdot)^{(r)}$ (point (d)), update the parameter vector with the Gauss–Newton formula:

$$\theta^{(r+1)} = \theta^{(r)} + \left(\sum_1^N \psi_t(t)^{(r)}\psi(t)^{(r)'} \right)^{-1} \left(\sum_1^N \epsilon_t(t)^{(r)}\psi(t)^{(r)} \right).$$

f) Iterate the procedure up to convergence.

The various filtering actions taking place in this procedure are summarized in Figure 5.4.

Remark 5.4 (Comparing two vectors: $\psi(t)$ vs $\phi(t)$) The special structure of the vector $\psi(t)$

$$\psi(t) = [\alpha(t - 1) \cdots \alpha(t - n_a) \; \vdots \; \beta(t - 1) \cdots \beta(t - n_b) \; \vdots \; \gamma(t - 1) \cdots \gamma(tn_c)]'$$

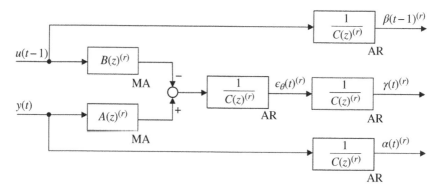

Figure 5.4 Data filtering for the estimation of ARMAX models.

is reminiscent of the structure of the *observation vector* introduced in the least squares method:

$$\phi(t) = [y(t-1) \cdots y(t-n_a) \vdots u(t-1) \cdots u(t-n_b)]'.$$

The passage from $\phi(t)$ to $\psi(t)$ can be outlined as follows. First, filter the original data $y(\cdot)$ and $u(\cdot)$ according to

$$C(z)^{(r)} \epsilon(t) = A(z)^{(r)} y(t) - B(z)^{(r)} u(t-1),$$

so as to work out the samples of the prediction error $\epsilon(\cdot)$ estimated at step r of the iterative procedure. Then construct the vector

$$\phi_{\mathrm{E}}(t) = [y(t-1) \cdots y(t-n_a) \vdots u(t-1) \cdots u(t-n_b) \vdots \epsilon(t-1) \cdots \epsilon(t-n_c)]',$$

which we may call the *extended observation vector*. We notice in passing that the previous expression for the computation of the prediction error can be rewritten as

$$\epsilon(t) = y(t) - \phi_{\mathrm{E}}(t)' \theta^{(r)}.$$

The generic entry of $\psi(t)$ is given by the corresponding entry of $\phi_{\mathrm{E}}(t)$ filtered through the same transfer function, namely $1/C(z)^{(r)}$

$$\psi(t) = \frac{1}{C(z)^{(r)}} \phi_{\mathrm{E}}(t).$$

Note that, while $\phi(t)$ depends upon data only, $\psi(t)$ also depends upon the estimated model.

Remark 5.5 (Numerical issues)   As we have seen, the computations required in the maximum likelihood algorithm call for various filtering actions, of MA type as well as AR type, as can be seen in Figure 5.4. Indeed, to work out the prediction error $\epsilon(\cdot)$, we have to filter $y(\cdot)$ through $A(z)/C(z)$ and

$u(\cdot)$ through $B(z)/C(z)$; moreover, in the computation of the entries α, β, and γ of vector ψ, $y(\cdot)$, $u(\cdot)$, and $\epsilon(\cdot)$ have to be filtered through $1/C(z)$.

Interestingly enough, all these transfer functions have the same denominator, polynomial $C(z)$. If, by any chance, at iteration r, any of such poles falls outside the unit disk, then the data filtering system becomes unstable and the results are not reliable. This is why, at each iteration, polynomial $C(z)$ is to be monitored, and, in case of instability, be replaced by a stable surrogate before proceeding to the subsequent step. This remedy is known as *projection* of polynomial $C(z)$.

Remark 5.6 (ML denomination)   The term *maximum likelihood* is borrowed from statistics, where the maximum likelihood estimation method has been conceived to find the best estimate of the parameters of a probabilistic model. The rationale of the method is: take as best parameters those maximizing the joint probability density function of observations. It can be seen that, under the Gaussian assumption, this principle applied to an ARMAX model leads to the prediction error technique introduced above.

Example 5.6 (Identification of an ARX and ARMAX model from simulated data)   The data generation mechanism

$$\mathscr{S} : y(t) = 0.3y(t-1) + u(t-1) + \eta(t) + 0.5\eta(t-1), \quad \eta(\cdot) \sim \mathrm{WN}(0,1)$$

is subject to the exogenous input $u(\cdot)$, which is also a white noise of unit variance, $u(\cdot) \sim \mathrm{WN}(0,1)$, uncorrelated with $\eta(\cdot)$. With such an input, the system is simulated so as to generate 1000 output snapshots. The pairs $\{u(t), y(t) | t = 1, 2, \ldots, 1000\}$ constitute the set of data for the estimation process. First, we identify an ARX(1,1) model:

$$\mathscr{M} : \hat{y}(t) = ay(t-1) + bu(t-1) + \xi(t), \quad \xi(\cdot) \sim \mathrm{WN}.$$

By means of the least squares method, we obtain

$$\hat{a} = 0.499,$$
$$\hat{b} = 1.000.$$

To assess the validity of the estimated model, we investigate the features of its prediction error, $\epsilon(t) = y(t) - \hat{y}(t)$. To this end, we estimate the covariance function of $\epsilon(t)$ as

$$\hat{\gamma}(r) = \frac{1}{N} \sum_{t}^{N-\tau} \epsilon(t)\epsilon(t+\tau),$$

for r ranging from 1 to 30, and then compute the estimated normalized covariance function

$$\hat{\rho}(\tau) = \frac{\hat{\gamma}(\tau)}{\hat{\gamma}(0)}$$

Table 5.1 Iterative ML identification of an ARMAX(1, 1, 1) model.

Iteration (r)	$\hat{a}^{(r)}$	$\hat{b}^{(r)}$	$\hat{c}^{(r)}$	$\hat{\lambda}^2$
1*	−0.400	−0.400	1.500	—
2	0.244	0.961	0.636	3.551
3	0.2815	1.014	0.530	1.021
4	0.290	1.019	0.509	1.000
5	0.292	1.019	0.507	1.000

again for r ranging from 1 to 30. The Anderson test at 5% is then applied. For $\alpha = 0.05$, the corresponding confidence strip is $\pm\beta = \pm0.0629$. In the performed simulation, in the range $r = 1, 2, \dots, 30$, one can count four samples of $\hat{\rho}(r)$ outside such confidence strip, a number exceeding 5% of 30, meaning that $\epsilon(\cdot)$ should not be seen as a white process.

The same data $\{u(t), y(t)|t = 1, 2, \dots, 1000\}$ have been used to fit an ARMAX(1,1,1) model:

$$\mathcal{M} : y(t) = ay(t-1) + bu(t-1) + \xi(t) + \xi(t-1), \quad \xi(\cdot) \sim \text{WN},$$

whose parameters have been estimated by the iterative (maximum likelihood) method above seen. With a null initialization, i.e. $a^{(0)} = 0$, $b^{(0)} = 0$, $c^{(0)} = 0$, the obtained estimates are shown in Table 5.1.

It turns out that the polynomial $C(z)$ estimated at the first iteration, $r = 1$, is unstable (the absolute value of $c^{(1)}$ exceeds 1). At iteration 1, therefore, a projection mechanism has been activated before proceeding to the filtering procedure for the determination of the entries of vector $\psi(t)$ from data. To be precise, we have replaced the estimated parameter with its inverse, so as to keep the same spectrum. In the table, the symbol $*$ is used to denote the intervention of the projection device. From the analysis of the obtained results, one can see that convergence is reached in a few iterations.

In the last column of the table, the variance of the estimated prediction error is also reported for completeness. Finally, we report that, with the estimated model, the whiteness test is passed.

5.5 Asymptotic Analysis

In the predictive approach, the optimal model in a given family

$$\{\mathcal{M}(\theta)|\theta \in \Theta\}$$

is obtained by minimizing the criterion

$$J_N(\theta) = \frac{1}{N} \sum_{\tau}^{N} \epsilon_\theta(t)^2,$$

where

$$\epsilon_\theta(t) = y(t) - \hat{y}_\theta(t)$$

is the prediction error of model $\mathcal{M}(\theta)$. As usual, we use the symbol $\hat{\theta}$ or $\hat{\theta}_N$ to denote the point of minimum of the criterion.

In the last sections above, we have seen how to elaborate data to find the parameter estimate. We address now the following question: Can we assess the properties of the obtained estimate? To this purpose, we change our paradigm of study: Rather than considering a specific sequence of data, we have to figure out what happens in general, by taking into consideration *all possible sequences.* This is why we now look at data as the effect of some *random event*, the outcome of which is denoted by s. This means that there is a sequence of data for a certain outcome, a different sequence for another outcome, and so on. Correspondingly, the prediction error is a random sequence too: $\epsilon_\theta(t) = \epsilon_\theta(t,s)$. In this framework, $J_N(\theta)$, as a function of θ, is not a single curve; it is a *set of curves* each of which is associated to the outcome s of the random event: $J_N(\theta) = J_N(\theta, s)$. In Figure 5.5, we see this set forming a pencil of functions for a certain number of data points. If we assume that the set of possible sequences of data form a stationary process, then we expect that

$$J_N(\theta) = J_N(\theta, s) \rightarrow \bar{J}(\theta) = E[\epsilon^2(t)].$$

Note that, the operator $E[\cdot]$ performs an average over all possible outcomes, so that $\bar{J}(\theta)$ is a deterministic function. This means that the pencil of performance index curves becomes more and more thin as the number of data grows; in the long run, when $N \rightarrow \infty$, the pencil collapses into a *unique deterministic curve*, as indicated in Figure 5.5.

Focus now on the set of points of minimum of the curves. Each curve has its own minimum, so that we have the set

$$\hat{\theta}_N = \hat{\theta}_N(s).$$

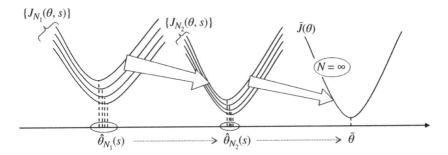

Figure 5.5 Performance index convergence.

This set is nothing else than a random vector. What happens to the sequence $\hat{\theta}_N$ as the number of data tends to infinity? Under mild assumptions, since the pencil tends to $\bar{J}(\theta)$, the random vector $\hat{\theta}_N$, minimizing $J_N(\theta)$, will tend to $\bar{\theta}$, the minimum point of the deterministic curve $\bar{J}(\theta)$; in case $\bar{J}(\theta)$ has multiple global minima, then $\hat{\theta}_N$ will tend to the set Δ of points of minimum of $\bar{J}(\theta)$. See again Figure 5.5.

We are led to the following conclusion: To analyze the properties of the estimate, we can study its asymptotic features, namely we analyze the asymptotic estimated model, $\mathcal{M}(\bar{\theta})$, or the set of possible asymptotic estimated models, namely $\{\mathcal{M}(\theta) | \theta \in \Delta\}$.

Example 5.7 (Yule–Walker equations revisited) Consider a zero-mean stationary process $y(t)$ characterized by a covariance function $\gamma_{yy}(\tau)$. To fit an AR(2) model, a possibility is to resort to the Yule–Walker equations as described in Section 1.9.2. We adopt now the prediction error identification rationale. For example, write the model in predictive form:

$$\hat{y}(t|t-1) = a_1 y(t-1) + a_2 y(t-2).$$

The parameters are determined by minimizing

$$\bar{J} = E[(y(t) - a_1 y(t-1) - a_2 y(t-2))^2].$$

A simple computation shows that

$$\bar{J} = (1 + a_1^2 + a_2^2)\gamma_{yy}(0) + 2(a_1 a_2 - a_1)\gamma_{yy}(1) - 2a_2\gamma_{yy}(2).$$

By deriving $\bar{J}$ with respect to a_1 and a_2 and setting the derivative to 0, we obtain

$$\gamma_{yy}(0)a_1 + \gamma_{yy}(1)a_2 - \gamma_{yy}(1) = 0,$$
$$\gamma_{yy}(0)a_2 + \gamma_{yy}(1)a_1 - \gamma_{yy}(2) = 0.$$

We let our readers compute the second derivatives of $\bar{J}$ with respect to the parameters and show that $(\hat{a}_1, \hat{a}_2)$ is indeed a point of minimum of J.

Note that these equations coincide with the Yule–Walker equations of Section 1.9.2. The solution is still given by formulas (1.12) and (1.13).

5.5.1 Data Generation System Within the Class of Models

An interesting case of study arises when the data generation mechanism $\mathcal{S}$ belongs to the family of models adopted in the identification process. This means that, within the family of models, there is a parameter vector, say θ°, such that $\mathcal{M}(\theta^\circ) = \mathcal{S}$. The question is then: If one resorts to a prediction error method, does the estimated model tend to system $\mathcal{S}$? This is equivalent to asking whether $\hat{\theta}_N$ asymptotically tends to θ°.

As an example, assume that the data generation mechanism is

$$\mathcal{S} : A^\circ(z)y(t) = B^\circ(z)u(t-1) + C^\circ(z)\eta(t),$$
$$\eta(\cdot) \sim WN(0,1),$$

where $A^\circ(z)$, $B^\circ(z)$, and $C^\circ(z)$ are polynomials evaluated in the system parameter vector θ°, while the family of models is given by

$$\mathcal{M} : A(z)y(t) = B(z)u(t-1) + C(z)\xi(t),$$
$$\xi(\cdot) \sim WN(0,\lambda^2),$$

with polynomials $A(z)$, $B(z)$, and $C(z)$ of the same orders as $A^\circ(z)$, $B^\circ(z)$, and $C^\circ(z)$, respectively. Note that, in view of expression (5.19), the prediction error $\epsilon_\theta(t)$ of the model is given by

$$\epsilon_\theta(t) = \frac{A(z)}{C(z)}y(t) - \frac{B(z)}{C(z)}u(t-1).$$

In particular, when we use here polynomials $A^\circ(z)$, $B^\circ(z)$, and $C^\circ(z)$, then we obtain the prediction error of the system generating data, namely $\epsilon_{\theta^\circ}(t) = \eta(t)$; otherwise, $\epsilon_\theta(t)$ is not, in general, a white process.

Let's now return to the general question we posed: Does the estimated model $\mathcal{M}(\hat{\theta}_N)$ asymptotically tend to $\mathcal{S}$?

To reply, consider the definition of prediction error

$$\epsilon_\theta(t) = y(t) - \hat{y}_\theta(t)$$

and add and subtract $\hat{y}_\theta^\circ(t)$

$$\epsilon_\theta(t) = (y(t) - \hat{y}_{\theta^\circ}(t)) + (\hat{y}_{\theta^\circ}(t) - \hat{y}_\theta(t)).$$

Consequently,

$$\bar{J}(\theta) = E[\epsilon_\theta(t)^2]$$
$$= E[(y(t) - \hat{y}_{\theta^\circ}(t))^2] + E[(\hat{y}_{\theta^\circ}(t) - \hat{y}_\theta(t))^2]$$
$$+ 2E[(y(t) - \hat{y}_{\theta^\circ}(t))((\hat{y}_{\theta^\circ}(t) - \hat{y}_\theta(t))].$$

Signal $y(t)$ is the output of the mechanism generating the data; moreover, $\hat{y}_{\theta^\circ}(t)$ is the predictor of the same system. When such mechanism is known, the corresponding prediction error $y(t) - \hat{y}_{\theta^\circ}(t)$ is uncorrelated with past data. For instance, in the above example, $y(t) - \hat{y}_{\theta^\circ}(t) = \eta(t) \sim WN$. Second, we observe that both $\hat{y}_{\theta^\circ}(t)$ and $\hat{y}_\theta(t)$ are predictors, and hence depend upon past data only. Therefore, $(y(t) - \hat{y}_\theta^\circ(t))$ and $(\hat{y}_{\theta^\circ}(t) - \hat{y}_\theta(t))$ are uncorrelated, so that $E[(y(t) - \hat{y}_{\theta^\circ}(t))(\hat{y}_{\theta^\circ}(t) - \hat{y}_\theta(t))] = 0$. Hence,

$$\bar{J}(\theta) = E[\epsilon_\theta(t)^2]$$
$$= E[(y(t) - \hat{y}_{\theta^\circ}(t))^2] + E[(\hat{y}_{\theta^\circ}(t) - \hat{y}_\theta(t))^2].$$

$E[(\hat{y}_{\theta^\circ}(t) - \hat{y}_\theta(t))^2]$ being non-negative, we have

$$\bar{J}(\theta) \geq E[(y(t) - \hat{y}_{\theta^\circ}(t))^2] = \bar{J}(\theta^\circ).$$

Therefore, θ° is a point of minimum of $\bar{J}(\theta)$. If $\bar{J}(\theta)$ has a unique global minimum, then we can claim that θ° is that point, so that the prediction error identification rationale leads to the correct estimate. If instead $\bar{J}(\theta)$ has a multiplicity of points of global minimum, then the estimate can converge either to θ° or to another point, say $\bar{\theta}$. However, the predictive performance is always the same, namely $\bar{J}(\bar{\theta}) = \bar{J}(\theta^\circ)$.

5.5.2 Data Generation System Outside the Class of Models

The data generator may not belong to the family of models. In practical problems, this happens very often, since a simple linear discrete-time model can provide only an approximated description of a real-world process. For example, the main variables of a heat exchanger are related to each other via nonlinear partial differential equations, so that an ARX or an ARMAX model can provide a rough description only. Notwithstanding this (obvious) limitation, such models are frequently used for two main reasons: first, they are apt for design purposes, e.g. to set up a suitable controller; second, they can be easily worked out from data via a simple identification procedure.

We illustrate this case by analyzing a few simple situations.

Example 5.8 (Estimating an AR model from an MA time series) A time series is generated by the MA system

$$\mathscr{S} : y(t) = \eta(t) + c\eta(t-1),$$
$$\eta(\cdot) \sim \text{WN}(0, \lambda^2), \quad \lambda^2 \neq 0, \quad c \neq 0.$$

Whatever the value of parameter c, the sequence $y(\cdot)$ is a zero-mean stationary process with covariance function

$$\gamma_{yy}(0) = (1 + c^2)\lambda^2,$$
$$\gamma_{yy}(1) = c\lambda^2.$$

As model we take an AR(1), namely, in prediction form

$$\mathscr{M}(\theta) = \mathscr{M}(a) : \quad \hat{y}(t) = ay(t-1).$$

To estimate the best parameter, we refer to the criterion

$$\bar{J} = E[(y(t) - \hat{y}(t))^2].$$

Now

$$\bar{J} = E[(y(t) - ay(t-1))^2]$$
$$= E[y(t)^2] + a^2 E[y(t-1)^2] - 2aE[y(t)y(t-1)];$$

being $E[y(t)^2] = E[y(t-1)^2] = \gamma_{yy}(0)$ and $E[y(t)y(t-1)] = \gamma_{yy}(1)$, we have

$$\bar{J} = (1 + a^2)\gamma_{yy}(0) - 2a\gamma_{yy}(1)$$

the minimum of which is

$$\bar{a} = \frac{\gamma_{yy}(1)}{\gamma_{yy}(0)}.$$

The data are constituted by the MA(1) time series, so that

$$\bar{a} = \frac{\gamma_{yy}(1)}{\gamma_{yy}(0)} = \frac{c}{1 + c^2}.$$

Note that $\bar{a}$ depends upon parameter c but is not influenced by variance λ^2.

The model associated with parameter $\bar{a}$ has to be seen as the optimal one within the AR(1) family, in that it provides the best data fitting in terms of prediction error variance.

Is there a way to understand that the best AR(1) model is not the "true" data generation mechanism? A possibility is to analyze the prediction error

$$\epsilon(t) = y(t) - \hat{y}(t) = y(t) - \bar{a}y(t-1).$$

Here, $y(t)$ is a linear combination of $\eta(t)$ and $\eta(t-1)$, while $y(t-1)$ is a linear combination of $\eta(t-1)$ and $\eta(t-2)$. Hence, $\epsilon(t)$ is a linear combination of $\eta(t), \eta(t-1)$, and $\eta(t-2)$, namely it is a moving average process of order 2, not a white process.

Remark 5.7 (On the stability of the identified model) It is easy to verify that, for each value of c, the estimated parameter $\bar{a}$ is in modulus less than 1. In other words, the optimal AR(1) model is stable. This fact is indeed expected as the given process $y(\cdot)$ is stationary. If the estimated model were unstable, it would generate a nonstationary process, so providing a fully unrealistic description of the given data.

Remark 5.8 (Identification and spectral factorization) Suppose that the original time series is generated by

$$\mathcal{S} : y(t) = \eta(t) + \frac{1}{c}\eta(t-1),$$
$$\eta(\cdot) \sim \text{WN}(0, \lambda^2), \qquad \lambda^2 \neq 0, \quad c \neq 0.$$

Then

$$\bar{a} = \frac{\gamma_{yy}(1)}{\gamma_{yy}(0)} = \frac{\frac{1}{c}}{1 + (\frac{1}{c})^2} = \frac{c}{1 + c^2}.$$

This means that the optimal AR model coincides with the one previously obtained in Example 5.8. The reason is that processes $\eta(t) + \frac{1}{c}\eta(t-1)$ and

$\eta(t) + c\eta(t-1)$ are indeed the same process in that they have the same spectrum up to a proportionality coefficient.

Example 5.9 (Estimating ARX for ARMAX) In this example, we probe the performance of the ARX(1,1) family of models in the estimation of two different types of systems: an ARX(1,1) and an ARMAX(1,1,1).

We start by writing the model in predictive form

$$\hat{\mathcal{M}}(\theta) : \hat{y}_\theta(t) = ay(t-1) + bu(t-1),$$
$$\theta = [ab]'.$$

To estimate the parameters, we resort to least squares, the prediction error minimization method tailored to ARX models. Assuming that processes $u(\cdot)$ and $y(\cdot)$ are stationary with zero mean, we can apply the asymptotic rationale presented above and conclude that the estimate, $\hat{\theta}_N = [\hat{a}_N \hat{b}_N]'$, will tend to one of the points of minimum of

$$\bar{J}(\theta) = E[(y(t) - \hat{y}_\theta(t))^2]$$
$$= E[y(t)^2] + E[\hat{y}_\theta(t)^2] - 2E[y(t)\hat{y}_\theta(t)].$$

By introducing the covariance functions

$$\gamma_{uu}(\tau) = E[u(t)u(t+\tau)],$$
$$\gamma_{yy}(\tau) = E[y(t)y(t+\tau)],$$
$$\gamma_{uy}(\tau) = E[u(t)y(t+\tau)],$$

and assuming that $E[y(t)u(t)] = 0$, we have

$$E[\hat{y}_\theta(t)^2] = a^2 E[y(t-1)^2] + b^2 E[u(t-1)^2] = a^2\gamma_{yy}(0) + b^2\gamma_{uu}(0).$$

Moreover,

$$E[y(t)\hat{y}_\theta(t)] = a\ E[y(t)y(t-1)] + bE[y(t)u(t-1)]$$
$$= a\ \gamma_{yy}(1) + b\gamma_{uy}(1).$$

Therefore,

$$\bar{J}(\theta) = (1+a^2)\ \gamma_{yy}(0) + b^2\ \gamma_{uu}(0) - 2a\gamma_{yy}(1) - 2b\ \gamma_{uy}(1),$$

the point of minimum of which is

$$\bar{a} = \frac{\gamma_{yy}(1)}{\gamma_{yy}(0)}, \quad \bar{b} = \frac{\gamma_{uy}(1)}{\gamma_{uu}(0)}.$$

We now apply this result to the following cases:

(a) Data generated by an ARX(1,1) system
(b) Data generated by ARMAX(1,1,1) system

(a) If the data generation mechanism is the ARX(1,1) system

$$\mathscr{S} : y(t) = a^\circ y(t-1) + b^\circ u(t-1) + \eta(t),$$
$$u(\cdot) \sim \text{WN}(0, \mu^2), \quad \mu^2 \neq 0,$$
$$\eta(\cdot) \sim \text{WN}(0, \lambda^2), \quad \lambda^2 \neq 0,$$
$$u(\cdot) \quad \text{and} \quad e(\cdot) \quad \text{uncorrelated},$$

with $|a^\circ| < 1$, then the output is a stationary process (with zero mean). In particular, $y(\cdot)$ at time t depends upon the input $u(\cdot)$ up to time $t-1$. $u(\cdot)$ being a white noise, the condition $E[y(t)u(t)] = 0$ is satisfied and the formulas above derived for $\bar{a}$ and $\bar{b}$ are valid.

We let our readers show that $\gamma_{uy}(1)$ and $\gamma_{yy}(1)$ are related to each other as

$$\gamma_{uy}(1) = b^\circ \gamma_{uu}(0),$$
$$\gamma_{yy}(1) = a^\circ \gamma_{yy}(0).$$

Therefore,

$$\bar{a} = \frac{\gamma_{yy}(1)}{\gamma_{yy}(0)} = a^\circ, \quad \bar{b} = \frac{\gamma_{uy}(1)}{\gamma_{uu}(0)} = b^\circ.$$

Thus, asymptotically, $\hat{a}_N$ converges to a° and $\hat{b}_N$ converges to b°, namely the estimate is *consistent*. This conclusion is in agreement with the previous analysis regarding the case when the data generator belongs to the family of models. We also note that the prediction error of the asymptotically estimated model obviously coincides with $\eta(t)$ and is therefore a white process.

(b) If the data generation mechanism is the ARMAX(1,1,1) system

$$\mathscr{S} : y(t) = a^\circ y(t-1) + b^\circ u(t-1) + \eta(t) + c^\circ \eta(t-1),$$

$$u(\cdot) \sim \text{WN}(0, \mu^2), \quad \mu^2 \neq 0,$$

$$\eta(\cdot) \sim \text{WN}(0, \lambda^2), \quad \lambda^2 \neq 0,$$

$$u(\cdot) \text{ and } \eta(\cdot) \text{ uncorrelated},$$

with $|a^\circ| < 1$, then the output constitute a zero-mean stationary process; moreover, the condition $E[y(t)u(t)] = 0$ is satisfied.

We again leave to our readers the computation of covariance functions $\gamma_{uy}(1)$ and $\gamma_{yy}(1)$:

$$\gamma_{uy}(1) = b^\circ \gamma_{uu}(0),$$

$$\gamma_{yy}(1) = a^\circ \gamma_{yy}(0) + c^\circ \text{Var}[\eta].$$

In conclusion,

$$\bar{a} = \frac{\gamma_{yy}(1)}{\gamma_{yy}(0)} = a^\circ + c^\circ \frac{\text{Var}[\eta]}{\text{Var}[y]}, \quad \bar{b} = b^\circ.$$

We see that, while the estimate of $\hat{b}_N$ is consistent, $\hat{a}_N$ does not tend to a°, the difference being proportional to the noise-to-signal ratio $\text{Var}[\eta]/\text{Var}[y]$ with a weighting coefficient given by c°. Hence, the estimate $\hat{a}_N$ of parameter a is consistent only if $c^\circ = 0$, a case when the ARMAX data generator is indeed an ARX system (so that the generator belongs to the family of models).

The detailed expression of $\bar{a}$ can be worked out by computing covariances $\gamma_{yy}(0)$ and $\gamma_{yy}(1)$:

$$\gamma_{yy}(0) = \frac{(b^\circ)^2 \mu^2 + (1 + (c^\circ)^2 + 2a^\circ c^\circ)\lambda^2}{1 - (a^\circ)^2},$$

$$\gamma_{yy}(1) = a^\circ \gamma_{yy}(0) + c^\circ \lambda^2.$$

Consequently,

$$\bar{a} = \frac{\gamma_{yy}(1)}{\gamma_{yy}(0)} = a^\circ + \frac{c^\circ \lambda^2}{\gamma_{yy}(0)} = \alpha \mu^2 + \beta \lambda^2.$$

where

$$\alpha = \frac{a^\circ (b^\circ)^2}{(b^\circ)^2 \mu^2 + (1 + (c^\circ)^2 + 2a^\circ c^\circ)\lambda^2}$$

$$\beta = \frac{a^\circ + c^\circ + a^\circ (c^\circ)^2 + (a^\circ)^2 c^\circ}{(b^\circ)^2 \mu^2 + (1 + (c^\circ)^2 + 2a^\circ c^\circ)\lambda^2}$$

5.5.2.1 Simulation Trial

With reference to an ARMAX(1, 1, 1) with $a^\circ = 0.3$, $b^\circ = 0.5$, $c^\circ = 1$, subject to the exogenous input $u(\cdot) \sim \text{WN}(0, 0.1)$ and disturbance $\eta(\cdot) \sim \text{WN}(0, 1)$, generate 300 samples of $u(\cdot)$ and $y(\cdot)$. With least squares, identify an ARX(1, 1) model and then test the prediction error whiteness.

5.5.3 General Considerations on the Asymptotics of Predictive Identification

For the sake of clarity, we now summarize the asymptotic behavior of prediction error identification methods. As discussed in the previous sections of this chapter, according to such methods, once a family of models $\mathfrak{M} = \{\mathcal{M}(\theta)|\theta \in$

Θ} has been selected, the optimal estimate with N data is given by the point of minimum of $J_N(\theta) = (1/N)\sum_t \epsilon_\theta(t)^2$, where $\epsilon_\theta(t)$ is the prediction error associated with model $\mathcal{M}(\theta)$ at time t. Asymptotically, such criterion will tend to $\bar{J}(\theta) = E[\epsilon^2(t)]$ and therefore we expect that the estimate will tend to the point of minimum of $\bar{J}(\theta)$ (see Figure 5.5).

Having denoted by Δ the set of points of minimum of $\bar{J}(\theta)$, we can distinguish four conceptual situations.

(a) $\mathcal{S} \in \mathfrak{M}$, *and Δ consists of a unique point $\bar{\theta}$:* In such a case, as seen in Section 5.5.1, $\bar{\theta}$ is just vector of the parameters $\theta°$ of the system generating data $\mathcal{S} = \mathcal{M}(\theta°)$. Therefore, the prediction error method provides an estimate $\hat{\theta}_N$ asymptotically converging to $\theta°$.

(b) $\mathcal{S} \notin \mathfrak{M}$, *and Δ consists of a unique point $\bar{\theta}$:* This case has been discussed in Section 5.5.2: The parameter estimate will asymptotically tend to a point $\bar{\theta}$, with the corresponding model $\mathcal{M}(\bar{\theta})$ to be seen as the best proxy of the true system in the selected family of models.

(c) $\mathcal{S} \in \mathfrak{M}$, *and Δ is constituted by a multiplicity of points:* Then, in Δ there exists a vector $\theta°$ such that $\mathcal{S} = \mathcal{M}(\theta°)$. The estimate will asymptotically tend to a point in Δ, possibly $\theta°$ or a different point. From a predictive point of view, all models associated to parameters belonging to Δ are equivalent in the sense that their predictive performance is the same.

(d) $\mathcal{S} \notin \mathfrak{M}$, *and Δ is constituted by a multiplicity of points:* The models defined by vectors in Δ are the best proxy of the data generation mechanism in the family. The estimate converges to a point in Δ.

These four situations are schematically depicted in Figure 5.6.

5.5.4 Estimating the Uncertainty in Parameter Estimation

Up to now, we have discussed the asymptotic behavior of the point estimate obtained by prediction error methods. Another important issue concerns the uncertainty affecting the identified parameters. We address now such issue, by assuming that $\mathcal{S} \in \mathfrak{M}$, and Δ consists of a unique point $\bar{\theta}$. Then, we know that $\bar{\theta}$ coincides with parameter vector $\theta°$ associated with the system generating the data ($\mathcal{S} = \mathcal{M}(\theta°)$). Having said that, consider the prediction error $\epsilon(t, \theta)$ of model $\mathcal{M}(\theta)$, and its gradient with respect to θ

$$\psi(t, \theta) = -\frac{d}{d\theta}\epsilon(t, \theta)'. \tag{5.23}$$

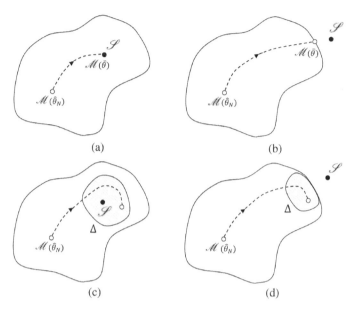

Figure 5.6 Asymptotic behavior of prediction error identification methods.

This is a column vector of dimension given by the number of parameters to be estimated. We suppose that such vector be stationary, so that matrix

$$\bar{R} = E[\psi(t, \theta^\circ)\psi(t, \theta^\circ)'] \tag{5.24}$$

does not depend upon time. It can be shown that, for high values of N, the variance of the estimator $\hat{\theta}_N$ obtained with a prediction error identification method is given by

$$\frac{1}{N}\text{Var}[\epsilon(t, \theta^\circ)]\bar{R}^{-1}. \tag{5.25}$$

The elements along the diagonals of such matrix provide the variance of the estimates of the single parameters entering vector θ. For instance, element $(1, 1)$ of the matrix is the variance of $\hat{\theta}_{1N}$ as N tends to infinity.

The above conclusion can be more precisely stated as follows:

$$\sqrt{N}(\hat{\theta}_N - \theta^\circ) \sim As \quad (0, \bar{P}), \tag{5.26}$$

where

$$\bar{P} = \mathrm{Var}[\epsilon(t, \theta^\circ)]\bar{R}^{-1}.$$

Obviously, formula (5.26) means that, when the number of data tends to ∞, the random vector $\sqrt{N}(\hat{\theta}_N - \theta^\circ)$ has zero mean and variance matrix $\bar{P}$.

In practice, matrix $\bar{R}$ and scalar $\mathrm{Var}[\epsilon(t, \theta^\circ)]$ are replaced by their sample surrogates, namely

$$\frac{1}{N}\sum_1^N \psi(t, \hat{\theta}_N)\psi(t, \hat{\theta}_N)',$$

$$\frac{1}{N}\sum_1^N \epsilon^2(t, \hat{\theta}_N),$$

respectively.

5.5.4.1 Deduction of the Formula of the Estimation Covariance

In search of an expression for $\hat{\theta}_N - \theta^\circ$, consider the vector $\psi_N(t, \theta)$ computed from $J_N(t, \theta)$, as

$$\psi_N(t, \theta) = -\frac{\mathrm{d}}{\mathrm{d}\theta}J_N(t, \theta).$$

The truncated Taylor expansion of $\psi_N(t, \theta)$, as a function of θ, about $\theta = \theta^\circ$ is

$$\psi_N(t, \theta) = \psi_N(t, \theta^\circ) + M_N(t, \theta^\circ)(\theta - \theta^\circ), \tag{5.27}$$

where $M_N(t, \theta)$ denotes the second derivative of $J_N(t, \theta)$ with respect to θ.

Evaluating (5.27) in $\theta = \hat{\theta}_N$, we obtain

$$\psi_N(t, \hat{\theta}_N) = \psi_N(t, \theta^\circ) + M_N(t, \theta^\circ)(\hat{\theta}_N - \theta^\circ).$$

By definition, $\hat{\theta}_N$ is the point of minimum of $J_N(t, \theta)$. Hence, $\psi_N(t, \hat{\theta}_N) = 0$, so that

$$M_N(t, \theta^\circ)(\hat{\theta}_N - \theta^\circ) = -\psi_N(t, \theta^\circ).$$

In light of Section 5.5, we expect that $J_N(\theta)$ converges to $\bar{J}(\theta)$ and that $\hat{\theta}_N$ tends to the unique point of minimum $\bar{\theta} = \theta^\circ$ of $\bar{J}(\theta)$. Under regularity conditions, the second derivative of $J_N(\theta)$ evaluated in $\hat{\theta}_N$ tends to the second derivative of $\bar{J}(\theta)$ in $\bar{\theta} = \theta^\circ$, which we will denote by $\bar{M}(\theta^\circ)$. The minimum being unique by assumption, such matrix will be non-singular. Therefore, for large values of N, we can write

$$(\hat{\theta}_N - \theta^\circ) = -\bar{M}(\theta^\circ)^{-1}\psi_N(t, \theta^\circ).$$

Note that $\bar{M}(\theta^\circ)$ is a symmetric deterministic matrix, so that

$$\lim_{N\to\infty} \mathrm{Var}[\sqrt{N}(\hat{\theta}_n - \theta^\circ)] = \bar{M}(\theta^\circ)^{-1}\bar{K}\bar{M}(\theta^\circ)^{-1}, \tag{5.28}$$

where

$$\bar{K} = \lim_{N \to \infty} \mathrm{Var}[\sqrt{N}\Psi_N(t,\theta^\circ)] = \lim_{N \to \infty} E[N\psi_N(t,\theta^\circ)\psi_N(t,\theta^\circ)'].$$

We now compute matrices $\bar{M}(\theta^\circ)$ and $\bar{K}$. As for $\bar{M}(\theta^\circ)$, we have

$$\bar{M}(\theta^\circ) = \frac{\mathrm{d}^2}{\mathrm{d}\theta^2}\bar{J}(\theta^\circ) = \frac{\mathrm{d}^2}{\mathrm{d}\theta^2}E[\epsilon(t,\theta^\circ)^2]$$

$$= 2E[\psi(t,\theta^\circ)\psi(t,\theta^\circ)'] + 2E\left[\epsilon(t,\theta^\circ)\frac{\mathrm{d}^2}{\mathrm{d}\theta^2}\epsilon(t,\theta^\circ)\right].$$

The second term in this sum can be neglected; indeed, the second derivative of $\epsilon(t,\theta^\circ)$ depends upon $u(\cdot)$ and $y(\cdot)$ only up to $t-1$, whereas $\epsilon(t,\theta^\circ)$ is independent of past. Hence,

$$E\left[\epsilon(t,\theta^\circ)\frac{\mathrm{d}^2}{\mathrm{d}\theta^\circ}\epsilon(t,\theta^\circ)\right] = E[\epsilon(t,\theta^\circ)]E\left[\frac{\mathrm{d}^2}{\mathrm{d}\theta^2}\epsilon(t,\theta^\circ)\right] = 0.$$

Therefore,

$$\bar{M}(\theta^\circ) = 2\bar{R}, \tag{5.29}$$

where

$$\bar{R} = E[\psi(t,\theta^\circ)\psi(t,\theta^\circ)']. \tag{5.30}$$

Passing to $\bar{K}$, we start by reconsidering the expression of performance index

$$J_N(\theta) = \frac{1}{N}\sum_{t}^{N}\epsilon(t,\theta)^2,$$

so that $\Psi_N(t,\theta)$ is given by

$$\Psi_N(t,\theta) = -\frac{\mathrm{d}}{\mathrm{d}\theta}J_N(t,\theta) = \frac{2}{N}\sum_{t}^{N}\epsilon(t,\theta)\psi(t,\theta),$$

and therefore,

$$\psi_N(t,\theta^\circ)\psi_N(t,\theta^\circ)' = \frac{4}{N^2}\left[\sum_{ij}^{N}\epsilon(i,\theta)\epsilon(j,\theta)\psi(i,\theta)\psi(j,\theta)'\right]_{\theta=\theta^\circ}.$$

Since

$$\bar{K} = \lim_{N \to \infty} E[N\psi_N(t,\theta^\circ)\psi_N(t,\theta^\circ)'],$$

we have

$$\bar{K} = \lim_{N \to \infty}\frac{4}{N}E\left[\sum_{ij}^{N}\epsilon(i,\theta)\epsilon(j,\theta)\psi(i,\theta)\psi(j,\theta)'\right]_{\theta=\theta^\circ}.$$

Thus,

$$\bar{K} = 4\text{Var}[\epsilon(t,\theta^\circ)]\bar{R}. \tag{5.31}$$

Combining this expression with (5.29), (5.30), and (5.28), we come to the conclusion that

$$\lim_{N\to\infty} \text{Var}[\hat{\theta}_N - \theta^\circ] = \frac{1}{N}\text{Var}[\epsilon(t,\theta^\circ)]\bar{R}^{-1} \tag{5.32}$$

or, equivalently, formula (5.26).

Remark 5.9 This result implies that the variance of the estimate tends to zero as $1/N$ (and the standard deviation as $1/\sqrt{N}$).

Remark 5.10 (Uncertainty in least squares estimation) In a least squares problem, the predictor is linear in the parameter and takes the expression

$$\hat{y}(t,\theta) = \phi(t)'\theta,$$

so that the prediction error is given by

$$\epsilon(t,\theta) = y(t) - \phi(t)'\theta.$$

Therefore,

$$\psi(t,\theta) = -\frac{d}{d\theta}\epsilon(t,\theta)' = \phi(t). \tag{5.33}$$

Remarkably, this derivative is independent of θ, so that matrix $\bar{R}$ can be determined without knowledge of parameter θ°:

$$\bar{R} = E[\psi(t,\theta^\circ)\psi(t,\theta^\circ)'] = E[\phi(t)\phi(t)']. \tag{5.34}$$

In view of the above result, for a large number of data, $\text{Var}[\hat{\theta}_N - \theta^\circ]$ is given by

$$\frac{1}{N}\text{Var}[\epsilon(t,\theta^\circ)]E[\phi(t)\phi(t)']^{-1}.$$

In practice, this formula is used in its sampled surrogate, namely

$$\frac{1}{N}\left(\frac{1}{N}\sum_1^N \epsilon^2(t,\hat{\theta}_N)\right)\left(\frac{1}{N}\sum_1^N \phi(t)\phi(t)'\right)^{-1}.$$

Example 5.10 (Parameter uncertainty in AR estimation) Let the data be generated by the AR(1) system:

$$\mathcal{S} : y(t) = a^\circ y(t-1) + c\eta(t), \quad \eta(\cdot) \sim \text{WN}(0,\lambda^2), \tag{5.35}$$

with $|a^\circ| < 1$. As model family, we also take the AR(1), so that

$$\mathcal{M} : \hat{y}(t) = ay(t-1)$$

and the associated prediction error is

$$\epsilon(t) = y(t) - \hat{y}(t) = y(t) - ay(t-1).$$

For the estimation of parameter a, we minimize

$$J_N = \frac{1}{N}\sum_{t}^{N}\epsilon(t)^2 = \frac{1}{N}\sum_{t}^{N}(y(t) - ay(t-1))^2.$$

From this formula, we have

$$\frac{d}{da}J_N = -\frac{2}{N}\sum_{t}^{N}(y(t) - ay(t-1))y(t-1).$$

By zeroing this gradient, we obtain

$$\hat{a}_N = \frac{1/N\sum_{t1}^{N}y(t)y(t-1)}{1/N\sum_{t1}^{N}y(t-1)^2}.$$

The denominator of such expression, $1/N\sum_{t1}^{N}y(t-1)^2$, is an estimate $\hat{\gamma}(0)$ of the variance $\gamma(0)$ of $y(\cdot)$, whereas the numerator, $1/N\sum_{t1}^{N}y(t)y(t-1)$, is an estimate $\hat{\gamma}(1)$ of the one-step covariance function $\gamma(1)$ of $y(\cdot)$:

$$\hat{a}_N = \frac{\hat{\gamma}(1)}{\hat{\gamma}(0)}.$$

Asymptotically, we expect that $\lim_{N\to\infty}\hat{\gamma}(0) = \gamma(0)$ and $\lim_{N\to\infty}\hat{\gamma}(1) = \gamma(1)$, so that

$$\lim_{N\to\infty}\hat{a}_N = \frac{\gamma(1)}{\gamma(0)}.$$

On the other hand, we know that in an AR(1) process, $\gamma(1) = a^\circ\gamma(0)$. Therefore, the estimate of parameter a will tend to θ°:

$$\lim_{N\to\infty}\hat{a}_N = a^\circ.$$

This is expected in view of what we have seen in Section 5.5.3.

According to the results presented above on the variance of the estimate, we can now assess the uncertainty in estimation. Since

$$\phi(t) = y(t-1),$$
$$\bar{R} = E[\phi(t)\phi(t)'] = E[y(t-1)^2] = \gamma(0),$$

we have

$$\text{Var}[\hat{a}_N] = \frac{1}{N}\frac{\lambda^2}{\gamma(0)} = \frac{1}{N}\frac{\text{Var}[e(t)]}{\text{Var}[y(t)]},$$

which can be replaced by

$$\frac{1}{N}\left(\frac{1}{N}\sum_{1}^{N}{}_{t}\,\epsilon(t,\hat{\theta}_N)^2\right)\left(\frac{1}{N}\sum_{1}^{N}{}_{t}\,y(t-1)^2\right)^{-1}$$

for the evaluation from data. From this expression, we learn that the uncertainty in estimation is proportional to the noise-to-signal ratio.

5.6 Recursive Identification

In the identification methods seen so far, all snapshots constituting the available set of data, from the first one to the last one, are used to form the estimate. Notice in passing that this holds true even in the case of the *maximum likelihood* method for the estimation of ARMAX models. Indeed, the ML algorithm is iterative, not recursive, in that within each iteration *all snapshots* are elaborated.

There are, however, situations in which the estimate must be done in real time, gradually as the data come in. For instance, in space missions, a main problem is trajectory control; for this purpose, the knowledge of the position of the spacecraft is a basic prerequisite. The position can be determined by the observation of the so-called fixed stars with an on-board camera. Such estimation requires continuous real-time updating to track the changes in the spacecraft location, i.e. it must be performed *online* as new observations become available. By contrast, when all snapshots are used in a batch way, we speak of *off-line* methods or *batch* methods.

For *online* estimation, we resort to *recursive methods*, enabling the formation of the estimate by updating the previous estimate in light of the last available snapshot. We shall introduce the following methods: *recursive least squares* (RLS), *recursive maximum likelihood* (RML), and *extended least squares* (ELS). The first one is related to ARX models; the remaining ones to ARMAX models.

5.6.1 Recursive Least Squares

The LS batch formula is

$$\hat{\theta}_t = S(t)^{-1}\sum_{1}^{t}{}_{i}\,\phi(i)y(i),$$

where the auxiliary matrix $S(t)$ is given by

$$S(t) = \sum_{1}^{t}{}_{i}\,\phi(i)\phi(i)'.$$

Here, the sum of terms $\phi(i)y(i)$ from $i = 1$ to $i = t$ appears. We split it into the sum up to instant $t - 1$, plus the remaining term containing the last snapshot $y(t)$:

$$\sum_{1}^{t}{}_i \phi(i)y(i) = \sum_{1}^{t-1}{}_i \phi(i)y(i) + \phi(t)y(t).$$

On the other hand, if we write the LS batch formula at time $t - 1$, we have

$$\sum_{1}^{t-1}{}_i \phi(i)y(i) = S(t - 1)\hat{\theta}_{t-1}.$$

Thanks to this expression, we obtain

$$\sum_{1}^{t}{}_i \phi(i)y(i) = S(t - 1)\hat{\theta}_{t-1} + \phi(t)y(t).$$

We now focus on the auxiliary matrix, where the sum of matrices $S(i)$ from $i = 1$ to $i = t$ appears. We split it into the same sum up to instant $t - 1$, plus the remaining term $\phi(t)\phi(t)'$:

$$S(t) = S(t - 1) + \phi(t)\phi(t)'.$$

Thus, we can write $S(t - 1)$ as $S(t - 1) = S(t) - \phi(t)\phi(t)'$, so that the sum of terms $\phi(i)y(i)$ from $i = 1$ to $i = t$ becomes

$$\sum_{1}^{t}{}_i \phi(i)y(i) = (S(t) - \phi(t)\phi(t)')\hat{\theta}_{t-1} + \phi(t)y(t).$$

Substituting into the LS batch formula, we obtain the RLS formula:

RLS (I form)

$$\hat{\theta}_t = \hat{\theta}_{t-1} + K(t)\epsilon(t),$$
$$K(t) = S(t)^{-1}\phi(t),$$
$$\epsilon(t) = y(t) - \phi(t)'\hat{\theta}_{t-1},$$
$$S(t) = S(t - 1) + \phi(t)\phi(t)'.$$

Here, the first and the last equations are the two basic recursions, providing the updating for the parameter estimate $\hat{\theta}_t$ and for the auxiliary matrix $S(t)$.

As we see from the first equation, the variation $\hat{\theta}_t - \hat{\theta}_{t-1}$ is given by $K(t)\epsilon(t)$. Hence, if the prediction error at time t is null, passing from time $t - 1$ to time t, the estimate keeps unchanged. This is rather intuitive: When the model estimated at time $t - 1$ perfectly predicts $y(t)$, there is no reason to modify the previous estimate of the parameter vector. If instead $\epsilon(t) \neq 0$, the estimate is modified by a quantity proportional to the value taken by $\epsilon(t)$, premultiplied by vector $K(t)$. Therefore, $K(t)$, named *algorithm gain*, determines how a non-null prediction error influences the various entries of the parameter vector.

Passing to the recursion of matrix $S(t)$, we observe that $\phi(t)\phi(t)'$ is positive semi-definite, so that $S(t) \geq S(t-1)$ (meaning that $S(t) - S(t-1)$ is positive semi-definite, see Appendix B). Therefore, if at a time point, for instance k, matrix $S(k)$ is invertible, then $S(t)$ is invertible at any time point $t \geq k$.

The initialization of the RLS algorithm can be made by means of a bunch of initial data, say from $t = 1$ up to a certain instant t_0. With such set of snapshots, one computes $S(t_0)$, thanks to the batch LS formula:

$$S(t_0) = \sum_{1}^{t_0} {}_i \phi(i)\phi(i)'.$$

This provides the initial value for the recursion in $S(t)$. Then, the recursion on $\hat{\theta}_t$ is initialized as

$$\hat{\theta}_{t_0} = S(t_0)^{-1} \sum_{1}^{t_0} {}_i \phi(i)y(i).$$

Remark 5.11 (Invertibility of $S(t)$)   Above we have implicitly assumed $S(t_0)$ to be invertible. In this regard, we note that the rank of matrix $\phi(i)\phi(i)'$ is equal to 1 (see Appendix B). So, indicating by n the total number of unknown parameters, $S(t_0)$ is an $n \times n$ matrix constructed as the sum of t_0 matrices of rank 1. In order to be invertible, or equivalently of rank n, a necessary condition is that $t_0 \geq n$.

RLS can be given various forms. Another useful expression is obtained by referring to matrix

$$R(t) = (1/t)S(t)$$

in place of $S(t)$. From the auxiliary equation in $S(t)$, we have

$$(1/t)S(t) = (1/t)S(t-1) + (1/t)\phi(t)\phi(t)'$$
$$= (t-1/t)(1/t-1)S(t-1) + (1/t)\phi(t)\phi(t)',$$

so that

$$R(t) = (t-1/t)R(t-1) + (1/t)\phi(t)\phi(t)'.$$

Therefore,

RLS (II form)

$$\hat{\theta}_t = \hat{\theta}_{t-1} + K(t)\epsilon(t),$$
$$K(t) = (1/t)R(t)^{-1}\phi(t),$$
$$\epsilon(t) = y(t) - \phi(t)'\hat{\theta}_{t-1},$$
$$R(t) = R(t-1) + (1/t)(\phi(t)\phi(t)' - R(t-1)).$$

Both the first and second form of RLS suffer from a numerical drawback: at each time point, it is necessary to invert a square matrix of the same dimension of the parameter vector, the matrix $S(t)$ in the first form and $R(t)$ in the second form. Fortunately, it is possible to rewrite the algorithm with an inversion limited to a scalar quantity instead of a matrix. This possibility is offered by the following result (see Appendix B for more details).

Matrix inversion lemma

Consider four matrices F, G, H, and K of suitable dimensions such that the square matrix $F + GHK$ can be formed according to the matrix multiplication rule. Suppose that F and H are invertible. Then

$$(F + GHK)^{-1} = F^{-1} - F^{-1}G(H^{-1} + KF^{-1}G)^{-1}KF^{-1}.$$

We now apply this lemma to the expression $S(t) = S(t-1) + \phi(t)\phi(t)'$ of the auxiliary matrix recursion, under the assumption that $S(t-1)$ is invertible. To this aim, we set $F = S(t-1)$, $G = \phi(t)$, $H = 1$, and $K = \phi(t)'$. In this way,

$$S(t)^{-1} = S(t-1)^{-1} + S(t-1)^{-1}\phi(t)(1 + \phi(t)'S(t-1)^{-1}\phi(t))^{-1}\phi(t)'S(t-1)^{-1}.$$

Then, by introducing

$$V(t) = S(t)^{-1},$$

as new auxiliary matrix, we can write

RLS (III form)

$$\hat{\theta}_t = \hat{\theta}_{t-1} + K(t)\epsilon(t),$$
$$K(t) = V(t)\phi(t),$$
$$\epsilon(t) = y(t) - \phi(t)'\hat{\theta}_{t-1},$$
$$V(t) = V(t-1) - \beta_{t-1}^{-1}V(t-1)\phi(t)\phi(t)'V(t-1),$$
$$\beta_{t\ 1} = 1 + \phi(t)'V(t-1)\phi(t).$$

Assuming that, at certain t_0, matrix $S(t_0)$ is invertible, the appropriate initialization of this last algorithm is

$$V(t_0) = \left(\sum_{i}^{t_0} \phi(i)\phi(i)'\right)^{-1},$$

$$\hat{\theta}_{t_0} = V(t_0)\sum_{i}^{t_0} \phi(i)y(i).$$

In the same way that the inverse of a positive real number is positive too, the inverse of a positive definite matrix is positive definite as well. Therefore, $V(t)$ is also symmetric and positive definite.

While **RLS (I form)** and **RLS (II form)** require the inversion of an $n \times n$ matrix at each step, in **RLS (III form)**, only the inversion of scalar $\beta_{t-1} = 1 + \phi(t)'V(t-1)\phi(t)$ is necessary.

Note that $\beta_{t-1} = 1 + \phi(t)'V(t-1)\phi(t) \geq 1$. Hence, the scalar β_{t-1} is invertible at any time point.

Remark 5.12 (Auxiliary matrices in RLS) In the expressions of RLS we have introduced three auxiliary matrices: $S(t)$, $R(t)$, and $V(t)$ (all of dimension $n \times n$). For the sake of clarity, in this remark, we summarize their main properties.

Matrix $S(t)$ is a sum of terms of the type $\phi(i)\phi(i)'$, each of which is symmetric and positive semi-definite. Hence, $S(t)$ is symmetric and positive semi-definite as well for each t. If at a time point t_0 the matrix is invertible, then it is invertible for each t, $t \geq t_0$. Invertibility and positive semi-definiteness entails that $S(t)$ is indeed positive definite $\forall t \geq t_0$.

Being $R(t) = (1/t)S(t)$, matrix $R(t)$ is symmetric and positive definite too.

As already noted, $V(t)$ is also symmetric and positive definite.

As for the time evolution of these matrices, we note that $S(t) - S(t-1) = \phi(t)\phi(t)'$. Hence, $S(t)$ is monotonically increasing (in the sense that $S(t) - S(t-1)$ is a positive semi-definite matrix). Consequently, $V(t)$ is monotonically decreasing, i.e. $V(t-1) - V(t)$ is positive semi-definite.

If $\phi(t)$ is a stationary vector process, we expect that

$$R(t) = \frac{1}{t}S(t) = \frac{1}{t}\sum_{1}^{t} \phi(i)\phi(i)' \to \bar{R} = E[\phi(t)\phi(t)'].$$

Correspondingly, since $S(t) = tR(t)$, in the long run

$$\|S(t)\| \to \infty.$$

Finally,

$$V(t) = S(t)^{-1} \to 0.$$

We end this remark by providing an interesting interpretation of matrix $V(t)$. If the data generation mechanism belongs to the family of regression models, i.e. $\mathcal{S} = \mathcal{M}(\theta°)$, then, as seen in Section 5.5.4, the uncertainty in parameter estimation with t snapshots is given by $(1/t)\mathrm{Var}[\epsilon(t,\theta°)]E[\phi(t)\phi(t)']^{-1}$. Therefore, matrix $V(t)$ is proportional to the covariance matrix of the parameter estimate.

Remark 5.13 (RLS gain) In agreement with the three forms of the RLS algorithm, we have three expressions for the gain

$$K(t) = S(t)^{-1}\phi(t),$$

$$K(t) = (1/t)R(t)^{-1}\phi(t),$$

$$K(t) = V(t)\phi(t).$$

As seen in the previous remark, we expect that $V(t) \to 0$. Therefore, we also expect that gain $K(t) \to 0$, so that

$$\hat{\theta}_t - \hat{\theta}_{t-1} = K(t)\epsilon(t) \to 0.$$

This corresponds to the fact that, asymptotically, $\hat{\theta}_t$ is convergent, $\hat{\theta}_t \to \bar{\theta}$.

Remark 5.14 (Conventional initialization)   The RLS algorithms are a rigorous recursive version of the batch least squares formula, provided that they are suitably initialized, as discussed above. Sometimes, however, one resorts to a conventional initialization. With reference to **RLS (III form)**, such initialization consists in posing

$$V(0) = \alpha I,$$

$$\hat{\theta}_0 = \theta(0).$$

Obviously, in order to preserve the sign-definiteness of $V(\cdot)$, we have to assume that $\alpha \geq 0$. As for the tuning of parameter α, recall that $V(t)$ is proportional to the covariance matrix of the parameter estimator. Therefore, selecting a small α (high α) means that the initial parametrization is subject to small (high) uncertainty. Correspondingly, the estimated parameter will slowly (quickly) drift away from $\hat{\theta}(0)$. As for the choice of $\theta(0)$, it is dictated by the *a priori* knowledge on the system characteristics.

5.6.2   Recursive Maximum Likelihood

For the estimation of ARMAX models, we have seen the so-called *maximum likelihood* method in Section 5.4. In that method, a main role was played by the *extended observation vector*

$$\phi_E(t)$$
$$= [y(t-1)\cdots y(t-n_a) \;\vdots\; u(t-1)\cdots u(t-n_b) \;\vdots\; \epsilon(t-1)\cdots \epsilon(t-n_c)]'.$$

Here, as usual, n_a, n_b, and n_c are the orders of polynomials $A(z)$, $B(z)$, and $C(z)$; $u(\cdot)$ and $y(\cdot)$ are the measured input and output data, while

$$\epsilon(t) = y(t) - \phi_E(t)'\theta^{(r)}$$

is the prediction error computed on the basis of the parameter estimate at a certain iteration of the algorithm, the rth iteration.

We also introduced the *filtered extended observation vector*:

$$\psi(t) = [\alpha(t-1)\cdots\alpha(t-n_a) \;\vdots\; \beta(t-1)\cdots\beta(t-n_b) \;\vdots\; \gamma(t-1)\cdots\gamma(tn_c)]',$$

where the various entries are given by the corresponding entry of $\phi_E(t)$ filtered through the estimated polynomial at the considered iteration $1/C(z)^{(r)}$

$$\psi(t) = \frac{1}{C(z)^{(r)}} \phi_E(t).$$

We also recall that the basic recursion in the ML algorithm is

$$\theta^{(r+1)} - \theta^{(r)} = \left(\sum_1^N {}_t \psi(t)\psi(t)' \right)^{-1} \left(\sum_1^N {}_t \psi(t)\epsilon(t) \right),$$

a formula that is reminiscent of the batch LS expression

$$\hat{\theta} = \left(\sum_1^N {}_t \phi(t)\phi(t)' \right)^{-1} \left(\sum_1^N {}_t \phi(t)y(t) \right).$$

Thus, one might be led to believe that, in the same way as it was possible to write LS in its recursive form RLS, it should also be possible to bring ML into a recursive version. Actually, there is a basic difference between these two formulas: While vector $\phi(t)$ in the LS formula depends upon data $u(\cdot)$ and $y(\cdot)$ only, vector $\psi(t)$ is obtained via filtering procedures based on systems depending upon the estimated parameters $\hat{\theta}_r$. Correspondingly, if we define

$$S(t) = \sum_1^t {}_i \psi(i)\psi(i)',$$

the expression

$$S(t) = S(t-1) + \psi(t)\psi(t)' \tag{5.36}$$

is not a recursive formula since both $S(t)$ and $S(t-1)$ depend upon the same parameter vector, $\hat{\theta}_t$. This expression would be recursive only if the left-hand member depends upon $\hat{\theta}_t$ and the right-hand member upon $\hat{\theta}_{t-1}$. To obtain a recursive formula, we have to accept an approximation, consisting in evaluating the right-hand member in $\hat{\theta}_{t-1}$. In this way, one can work out the RML algorithm as

RML

$$\hat{\theta}_t = \hat{\theta}_{t-1} + K(t)\epsilon(t),$$
$$K(t) = S(t)^{-1}\phi(t),$$
$$\epsilon(t) = y(t) - \phi_E(t)'\hat{\theta}_{t-1},$$
$$S(t) = S(t-1) + \psi(t)\psi(t)',$$
$$C(z)^{(t-1)}\psi(t) = \phi_E(t).$$

As for RLS, RML also can be written in various equivalent forms characterized by different auxiliary matrices.

5.6.3 Extended Least Squares

A simple recursive algorithm to identify an ARMAX model can be worked out according to the following empirical rationale. In the usual ARMAX model

$$y(t) = a_1 y(t-1) + a_2 y(t-2) + \cdots + a_{n_a} y(t - n_a)$$
$$+ b_1 u(t-1) + b_2 u(t-2) + \cdots + b_{n_b} u(t - n_b)$$
$$+ \xi(t) + c_1 \xi(t-1) + c_2 \xi(t-2) + \cdots + c_{n_c} \xi(t - n_c).$$

$u(\cdot)$ and $y(\cdot)$ are measurable, while $\xi(\cdot)$ is a remote signal. However, assume for a moment that, at time t, the values $\epsilon(t-1), \epsilon(t-2), \ldots, \epsilon(t - n_c)$ taken by $\xi(\cdot)$ at instants $t-1, t-2, \ldots, t - n_c$, were available. Then we could construct the *extended observation vector* as

$$\phi_E(t) = [y(t-1) \cdots y(t - n_a) \vdots u(t-1) \cdots u(t - n_b) \vdots \epsilon(t-1) \cdots \epsilon(t - n_c)]'.$$

Correspondingly, the ARMAX model could be written in the linear regression form:

$$y(t) = \phi_E(t)'\theta + \xi(t),$$

where

$$\theta = [a_1 \cdots a_{n_a} \vdots b_1 \cdots b_{n_b} \vdots c_1 \cdots c_{n_c}]'.$$

Since $\xi(\cdot)$ is white, the optimal predictor would be

$$\hat{y}(t) = \phi_E(t)'\theta.$$

We could then apply the least squares method, for example in the third form

$$\hat{\theta}_t = \hat{\theta}_{t-1} + K(t)\epsilon(t),$$
$$K(t) = V(t)\phi_E(t),$$
$$V(t) = V(t-1) - \beta_{t-1}^{-1} V(t-1)\phi_E(t)\phi_E(t)'V(t-1),$$
$$\beta_{t-1} = 1 + \phi_E(t)'V(t-1)\phi_E(t).$$

Actually, this is not an applicable algorithm as various items are unknown. To be precise, the last n_c elements, $\epsilon(t-1), \epsilon(t-2), \ldots, \epsilon(t - n_c)$, of the extended observation vector $\phi_E(t)$ are not known, as well as the term $\epsilon(t)$ appearing in the parameter updating equation $\hat{\theta}_t = \hat{\theta}_{t-1} + K(t)\epsilon(t)$. An empirical idea is to replace these unknowns with a surrogate: the prediction errors at various time points as recursively computed by means of the last estimate of the parameter vector, i.e.

$$\epsilon(t) = y(t) - \phi_E(t)'\hat{\theta}_{t-1}.$$

Then the algorithm operation requires a continuous interplay between the updating equations for the parameter vector and the updating equation for the prediction error. From one side, once the observation vector $\phi_E(t)$ is available, one can update the auxiliary matrix $V(t)$ and compute the gain $K(t)$..From the other side, once the parameter vector estimate $\hat{\theta}_{t-1}$ is available, the prediction error $\epsilon(t)$ can be computed by means of the last snapshot $y(t)$ and the last extended observation vector $\phi_E(t)$ as $\epsilon(t) = y(t) - \phi_E(t)'\hat{\theta}_{t-1}$; with the new estimate $\epsilon(t)$ of the prediction error, it is possible to update the parameter estimate as $\hat{\theta}_t = \hat{\theta}_{t-1} + K(t)\epsilon(t)$, and also one can update the extended observation vector by introducing $\epsilon(t)$ in place of $\epsilon(t-1)$, shifting all remaining terms in the sequence of the last n_c elements of $\phi_E(t)$, and dropping out the last entry $\epsilon(t - n_c)$ of $\phi_E(t)$.

The so-obtained algorithm is named ELS. Note that ELS does not admit a batch version, and it must be initialized in a purely conventional way.

Though empirical, ELS has the advantage of being a simple method to estimate ARMAX models, obtained by adapting (extending) the popular LS algorithm. It coincides with RML if the two vectors $\psi(t)$ and $\phi_E(t)$ are set coincident. In ELS, the filtering actions for the passage from $\phi_E(t)$ to $\psi(t)$ are not required, a fact leading to a numerical simplification with respect to RML. On the other hand, the empirical nature of ELS may lead to erroneous estimates of the parameters, even with a large number of data.

Example 5.11 (Counterexample to ELS convergence)   In the paper (Ljung et al. 1975), the attention is focused on the ARMA process

$$\mathcal{S} : y(t) = -0.9y(t-1) - 0.95y(t-2)$$
$$+ e(t) + 1.5e(t-1) + 0.75e(t-2), \quad e(\cdot) \sim \text{WN}(0,1)$$

as an example of the possible non-convergence of the ELS algorithm. We suggest our readers to proceed as follows. Generate 2000 snapshots $y(\cdot)$ by simulation. Then identify the parameters with the ELS method applied to the ARMA (2,2) family of models with parameters

$$\theta = [a_1 \ a_2 \ c_1 \ c_2]'.$$

As for the initialization of the recursive identification algorithm, take $\hat{\theta}_0 = 0$ and $V(0) = 100I$. Observe the evolution in time of the estimated parameter $\hat{c}_2$. Notwithstanding the large number of data, one will note that $\hat{c}_2$ does not converge to the true value $c_2^\circ = 0.75$.

Second, repeat the same estimation trial with the initialization

$$\hat{\theta}_0 = [-0.9 - 0.95 \ 1.5 \ 0.75]'$$

and $V(0) = 0.01I$. This means that we use the true parametrization to start with; moreover, with such a small $V(0)$, we implicitly invite the algorithm to keep close to $\hat{\theta}_0$ (recall that $V(\cdot)$ can be interpreted as a variance in parameter estimation matrix). Notwithstanding all that, in the long run, $\hat{c}_2$ will diverge from $\hat{c}_2^\circ$.

Finally, repeat the identification exercise by using RML in place of ELS and compare with the previous results.

5.7 Robustness of Identification Methods

5.7.1 Prediction Error and Model Error

A main objective of identification is to assess the mismatch between model and reality. Can we say something in this respect when the identification is performed with a prediction error minimization method? In other words, can we pass from the *prediction error* to the *model error*?

To reply to this question, we assume that data are generated by a single-input single-output system $\mathcal{S}$ with transfer function $G_0(z)$ subject to an output disturbance described by an ARMA process characterized by a spectral factor $W_0(z)$, i.e.

$$\mathcal{S} : y(t) = G_0(z)u(t-1) + W_0(z)e(t), \quad e(t) \sim \mathrm{WN}(0, \lambda^2). \tag{5.37}$$

As model, we take

$$\mathcal{M}(\theta) : y_\theta(t) = G(z, \theta)u(t-1) + W(z, \theta)\xi(t), \quad e(t) \sim \mathrm{WN}(0, \mu^2). \tag{5.38}$$

Such model was already studied in Section 4.5.1 and named Box and Jenkins model. The quantity

$$\Delta G(z, \theta) = G_0(z) - G(z, \theta)$$

is the *model error*. In many fields, in particular in control engineering, this mismatch is more significant than the prediction error by itself. It is therefore important to see how $\Delta G(z, \theta)$ depends upon $\epsilon(\theta)$. With this objective in mind, we start recalling that the predictor associated with model (5.38) is

$$\hat{y}_\theta(t|t-1) = (1 - W(z, \theta)^{-1})y(t) + W(z, \theta)^{-1}G(z, \theta)u(t-1),$$

so that the prediction error is given by

$$\begin{aligned}
\epsilon_\theta(t) &= y(t) - \hat{y}_\theta(t|t-1) \\
&= y(t) - (1 - W(z, \theta)^{-1})y(t) - W(z, \theta)^{-1}G(z, \theta)u(t-1) \\
&= W(z, \theta)^{-1}y(t) - W(z, \theta)^{-1}G(z, \theta)u(t-1).
\end{aligned}$$

Summing up

$$\epsilon_\theta(t) = W(z, \theta)^{-1}(y(t) - G(z, \theta)u(t-1)). \tag{5.39}$$

Taking into account the expression of $y(\cdot)$ given by (5.37), we then have

$$\begin{aligned}
\epsilon_\theta(t) &= W(z, \theta)^{-1}[(G_0(z) - G(z, \theta))u(t-1) + W_0(z)e(t)] \\
&= W(z, \theta)^{-1}[\Delta G(z, \theta)u(t-1) + W_0(z)e(t)]. \tag{5.40}
\end{aligned}$$

Expression (5.40) links the model error to the prediction error, pointing out that $\epsilon_\theta(t)$ depends linearly upon the exogenous variable $u(\cdot)$ and upon white noise $e(\cdot)$. In particular, we see that the influence of $u(\cdot)$ on $\epsilon_\theta(\cdot)$ is determined by the model error $\Delta G(z, \theta)$.

5.7.2 Frequency Domain Interpretation

We now assume that parameter θ is estimated by a prediction error method, i.e. by minimizing

$$J_N(\theta) = \frac{1}{N} \sum_{t}^{N} \epsilon_\theta(t)^2.$$

We know that, asymptotically,

$$J_N(\theta) = \frac{1}{N} \sum_{t}^{N} \epsilon_\theta(t)^2 \xrightarrow{N\to\infty} \bar{J}(\theta) = E[\epsilon_\theta(t)^2].$$

$\bar{J}(\theta)$ can be given a useful frequency domain expression. To derive it, we see $\epsilon_\theta(t)$ as a stationary process. Its variance $\mathrm{Var}[\epsilon_\theta(t)]$ coincides with the area of the spectrum $\Gamma_{\epsilon\epsilon}(\omega)$ of $\epsilon_\theta(t)$ from $-\pi$ to $+\pi$ up to coefficient $1/2\pi$. Hence, we have

$$\bar{J}(\theta) = E[\epsilon_\theta(t)^2] = \mathrm{Var}[\epsilon_\theta(t)] = \frac{1}{2\pi} \int\limits_{-\pi}^{+\pi} \Gamma_{\epsilon\epsilon}(\omega)d\omega. \tag{5.41}$$

$\Gamma_{\epsilon\epsilon}(\omega)$ can be derived with the *fundamental theorem of spectral analysis* presented in Section 1.13. Under the assumption that $u(\cdot)$ and $e(\cdot)$ are uncorrelated, by denoting with $\Gamma_{uu}(\omega)$ the spectrum of $u(\cdot)$, we obtain

$$\Gamma_{\epsilon\epsilon}(\omega) = \left[|W(e^{i\omega}, \theta)^{-1} \Delta G(e^{i\omega}, \theta)|^2 \Gamma_{uu}(\omega) \right] + \left[|W(e^{i\omega}, \theta)^{-1} W_0(e^{i\omega})|^2 \lambda^2 \right].$$

If the variance λ^2 is small, we can neglect the second term and write

$$\Lambda_{\epsilon\epsilon}(\omega) = \left[|W(e^{i\omega}, \theta)^{-1} \Delta G(e^{i\omega}, \theta)|^2 \Gamma_{uu}(\omega) \right],$$

so that

$$\bar{J}(\theta) = \frac{1}{2\pi} \int\limits_{-\pi}^{+\pi} |\Delta G(e^{i\omega}, \theta)|^2 g(\omega, \theta)d\omega, \tag{5.42}$$

where

$$g(\omega, \theta) = \frac{\Gamma_{uu}(\omega)}{|W_0(e^{i\omega}, \theta)|^2}.$$

This expression has the merit of explaining the effect that the exogenous input has, frequency by frequency, on the predictive identification performance index. As can be seen, this effect is modulated by coefficient $g(\omega, \theta)$, therefore playing the role of a *frequency penalization*.

For instance, if $g(\omega, \theta)$ is constant (with respect to ω), then the model errors ΔG occurring at various frequencies would have the same weight in their contribution to the overall performance index. If instead $g(\omega, \theta) = 0$ for $\omega \geq \bar{\omega}$, then only the values taken by ΔG for $\omega \leq \bar{\omega}$ would matter. If one takes as input a white noise, then

$$u(t) \sim \mathrm{WN}(0, \sigma^2) \longrightarrow g(\omega, \theta) = \frac{\sigma^2}{|W_0(e^{i\omega}, \theta)|^2},$$

so that the frequency penalization is determined by the inverse of $W(z)$.

5.7.3 Prefiltering

The original sequences of input and output snapshots, $u(\cdot)$ and $y(\cdot)$, can be passed through a filter to generate new sequences $u_F(\cdot)$ and $y_F(\cdot)$ as follows:

$$u_F(t) = L(z)u(t),$$
$$y_F(t) = L(z)y(t),$$

where $L(z)$ is the transfer function of a stable dynamical system (stability preserve stationarity). If the identification is performed with $u_F(\cdot)$ and $y_F(\cdot)$, then expression (5.42) becomes

$$\bar{J}_F(\theta) = \frac{1}{2\pi} \int_{-\pi}^{+\pi} |\Delta G(e^{i\omega}, \theta)|^2 g_F(\omega, \theta) d\omega,$$

where

$$g_F(\omega, \theta) = \frac{\Gamma_{uu}(\omega)}{|W(e^{i\omega}, \theta)|^2} |L(e^{i\omega})|^2$$

is the new penalization function.

By a suitable choice of $L(z)$, one can modulate function $g_F(\omega, \theta)$, so as to force the identification method to improve the estimation accuracy in the band of frequencies of interest.

5.8 Parameter Tracking

In our approach to identification via black box models, we have assumed that the parameters were constant. If there is a drift in a parameter, the recursive methods introduced so far cannot provide a satisfactory tracking. Indeed, such algorithms are a recursive version of methods conceived to estimate *constant* parameters. In this section, we introduce a widely used variant to estimate parameters subject to time variations.

Consider again the least squares cost function

$$J(\theta) = \frac{1}{t} \sum_{1}^{t}{}_i \epsilon(i)^2 = \frac{1}{t} \sum_{1}^{t}{}_i (y(i) - \theta'\phi(i))^2.$$

In this formula, prediction errors $\epsilon(i) = y(i) - \theta'\phi(i)$ are equally treated, in the sense that they contribute with the same weight to $J(\theta)$. In this way, the error associated with the last time point is equally important as that of the first time point, even if this initial moment is very remote with respect to the current time t. This may be inappropriate when the system dynamics are subject to a change in time, as may happen when the interval elapsed between the last and first point is large. Therefore, in order to best estimate the recent dynamics, less weight should be given to past data. This may be achieved by modifying the cost as follows:

$$J(\theta) = \frac{1}{t} \sum_{1}^{t} \mu^{t-i} \epsilon(i)^2,$$

where μ, $\mu \in (0, 1]$, is named the *forgetting factor* (FF). With the forgetting factor, it is possible to reduce the weight given to past data in favor of fresh data. Indeed, the weight of the last snapshot at time t, $\epsilon(t)$, is $\mu^{t-i} = \mu^{t-t} = 1$, the weight of the last but one snapshot, $\epsilon(t)$, is $\mu^{t-i} = \mu^{t-(t-1)} = \mu$, and so on. The weight of past data is exponentially decreasing as $t - i$ becomes larger and larger.

For $\mu = 1$, the standard RLS is retrieved, with all snapshots, from the first to the last one, used to form the estimate. For $\mu < 1$, the quantity $1/(1 - \mu)$ is named *data window* as it provides rough estimate of the number of past data used in the recursive algorithm to form the estimate. For instance, for $\mu = 0.99$, the data window is 100.

It is easy to see that, with the FF, the RLS algorithm becomes

RLS I form with FF

$$\hat{\theta}_t = \hat{\theta}_{t-1} + K(t)\epsilon(t),$$
$$K(t) = S(t)^{-1}\phi(t),$$
$$\epsilon(t) = y(t) - \phi(t)'\hat{\theta}_{t-1},$$
$$S(t) = \mu S(t-1) + \phi(t)\phi(t)'.$$

By comparing these formulas with those of the standard RLS, we see that the only change concerns the equation of auxiliary matrix $S(\cdot)$, also called *information matrix*:

$$S(t) = S(t-1) + \phi(t)\phi(t)', \qquad \text{standard RLS,}$$
$$S(t) = \mu S(t-1) + \phi(t)\phi(t)', \qquad \text{RLS with FF.}$$

In the standard RLS, matrix $S(t)$ is built from $S(t-1)$ by adding $\phi(t)\phi(t)'$; in this addition, $S(t-1)$ represents the information coming from the past (up to $t-1$), while $\phi(t)\phi(t)'$ represents the fresh information (conveyed by the last observation vector $\phi(t)$). In the FF variant, such sum is performed by multiplying $S(t-1)$ by coefficient μ; this leads to an attenuation of the importance of past information with respect to the fresh one supplied by $\phi(t)\phi(t)'$.

Another way to look at the effect of FF in recursive algorithms is to consider the trace of matrix $S(t)$:

$$\text{trace}[S(t)] = \mu \ \text{trace}[S(t-1)] + \|\phi(t)\|^2. \tag{5.43}$$

$S(t)$ being positive semi-definite, its trace is a non-negative number. In the absence of a forgetting effect, namely for $\mu = 1$, we expect that trace$[S(t)]$ will diverge, so that the norm of gain $K(t)$ will become smaller and smaller. This means that the sensitivity of the estimation to information conveyed by fresh data will become negligible in the long run. On the contrary, under the action of a forgetting factor ($\mu < 1$), expression (5.43) is that of a scalar stable system fed by $\|\phi(t)\|$. If $\phi(t)$ is a stationary process, then trace$[S(t)]$ will tend to a stationary process as well, with variance $(1 - \mu^2)^{-1}\sigma^2$, where σ^2 is the variance of $\|\phi(t)\|^2$. In this way, gain $K(t)$ do not tend to zero anymore and the sensitiveness of the estimation to current data is guaranteed even in the long run.

Example 5.12 (Effect of FF in parameter tracking)   A system with two uncorrelated inputs $u_1(t) \sim \text{WN}(0, 0.01)$ and $u_2(t) \sim \text{WN}(0, 0.01)$ generates a signal $y(t)$ according to the rule

$$y(t) = \theta_1(t)u_1(t) + \theta_2(t)u_2(t) + \xi(t), \quad \xi(t) \sim \text{WN}(0, 0.001). \tag{5.44}$$

Parameter $\theta_2(t)$ is a constant:

$$\theta_2(t) = 1,$$

whereas $\theta_1(t)$ is initially constant and then, after instant 400, is sinusoidally time varying:

$$\theta_1(t) = \begin{cases} 1, & t < 400, \\ \sin\left(\dfrac{\pi t}{100} + \dfrac{\pi}{2}\right), & t > 400. \end{cases}$$

The purpose of this example is to study the tracking performance of recursive identification methods.

As shown in Figure 5.7, we have considered three cases, marked as (a), (b), and (c) in the drawing. In simulation (a), the estimate of θ_1 obtained with the standard RLS is depicted. As can be seen, after time 400, the tracking is poor. This is due to the saturation of matrix $S(t)$ leading to a very small gain $K(t)$ on the long run. Second, the estimate has been performed with the FF variant of RLS with $\mu = 0.95$ (b) and with a lower FF, $\mu = 0.75$ (c). By inspecting these diagrams, one can appreciate the improvement in the tracking performance when the FF is lower. There is, however, another side of the coin: In the initial phase, when $\theta_1(t)$ is constant, the estimate is more and more waiving, especially when the FF takes a lower value. The reason is reducing the FF results in a shorter data window, so that the estimate becomes more sensitive to the noise.

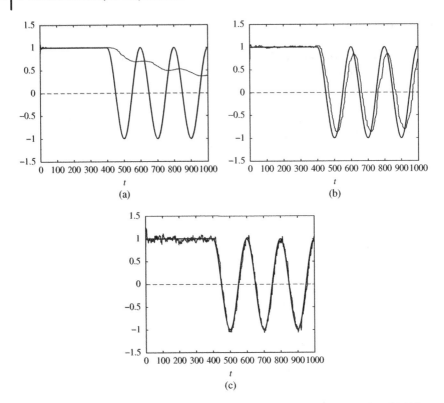

Figure 5.7 Estimate of parameter θ_1 in Example 5.12: (a) standard RLS algorithm; (b) RLS with forgetting factor $\mu = 0.95$; and (c) RLS with forgetting factor $\mu = 0.75$.

The performance of RLS with forgetting factor in the identification of time-varying systems is studied in a number of papers such as Guo et al. (1993) and Bittanti and Campi (1994).

We caution our readers that resorting to the forgetting factor may lead to drawbacks in case of long periods of data quiescence. For instance suppose that, for a long interval of time, $\phi(i) = 0$. Then, from Eq. (5.43), we see that the trace of $S(t)$ will progressively decrease and tend to 0. This implies that trace$[S(t)^{-1}]$ will become very large, a fact known as *wind up*. Now, gain $K(t)$ is the product of two terms: $S(t)^{-1}$ and $\phi(i)$. Until $\phi(i) = 0$, such product is null no matter how large is $K(t)$. But, at the end of the quiescence period, when $\phi(t)$ will be no more null, the gain $K(t) = S(t)^{-1}\phi(t)$ will have a sudden variation. This may produce a parameter jump, called *bursting phenomenon*, as illustrated in the following example.

Example 5.13 (Wind up and bursting)   Consider again system (5.44), with constant parameters,

$$\theta_1 = 1, \quad \theta_2 = 1,$$

and

$$u_1(t) = \begin{cases} \sim \text{WN}(0, 0.1), & t \le 200, \\ 0, & 200 < t \le 900, \\ \sim \text{WN}(0, 0.1), & t > 900, \end{cases}$$

$$u_2(t) \sim \text{WN}(0, 1),$$

$$\xi(t) \sim (0, 0.001),$$

$u_1(t)$, $u_2(t)$, and $\xi(t)$ uncorrelated to each other.

The estimate is performed with RLS with a forgetting factor $\mu = 0.9$.

In the data generation mechanism, there is a lack of excitation of parameter θ_1. Indeed, input $u_1(t)$ is quiescent over a long interval of time, from $t = 100$ to $t = 900$. On the other hand, at $t = 900$, $u_1(t)$ becomes exciting again, and the parameter estimate is subject to a jump at that point, as can be seen in Figure 5.8.

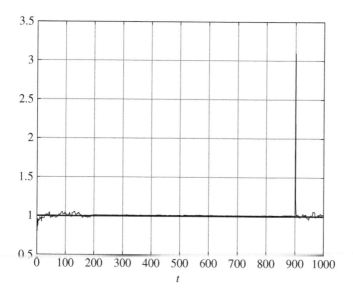

Figure 5.8 The estimate of parameter θ_1 in Example 5.13 exhibits a bursting phenomenon at time $t = 900$.

A remedy to such bizarre dysfunction responsible of a sudden burst consists in resorting to a time-varying forgetting factor: When the parameters are drifting, one resorts to the forgetting factor variant with a value of μ tuned according to the speed of variation of the parameters; if instead there is evidence that the parameters are constant, then resorting to the standard RLS ($\mu = 1$) is more appropriate. Another possibility is to analyze the distribution of the excitation of the incoming information in the parameter space and selectively adapt the forgetting factor direction by direction in such a space (*directional forgetting*). For details, we refer to the literature, for instance the papers Bittanti et al. (1990), or Ding and Chen (2005).

6

Model Complexity Selection

6.1 Introduction

In black box identification, one of the main problems is choosing the appropriate complexity of the model. This stimulating issue has been studied by scholars of various disciplines, including statisticians, computer scientists, control engineers, and many others. The essential terms of the problem can be outlined as follows. The basic identification criterion adopted so far is

$$J(\theta) = \frac{1}{N} \sum_{t=1}^{N} \epsilon_\theta(t)^2,$$

where N is the number of data, θ is the parameter vector, and $\epsilon_\theta(t)$ the prediction error at time t of the considered model $\mathcal{M}(\theta)$. $J(\theta)$ provides an index of fit of $\mathcal{M}(\theta)$ to data. Hence, if $\hat{\theta}_N$ is the point of minimum of the adopted criterion, $J(\hat{\theta}_N)$ is the index of fit associated with the optimal model $\mathcal{M}(\hat{\theta}_N)$. When a more complex class of models is considered, there are more degrees of freedom, so that we expect a lower index of fit of the optimal model. In short, the higher the model order, the better the data fitting.

The question is: To what extent is it advisable to enlarge the class of models to achieve a better fit? What is the appropriate balance between complexity and fit? This is illustrated in the following example.

Example 6.1 (Identification with ARX models of various orders) The ARX(2, 2) system

$$\mathcal{S} \ : \ y(t) = 1.2y(t-1) - 0.32y(t-2)$$
$$+ u(t-1) + 0.52u(t-2) + \eta(t),$$

where $\eta(t) \sim \text{WN}(0, 1)$, $u(t) \sim \text{WN}(0, 4)$, with $u(\cdot)$ and $\eta(\cdot)$ uncorrelated, is identified with the class of ARX(n, n) models:

$$\mathcal{M} \ : \ y(t) = a_1 y(t-1) + a_2 y(t-2) + \cdots + a_n y(t-n)$$
$$+ b_1 u(t-1) + b_2 u(t-2) + \cdots + b_n u(t-n) + \xi(t),$$

Model Identification and Data Analysis, First Edition. Sergio Bittanti.
© 2019 John Wiley & Sons, Inc. Published 2019 by John Wiley & Sons, Inc.

for $n = 1, 2, 3$. The parameter identification is performed via LS with 2000 pairs $\{u(t), y(t)\}$ generated by simulation.

The results of the estimation procedure are summarized in Table 6.1. In the table, one can find the following: the parameter estimates, the corresponding uncertainty in percent (in parenthesis), and the value taken by criterion $J = (\epsilon(1)^2 + \epsilon(2)^2 + \cdots + \epsilon(2000)^2)/2000$, where $\epsilon(t)$ is the prediction error of the identified model. We also report the outcome of the Anderson test of whiteness (number of points of the covariance function of $\epsilon(t)$ outside the acceptable range associated to a confidence level of 5% over a total of 30 points); see Section 2.7.

From the table, it is apparent that, passing from ARX(1,1) to ARX(2,2), there is a remarkable improvement; indeed, the value of the performance index decreases significantly, and the number of points violating the Anderson test passes from 7 to 0. On the opposite, the passage from ARX(2,2) to ARX(3,3) leads to a negligible reduction in J; on the top of that, parameters $\hat{a}_3$ and $\hat{b}_3$ are close to zero and have high standard deviation. All these outcomes strongly suggest that the proper order is $n = 2$.

In simulation examples, such as the one presented above, the choice of complexity may be relatively easy. In many other cases, however, the choice is more intriguing. In particular, it may happen that $J(\hat{\theta}_N)$ be monotonically decreasing with respect to the model complexity without a clear-cut point and that the whiteness test may not be satisfied even for a large number of parameters.

As a matter of fact, there is a fundamental limitation in such a type of analysis: It is not really true that the value taken by $J(\hat{\theta}_N)$ is a fair assessment of the model

Table 6.1 Identification of model ARX(n, n) for data generated by the system of Example 6.1.

ARX(1, 1)	$\hat{a} = 0.932$ (0.6%) $\hat{b} = 0.975$ (2.3%) $J = 3.864$ Anderson test 5%: 7		
ARX(2, 2)	$\hat{a}_1 = 1.204$ (1%) $\hat{b}_1 = 0.984$ (1%) $J = 0.998$ Anderson test 5%: 0	$\hat{a}_2 = -0.320$ (3%) $\hat{b}_2 = 0.485$ (3%)	
ARX(3, 3)	$\hat{a}_1 = 1.194$ (2%) $\hat{b}_1 = 0.984$ (1%) $J = 0.997$ Anderson test 5%: 0	$\hat{a}_2 = -0.299$ (10%) $\hat{b}_2 = 0.494$ (5%)	$\hat{a}_3 = -0.019$ (68%) $\hat{b}_3 = -0.016$ (120%)

performance in data fitting. Indeed, $J(\hat{\theta}_N)$ is a function of two basic ingredients: (i) parameter $\hat{\theta}_N$ and (ii) data $y(\cdot)$. The critical point is the fact that $\hat{\theta}_N$ is found by means of the same sequence $y(\cdot)$ used to compute index $J(\hat{\theta}_N)$. Therefore, the higher the model order, the better the data fitting. On the other hand, proceeding this way, by increasing the order without any limit, the model will also fit the noise corrupting data (a phenomenon known as *overfitting*). We come to the conclusion that $J(\hat{\theta}_N)$ is a *subjective* evaluation of model performance. For an *objective* assessment, one should rather consider how the model associated with $\hat{\theta}_N$ performs on data *different* than those used to determine $\hat{\theta}_N$.

We will now see how to proceed in order to find an objective criterion. For simplicity, in our analysis, we refer to a time series identified with an AR model of order n, so that the observation vector and the parameter vectors are

$$\phi(t) = [y(t-1)y(t-2) \quad \cdots \quad y(t-n)]',$$
$$\theta = [a_1 \quad a_2 \quad \cdots \quad a_n]'.$$

Correspondingly, the prediction error is

$$\epsilon_\theta(t) = y(t) - \phi(t)'\theta.$$

6.2 Cross-validation

If the number of data is high, one can partition them into two segments: one is used for identification purposes and the other to test the quality of the identified model. Correspondingly, we speak of *identification data* (first segment) and *validation data* (second segment).

To be precise, once an interval of possible values for the model order n, say $n_1 \leq n \leq n_2$, is selected, for each order, one finds the best model by using the identification data. This is the *model identification phase*. Then, the performance of each optimal model is assessed by probing its predictive performance over the validation data (*model validation phase*). The most appropriate order is that leading to the minimization of the performance index over this last segment.

In many cases however, the available data sequence is not as long, and the cross-validation rationale would lead to an excessively small number of identification data. Then the question becomes: Is it possible to construct an *objective* criterion to assess the performance of a model starting from a *subjective* one?

6.3 FPE Criterion

6.3.1 FPE Concept

Suppose we have a model, the model associated to a certain vector θ of parameters. We want to establish its capacity of interpreting data in an objective way.

To this purpose, it is fundamental to bear in mind that the sequence of snapshots at hand is just one of the many sequences we could extract from the given phenomenon. A way to capture this uncertainty is to see the data as part of a realization of a stochastic process, the realization associated with a certain outcome, say s, of an underlying random experiment. We write $y(t) = y(t, s)$. Correspondingly, the predictor is $\hat{y}(t, s, \theta)$.

To obtain an *objective* index of model accuracy, one must get rid of the particular sequence of data at hand and accounting for *all realizations*. This leads to consider the averaged index

$$\bar{J}(\theta) = E[(y(t, s) - \hat{y}(t, s, \theta))^2]. \tag{6.1}$$

The averaging provided by the expectation operator $E[\cdot]$ enables to see $\bar{J}(\theta)$ as the accuracy of model $\mathcal{M}(\theta)$ over *all possible sequences of data*.

Now, we cannot forget that the model is estimated from data. Therefore, the estimate, $\hat{\theta}_N$, depends upon data too:

$$\hat{\theta}_N = \hat{\theta}_N(s).$$

By evaluating $\bar{J}$ at $\hat{\theta}_N$, we obtain the quantity $\bar{J}(\hat{\theta}_N(s))$, which represents the accuracy of model $\mathcal{M}(\hat{\theta}_N(s))$ over all possible sequences of data. On the other side, the model associated with parameter vector $\hat{\theta}_N(s)$ is just one of the possible models that can be estimated. To achieve a fully *objective* index of accuracy, one can consider the average $E[\bar{J}(\hat{\theta}_N(s))]$ over all possible s. The so-obtained quantity is denoted by the acronym *FPE* (*final prediction error*):

$$FPE = E[\bar{J}(\hat{\theta}_N(s))].$$

FPE represents the average accuracy of all possible estimates associated with all possible sequences of data. By minimizing the FPE, we obtain the optimal complexity.

6.3.2 FPE Determination

We now evaluate FPE with reference to the problem of estimating an AR model for a given time series. To be precise, we assume that the data are generated by

$$\mathcal{S}: \quad y(t) = \phi(t, s)'\theta° + \eta(t),$$

where $\eta(\cdot)$ is a white noise of variance λ^2. The model, in predictive form, is

$$\hat{y}(t, s, \theta) = \phi(t)'\theta = \phi(t, s)'\theta.$$

Note that

$$\phi(t, s) = [y(t - 1, s) \ y(t - 2, s) \cdots y(t - n, s)]'.$$

Hence, $\phi(t,s)$ depends upon past data only. Consequently, $E[\phi(t,s)]\eta(t) = 0$. Therefore,

$$\bar{J}(\theta) = E[(y(t,s) - \hat{y}(t,s,\theta))^2] = E[(\phi(t,s)'(\theta° - \theta) + \eta(t))^2]$$
$$= (\theta° - \theta)'E[\phi(t,s)\phi(t,s)'](\theta° - \theta) + \lambda^2.$$

Letting

$$\bar{R} = E[\phi(t,s)\phi(t,s)'],$$

we have

$$\bar{J}(\theta) = (\theta° - \theta)'\bar{R}(\theta° - \theta) + \lambda^2.$$

Hence,

$$\text{FPE} = E[(\theta^0 - \hat{\theta}_N(s))'\bar{R}(\theta^0 - \hat{\theta}_N(s))] + \lambda^2.$$

On the other hand, in Section 5.5.4, we have seen that, for large values of N,

$$\text{Var}[(\theta° - \hat{\theta}_N(s))] = \frac{\lambda^2}{N}\bar{R}^{-1}.$$

By letting

$$v = \theta° - \hat{\theta}_N(s),$$

we can conclude that

$$\text{FPE} = E[v'\text{Var}[v]^{-1}v]\frac{\lambda^2}{N} + \lambda^2. \tag{6.2}$$

We now prove the following.

Proposition 6.1 *For any n-dimensional random vector v with zero mean value, $E[v'\text{Var}[v]^{-1}v] = n$.*

Proof: $v'\text{Var}[v]^{-1}v$ being a scalar, it coincides with its trace $\text{tr}(v'\text{Var}[v]^{-1}v)$. On the other side, given two matrices A and B, $\text{tr}(AB) = \text{tr}(BA)$, whenever both products AB and BA make sense. Then,

$$E[v'\text{Var}[v]^{-1}v] = E[\text{tr}\{v'\text{Var}[v]^{-1}v\}]$$
$$= E[\text{tr}\{\text{Var}[v]^{-1}vv'\}]$$
$$= \text{tr}\{E[\text{Var}[v]^{-1}vv']\}$$
$$= \text{tr}\{\text{Var}[v]^{-1}E[vv']\}$$
$$= \text{tr}\{\text{Var}[v]^{-1}\text{Var}[v]\}$$
$$= \text{tr}\ I = n.$$

Substituting this expression into formula (6.2), we obtain

$$\text{FPE} = \frac{n}{N}\lambda^2 + \lambda^2.$$

Here, the variance λ^2 is unknown and must be evaluated from data. A possibility is to replace it with a sample surrogate such as

$$\hat{\lambda}^2 = \frac{1}{N-n} \sum_1^N \epsilon(t)^2.$$

In this way,

$$\hat{\lambda}^2 = \frac{1}{N-n} \sum_1^N \epsilon(t)^2 = \frac{N}{N-n} \frac{1}{N} \sum_1^N \epsilon(t)^2 = \frac{N}{N-n} J(\hat{\theta}_N)^{(n)},$$

where $J(\hat{\theta}_N)^{(n)}$ is the usual *subjective* criterion for a model of order n. This leads to the celebrated formula

$$\text{FPE} = \frac{N+n}{N-n} J(\hat{\theta}_N)^{(n)}.$$

While $J(\hat{\theta}_N)^{(n)}$ is monotonically decreasing with n, for n, sufficiently large FPE will be increasing with n due to coefficient $N+n/N-n$ (note that, if $n \to N$, $\frac{N+n}{N-n} \to \infty$).

In conclusion, we expect that FPE will be initially decreasing and then, for large values of n, increasing. The optimal complexity is that for which FPE presents a minimum.

Although the criterion was here deduced for AR models, it is commonly used for any types of models by taking n as the total number of parameters.

Example 6.2 (Example 6.1 continued) We now reconsider the system of Example 6.1, where $N = 2000$ simulation snapshots were generated by simulation for an ARX(2,2) system, identified with ARX(n, n) models. By computing the FPE for $n = 1, 2, \ldots, 5$, we obtain

n	1	2	3	4	5
FPE	3.8765	1.0035	1.0041	1.0040	1.0040

The minimum is achieved for the order of the data generation mechanism ($n = n^\circ = 2$).

6.4 AIC Criterion

Another popular criterion for complexity selection is the *Akaike information criterion* or *AIC*. This criterion is deduced with statistical considerations, in the intent of minimizing the *discrepancy* between the probability density distribution of real data and the distribution that would have been produced by a certain model. As a measure of *discrepancy*, the so-called *Kullback distance*, $E[\ln(p_{\text{true}}/p_{\text{model}})]$, has been taken.

Following these ideas, Japanese statistician Hirotugu Akaike worked out this criterion

$$\text{AIC} = 2\frac{n}{N} + \ln J(\hat{\theta}_N)^{(n)}.$$

The optimal order is again determined by minimization with respect to the model order n.

Note that in the AIC expression, there is the sum of two terms. The second one is a logarithmic assessment of the fit to data of the optimal model of a given order n; this term is decreasing with n. The first one, instead, is a penalization of model complexity; this term is linearly increasing with n. Hence, the order leading to a minimum of AIC corresponds to the best trade-off between fitting and complexity.

6.4.1 AIC Versus FPE

For a large number of snapshots, FPE and AIC lead to the same estimates of the optimal order. Indeed, by recalling that $\ln(1 + x) \cong x$ if x is small, we have

$$\ln \text{FPE} = \ln \frac{1 + (n/N)}{1 - (n/N)} \hat{J}(\theta_N)^{(n)}$$

$$= \ln[1 + (n/N)] - \ln[1 - (n/N)] + \ln J(\hat{\theta}_N)^{(n)}$$

$$\cong 2(n/N) + \ln \hat{J}(\theta_N)^{(n)} = \text{AIC}.$$

6.5 MDL Criterion

The FPE and AIC criteria are worked out by statistical considerations. On the contrary, the *MDL* (*minimum description length*) criterion is based on the concept of *information complexity*. To be more precise, consider the problem of binary coding a set of data. One can proceed by coding the given time series of data. Alternatively, one can fit first a model and evaluate the corresponding prediction error. The same information provided with the original time series can be alternatively but equivalently given by (i) coding the model $\mathcal{M}(\theta)$ fitted to data and (ii) coding the series of prediction errors $y(\cdot) - \hat{y}_\theta(\cdot)$.

Now, it is apparent that a binary coding for the parameter vector requires an increasing amount of bits with the model complexity (increasing number of parameters). On the opposite, the prediction error signal becomes smaller and smaller when the complexity increases (the variance becomes smaller and smaller). Hence, with a given set of quantization levels, coding $y(\cdot) - \hat{y}_\theta(\cdot)$ is less demanding when the model is more complex. So, there is a trade-off between these two descriptions, corresponding to which the overall coding effort is minimum. This is the MDL principle. According to the above ideas, the computation of the appropriate complexity as the one for which the overall modeling effort requires a minimum number of bits leads to criterion:

$$MDL = (\ln N)\frac{n}{N} + \ln J(\hat{\theta}_N)^{(n)}.$$

Again, the optimal number n of parameters is achieved by minimizing the criterion.

6.5.1 MDL Versus AIC

As in AIC, in MDL, there is a complexity penalization term and a fitting term. The latter is the same in the two criteria, namely $\ln J(\hat{\theta}_N)^{(n)}$. The former is $(\ln N)\frac{n}{N}$ in MDL and $(2)\frac{n}{N}$ in AIC. In both cases, the complexity penalization is linear in n; however, the slope of the corresponding straight line is different. For AIC, the slope is $(\frac{2}{N})$, whereas for MDL, the slope is $(\frac{\ln N}{N})$. For instance, for $N = 10, 100, 1000$, the slope is $0.2, 0.02, 0.002$ for AIC and $0.23, 0.04, 0.069$ for MDL, respectively. This means that MDL leads to models of lower complexity, namely it is more parsimonious.

Example 6.3 (Order determination with FPE, AIC, MDL) Consider the AR(3) process generated by

$$\mathcal{S} : y(t) = 0.6y(t-1) - 0.1y(t-2) - 0.2y(t-3) + \eta(t),$$

where $\eta(t) \sim WN(0, 1)$.

$\mathcal{S}$ is stable, so that it generates a stationary process. With 30 simulation trials, 30 independent realizations of the process have been generated. To be precise, 30 realizations of a white noise $\eta(t)$ have been produced and the corresponding output signals $y(t)$ have been stored. By denoting with N the number of data of each sequence and with i the index of the sequence ($i = 1, 2, \ldots, 30$), the set of available data can be summarized as follows:

$$\{y^{(i)}(t)|t = 1, 2, \ldots, N\}$$

for the ith sequence, $i = 1, 2, \ldots, 30$.

The data have been processed as follows. With a sequence, say the ith one, we proceed to the identification of an AR(n) model, with order n ranging from 1 to 10. For each n, the model parameters were determined via least squares. Then, the sample variance of prediction errors associated with each identified model was computed, and from it, criteria FPE, AIC, and MDL were computed. The entire process was repeated for $n = 1, 2, \ldots, 10$. By minimizing FPE, AIC, or MDL with respect to n, the optimal order $\hat{n}$ is found for each of the 30 sequences. The obtained results are summarized in Table 6.2, where the number of times (over 30 sequences) for which the estimated optimal order $\hat{n}$ takes a certain value is reported.

As can be seen, if N is small, for example $N = 30$, then the order $n = 1$ is endowed by all criteria; to be precise, FPE indicates such order 8 times over 30,

Table 6.2 Choice of the optimal order (Example 6.3).

$\hat{n}$		1	2	3	4	5	6	7	8	9	10
$N = 30$	FPE	8	4	2	2	2	1	1	3	2	5
	AIC	5	4	3	2	2	1	1	3	2	7
	MDL	17	5	3			6	3	1	1	
$N = 50$	FPE	3	5	5	2	2	4	3	4	1	1
	AIC	3	5	5	2	2	4	3	4	1	1
	MDL	14	9	6		1					
$N = 100$	FPE	2	9	10	2	2			1	2	2
	AIC	2	9	10	2	2			1	2	2
	MDL	8	11	10		1					
$N = 250$	FPE		2	15	5	1		1	3	2	1
	AIC		2	15	5	1		1	3	2	1
	MDL	1	10	18	1						
$N = 500$	FPE			25	2		1	1		1	
	AIC			25	2		1	1		1	
	MDL			30							

This table reports the number of times a given criterion leads to the choice of the optimal order $\hat{n}$ in the 30 simulations. Empty boxes correspond to orders that are never chosen.

AIC 5 times over 30, and MDL 17 times over 30. If we pass to $N = 100$, the order designated by FPE and AIC is the correct one, namely $n = 3$, 10 times over 30. Criterion MDL, instead, prefers a more parsimonious model of order $n = 2$ (11 times over 30).

If N is further increased, then the correct choice of the order is more and more frequent. In particular, when $N = 500$, with AIC, the correct order $n = 3$ is estimated 25 times over 30, while $\hat{n} = 4$ is 2 times over 30. For the remaining $30 - 25 - 2 = 3$ trials, we find once $\hat{n} = 6$, once $\hat{n} = 7$, and once $\hat{n} = 8$. Always for $N = 500$, if we pass to MDL, the correct order $\hat{n} = 3$ is found in all 30 trials.

We finally observe that the results achieved with FPE and AIC are not coincident, but very similar.

Example 6.4 (Order determination with cross validation) For the sake of completeness, we have also applied the cross-validation procedure to the data of the previous example. Again we have considered AR(n) models of order $n = 1, 2, \ldots, 10$.

The model parameter is found by processing a segment of data and then the optimal model is tested with a different segment of data. We denote by N_{id} and N_{val} the number of identification and validation data, respectively.

Table 6.3 Choice of the optimal order via the cross-validation method (Example 6.4).

N_{val} \ $\hat{n}$	1	2	3	4	5	6	7	8	9	10
$N_{id} = 50$										
50	3	9	7	3	2	1	4			1
100	5	6	11	4	1	1	1			1
200	5	11	9	1	3	1				
$N_{id} = 100$										
50	1	7	8	2		1	6	2		3
100	1	6	7	6	4			1	4	1
200	1	5	13	4	2	2	2	1		
$N_{id} = 500$										
50	2	2	9	1	4	2	3		3	4
100		2	10	4	1	4		4	1	3
200	2	4	9	4	1	3		1	1	5

Empty boxes correspond to orders that are never chosen.

As optimal order, we consider the one for which the variance of the prediction error over the validation data is minimized. The results are summarized in Table 6.3, where we report the number of times (over a total of 30 trials) for which order n was chosen. As can be seen, the percentage of errors is non-negligible even for $N_{id} = 500$.

Remark 6.1 (Asymptotic analysis of model complexity criteria) The criteria previously introduced supply a useful tool for complexity selection. Under the assumption that the data generation mechanism belongs to the class of models, advanced theoretical studies have led to the conclusion that FPE and AIC have a non-null probability of overestimating the order of the model, while MDL leads to an asymptotically correct estimate of the order.

Above criteria for model complexity selection have been studied in a large number of papers. Among them, for the statistical-based methods such as AIC and FPE, we acknowledge the work by Akaike (1974). For MDL, we make reference to a Rissanen (1978).

6.6 Durbin–Levinson Algorithm

In predictive identification, we learned how to identify AR, MA, or ARMA models by means of least squares (AR) or maximum likelihood (MA and ARMA) techniques. We now complement what we have already seen with further observations providing useful hints on the class of models best suited to fit a given sequence of data. To this end, we first recall that we defined the covariance function and the normalized covariance function in Chapter 1.

Their estimation from data has been treated in Chapter 2. A distinctive feature of these functions is that, for an MA(n) process, they are null for $\tau \geq n + 1$:

$$\gamma(\tau) = E[y(t)y(t + \tau)] = 0, \ \tau \geq n + 1 \text{ for an MA}(n) \text{ process.}$$

So, if we have a time series whose covariance function (or normalized covariance function $\rho(\tau) = \gamma(\tau)/\gamma(0)$) is vanishing at a certain point, we can infer that the series can be interpreted as a moving average of a certain order, the order after which the covariance function begins to be equal to zero.

We now pose the following main question: Can a similar tool be set up for AR models?

Fortunately, the reply is affirmative. Indeed, it is possible to introduce another covariance function, the so-called *partial covariance function* or simply *PARCOV*, playing for AR processes a role analogous to that played by the covariance function for MA processes.

In order to define the PARCOV, we have to make a preliminary step by introducing the so-called *Durbin–Levinson algorithm*, an algorithm enabling the iterative estimation of AR(n) models by iterating over the order n.

The algorithm can be introduced by referring either to the least squares normal equations or to the Yule–Walker equations. Here, we adopt the latter option, by first discussing the passage from the AR(1) to the AR(2) model.

6.6.1 Yule–Walker Equations for Autoregressive Models of Orders 1 and 2

For an AR(1) model

$$y(t) = a_1 y(t - 1) + \eta(t), \quad \eta(t) \sim \text{WN}(0, \lambda^2),$$

the Yule–Walker equations are obtained by multiplying both members by $y(t - \tau)$ and then applying the expected value operator, so that

$$\gamma(0) = a_1 \gamma(1) + \lambda^2,$$
$$\gamma(\tau) = a_1 \gamma(\tau - 1), \quad \forall \tau \geq 1.$$

These formulas can be used in two ways: given the model, they enable evaluating the covariance function $\gamma(\cdot)$; alternatively, they enable the determination of a_1 and λ^2 given $\gamma(\cdot)$. The point of view of interest here is the latter one. To be precise, by solving the Yule–Walker equations, we can determine parameter a_1 as

$$a_1 = \frac{\gamma(\tau)}{\gamma(\tau - 1)} \tag{6.3}$$

and then we can find variance λ^2 as

$$\lambda^2 = \gamma(0) - a_1 \gamma(1). \tag{6.4}$$

The formula for a_1 holds for any τ. In real problems, however, $\gamma(\cdot)$ must be estimated from data, as seen in Section 2.5, for instance with the expression (2.10). From that formula, we learn that, when N data are available, only $N - \tau$ snapshots are used to compute the covariance at lag τ. Hence, when τ increases, the estimate is formed with a decreasing number of data and is therefore more and more uncertain. For this reason, one usually resorts to the Yule–Walker formulas

$$a_1 = \frac{\gamma(1)}{\gamma(0)}$$

and

$$\lambda^2 = \gamma(0) - \frac{\gamma(1)^2}{\gamma(0)}.$$

Turn now to the AR(2) model:

$$y(t) = A_1 y(t-1) + A_2 y(t-2) + \eta(t), \quad \eta(t) \sim \text{WN}(0, \Lambda^2),$$

where the parameters are indicated by capital symbols to avoid possible confusions with the AR(1) case. In other words, a_1 is the first parameter of the AR(1) model, while A_1 is the first parameter of the AR(2) model.

The Yule–Walker equations are again obtained by multiplying the two members of this relation by $y(t - \tau)$ and then applying the expected value operator. Thus, for $\tau = 0, 1, 2$,

$$\gamma(0) = A_1 \gamma(1) + A_2 \gamma(2) + \Lambda^2,$$
$$\gamma(1) = A_1 \gamma(0) + A_2 \gamma(1),$$
$$\gamma(2) = A_1 \gamma(1) + A_2 \gamma(0).$$

Given $\gamma(0), \gamma(1)$, and $\gamma(2)$, this is a linear system in the unknowns A_1, A_2, and Λ^2. From the last two equations, we compute A_1 and A_2 as

$$\begin{bmatrix} A_1 \\ A_2 \end{bmatrix} = \begin{bmatrix} \gamma(0) & \gamma(1) \\ \gamma(1) & \gamma(0) \end{bmatrix}^{-1} \begin{bmatrix} \gamma(1) \\ \gamma(2) \end{bmatrix}. \tag{6.5}$$

Then, by means of the first equation, Λ^2 can be determined.

6.6.2 Durbin–Levinson Recursion: From AR(1) to AR(2)

Let's compare the Yule–Walker equations previously obtained.
For AR(2):

$$\gamma(0) = A_1 \gamma(1) + A_2 \gamma(2) + \Lambda^2, \tag{6.6a}$$

$$\gamma(1) = A_1 \gamma(0) + A_2 \gamma(1), \tag{6.6b}$$

$$\gamma(2) = A_1 \gamma(1) + A_2 \gamma(0), \tag{6.6c}$$

and for AR(1):

$$\gamma(0) = a_1\gamma(1) + \lambda^2,$$ (6.7a)

$$\gamma(1) = a_1\gamma(0).$$ (6.7b)

When we use such equations to estimate the parameters, the covariance function is given. Hence, the values of $\gamma(\tau)$ appearing in these systems are the same, as they depend upon data only. Hence, from (6.6b) and (6.7b),

$$a_1\gamma(0) = A_1\gamma(0) + A_2\gamma(1),$$

so that

$$A_1 = a_1 - A_2\frac{\gamma(1)}{\gamma(0)}.$$ (6.8)

Moreover, (6.7a) and (6.6a) entail

$$a_1\gamma(1) + \lambda^2 = A_1\gamma(1) + A_2\gamma(2) + \Lambda^2.$$ (6.9)

In (6.9), replacing A_1 with expression (6.8), we obtain

$$a_1\gamma(1) + \lambda^2 = a_1\gamma(1) - A_2\frac{\gamma(1)^2}{\gamma(0)} + A_2\gamma(2) + \Lambda^2.$$

Hence,

$$\Lambda^2 = \lambda^2 - A_2\left[\gamma(2) - \frac{\gamma(1)^2}{\gamma(0)}\right].$$

We now prove that the quantity in the square brackets can be expressed as follows:

$$\gamma(2) - \frac{\gamma(1)^2}{\gamma(0)} = A_2\lambda^2.$$ (6.10)

Indeed, substituting expression (6.8) in (6.6c), we have

$$\gamma(2) = \left[a_1 - A_2\frac{\gamma(1)}{\gamma(0)}\right]\gamma(1) + A_2\gamma(0).$$

Here, we replace a_1 with the expression deduced from (6.7b):

$$\gamma(2) - \frac{\gamma(1)^2}{\gamma(0)} = A_2\left[\gamma(0) - \frac{\gamma(1)^2}{\gamma(0)}\right].$$

On the other hand,

$$\gamma(0) - \frac{\gamma(1)^2}{\gamma(0)} = \lambda^2,$$

so that (6.10) is proven. Therefore, the two variances are related as

$$\Lambda^2 = \lambda^2[1 - A_2^2].$$

Summing up, parameters A_1, A_2, and Λ^2 of the AR(2) can be obtained from parameters a_1 and λ^2 of the AR(1), thanks to the following relationships known as *Durbin–Levinson* formulas:

$$A_2 = \frac{1}{\lambda^2} [\gamma(2) - a_1\gamma(1)], \tag{6.11a}$$

$$A_1 = a_1 - A_2 \frac{\gamma(1)}{\gamma(0)}, \tag{6.11b}$$

$$\Lambda^2 = \lambda^2[1 - A_2^2]. \tag{6.11c}$$

According to them, to determine the AR(2) model once the AR(1) has been estimated, one has to evaluate first parameter A_2 with (6.11a) and then update the remaining parameters by means of (6.11b) and (6.11c).

Remark 6.2 (From λ^2 to Λ^2) Passing from the AR(1) to the AR(2) model, there are more degrees of freedom for data description. Therefore, anyone would expect a better fit, i.e.

$$\Lambda^2 \leq \lambda^2.$$

This is indeed correct, as the following observation shows. Consider the transfer function of the AR(2) model:

$$W(z) = \frac{1}{1 - A_1 z^{-1} - A_2 z^{-2}} = \frac{z^2}{z^2 - A_1 z - A_2}.$$

In order to have a stationary process, the stability condition has to be satisfied. This means that the roots of the denominator of $W(z)$ (written in positive powers of z), namely the roots of $z^2 - A_1 z - A_2$, must be inside the unit disk. Since parameter A_2 is the product of such roots, we conclude that

$$|A_2| \leq 1.$$

In turn, in view of (6.11c), this leads to the expected conclusion $\Lambda^2 \leq \lambda^2$.

Remark 6.3 (From a_1 to A_1 and A_2) Suppose that $\gamma(2) = a_1\gamma(1)$. Then, from (6.11a) and (6.11b), we see that $A_2 = 0$ and $A_1 = a_1$. Thus, the AR(2) model self-downgrades to the already estimated AR(1) model. This is no surprise since the condition $\gamma(2) = a_1\gamma(1)$ is equivalent to saying that model AR(1) perfectly fits $\gamma(2)$ besides $\gamma(0)$ and $\gamma(1)$. Hence, the AR(2) model (computed from $\gamma(0)$, $\gamma(1)$, and $\gamma(2)$) cannot provide a better performance of the AR(1). Correspondingly, $\Lambda^2 = \lambda^2$, as it appears from (6.11c).

When instead $A_2 \neq 0$, then $\Lambda^2 < \lambda^2$, i.e. passing from an AR(1) to an AR(2) leads to a real improvement in data fitting.

6.6.3 Durbin–Levinson Recursion for Models of Any Order

The Yule–Walker equations can be generalized to autoregressive models of any order. For an AR(n), we have

$$
\begin{bmatrix} a_1 \\ a_2 \\ \cdots \\ a_{n-1} \\ a_n \end{bmatrix}
=
\begin{bmatrix}
\gamma(0) & \gamma(1) & \gamma(2) & \cdots & \gamma(n-1) \\
\gamma(1) & \gamma(0) & \gamma(1) & \cdots & \gamma(n-2) \\
\cdots & \cdots & \cdots & \cdots & \cdots \\
\gamma(n-2) & \gamma(n-3) & \gamma(n-4) & \cdots & \gamma(1) \\
\gamma(n-1) & \gamma(n-2) & \gamma(n-3) & \cdots & \gamma(0)
\end{bmatrix}^{-1}
\begin{bmatrix} \gamma(1) \\ \gamma(2) \\ \cdots \\ \gamma(n-1) \\ \gamma(n) \end{bmatrix}. \quad (6.12)
$$

Note that this computation requires the inversion of a Toeplitz matrix of dimension equal to the number of parameters to be estimated, an inversion that can be avoided by the Durbin–Levinson algorithm.

In order to write the algorithm for models of any orders, we shall denote by

$$
a_i^{(n+1)}, \quad i = 1, 2, \ldots, n+1
$$

the parameters of the AR($n+1$) model obtained by solving the corresponding Yule–Walker equations and with

$$
a_i^{(n)}, \quad i = 1, 2, \ldots, n
$$

the parameters of the AR(n) model. Moreover, we indicate with $\lambda_{(n+1)}^2$ and $\lambda_{(n)}^2$ the prediction error variances for AR($n+1$) and AR(n), respectively.

The Durbin–Levinson algorithm for the passage from AR(n) $\rightarrow$ AR($n+1$) is

$$
a_{n+1}^{(n+1)} = \frac{1}{\lambda_{(n)}^2} \left[\gamma(n+1) - \sum_1^n a_i^{(n)} \gamma(n+1-i) \right], \quad (6.13a)
$$

$$
a_i^{(n+1)} = a_i^{(n)} - a_{n+1}^{(n+1)} a_{n+1-i}^{(n)}, \quad (6.13b)
$$

$$
\lambda_{(n+1)}^2 = \left[1 - (a_{n+1}^{(n+1)})^2 \right] \lambda_{(n)}^2. \quad (6.13c)
$$

By setting $n = 1$, these formulas return equations (6.11).

This algorithm allows the estimation of various autoregressive models of different orders, by the inversion of a scalar for each order. In this way, the inversion of a Toeplitz matrix of dimension equal to the number of parameters to be estimated is no more necessary.

Remark 6.4 (Reflection on the transition from AR(n) to AR($n+1$)) The previous considerations for the passage AR(1) > AR(2) can be extended to models of any order as follows:

(i) Suppose that

$$
\gamma(n+1) - \sum_1^n a_i^{(n)} \cdot \gamma(n+1-i) = 0.
$$

This means that model AR(n) fits the Yule–Walker equations of order $n + 1$ as well. Therefore, it is not surprising that

$$a_{n+1}^{(n+1)} = 0,$$

as claimed by Eq. (6.13a). This means that the AR($n + 1$) model downgrades to the AR(n) model.

(ii) Parameter

$$a_{n+1}^{(n+1)}$$

is the known term of a stable polynomial. Hence, we expect that

$$|a_{n+1}^{(n+1)}| < 1.$$

(iii) From the previous inequality and Eq. (6.13c), we conclude that

$$\lambda_{(n+1)}^2 \leq \lambda_{(n)}^2,$$

meaning that the fitting of the higher order model is more accurate. From (6.13c), we also see that $\lambda_{(n+1)}^2 = \lambda_{(n)}^2$, when $a_{n+1}^{(n+1)} = 0$.

Remark 6.5 (LS Versus YW) We worked out the Durbin–Levinson algorithm by making reference to the Yule–Walker (YW) equations. We will now see that the least squares estimator asymptotically coincides with is the Yule–Walker estimator. Indeed, the LS formula is

$$\hat{\theta}_{N_{(LS)}} = \left\{ \frac{1}{N} \sum_t \phi(t)\phi(t)' \right\}^{-1} \left\{ \frac{1}{N} \sum_t \phi(t)y(t) \right\},$$

where $\phi(t)$ is the observation vector. For a time series, this vector is

$$\phi(t) = [y(t-1)y(t-2)\cdots y(t-n)]'.$$

When $N \to \infty$,

$$\frac{1}{N} \sum_t \phi(t)\phi(t)' = \frac{1}{N} \begin{bmatrix} \sum_t y(t)^2 & \cdots & \sum_t y(t)y(t-n+1) \\ \cdots & \cdots & \cdots \\ \cdots & \cdots & \cdots \\ \sum_t y(t)y(t-n+1) & \cdots & \sum_t y(t)^2 \end{bmatrix}$$

$$\to \begin{bmatrix} \gamma(0) & \cdots & \gamma(n-1) \\ \gamma(1) & \cdots & \gamma(n-2) \\ \cdots & \cdots & \cdots \\ \gamma(n-1) & \cdots & \gamma(0) \end{bmatrix}$$

while

$$\frac{1}{N}\sum_{t}\phi(t)y(t) = \frac{1}{N}\begin{bmatrix}\sum_{t}y(t)y(t-1)\\ \dots \\ \dots \\ \sum_{t}y(t)y(t-n)\end{bmatrix} \rightarrow \begin{bmatrix}\gamma(1)\\ \gamma(2)\\ \dots \\ \gamma(n)\end{bmatrix}.$$

By comparing these expressions with (6.12), we see that, for a large number of data, the LS estimates coincide with the YW estimates.

6.6.4 Partial Covariance Function

By means of the Durbin–Levinson algorithm, one can easily estimate autoregressive models of increasing orders. For an AR(n), the corresponding parameters have been denoted with the symbols $a_i^{(n)}, i = 1, 2, \dots, n$. We now focus our attention on the last of such parameters, namely

$$a_n^{(n)}$$

and we introduce the *partial covariance function* as

$$\text{PARCOV}(n) = a_n^{(n)}, \quad n = 1, 2, \dots$$

Motivated by a search of an analogy with the covariance function $\gamma(\tau)$, the PARCOV is written in τ as well:

$$\text{PARCOV}(\tau) = a_\tau^{(\tau)}, \quad \tau = 1, 2, \dots$$

From previous considerations, we know that

$$|\text{PARCOV}(\tau)| < 1, \quad \tau = 1, 2, \dots.$$

Moreover, if the data are generated by an autoregressive model of order n, then

$$\text{PARCOV}(\tau) = 0, \quad \tau = n+1, n+2, \dots$$

A main issue of time series analysis is the choice of the type of model to best fit data. If the series can be seen as a realization of a stationary process, then we can proceed by estimating the normalized covariance function $\rho(\tau) = \gamma(\tau)/\gamma(0)$ and the partial covariance function $\text{PARCOV}(\tau)$. Note that both functions take values between -1 and $+1$.

The diagrams of $\rho(\tau)$ and $\text{PARCOV}(\tau)$ as functions of τ, which can be easily constructed from data, are very useful to have an hint on the best type of model to be adopted. Indeed, we know that, if $\rho(\tau) \simeq 0$ for $\tau > n$, then we can assume that data were generated by an MA model of order n. If instead $\text{PARCOV}(\tau) = 0$ for $\tau > n$, the data can be interpreted as an AR model of order n. If neither $\rho(\tau)$ nor $\text{PARCOV}(\tau)$ present a sharp zeroing point, then one can resort to an ARMA model whose orders can be decided by analyzing objective criteria such as the FPE, AIC, or MDL.

7

Identification of State Space Models

7.1 Introduction

In this chapter, we address the problem of black box identification of state space models, namely the problem of estimating the system matrices directly from input–output observations. To face it, we introduce a new estimation paradigm, different from the ideas around which the previous approach of parametric identification via the prediction error rationale is hinged. The new estimation paradigm is rooted in system notions such as reachability, observability, and system realization. The readers who are not familiar with such topics are referred to Appendix A.

Actually, many methods have been studied in this field, giving rise to a branch of system identification known in our days as *subspace identification methods*. Herein we present a basic technique, the roots of which go back to the celebrated paper by Ho and Kalman (1965). We focus on SISO deterministic models:

$$x(t + 1) = Fx(t) + Gu(t), \tag{7.1a}$$

$$y(t) = Hx(t), \tag{7.1b}$$

where

- input $u(\cdot)$ and output $y(\cdot)$ are scalar,
- state $x(\cdot)$ is a vector with n entries,
- F, G, and H are real matrices of suitable dimensions (compatible with the dimensions of u, x, y).

We address the following:

State space identification problem
Given the measurements of input $u(\cdot)$ and output $y(\cdot)$ over a certain time interval, whereas state $x(\cdot)$ in not measurable, find

- order n,
- matrices F, G, and H.

Model Identification and Data Analysis, First Edition. Sergio Bittanti.
© 2019 John Wiley & Sons, Inc. Published 2019 by John Wiley & Sons, Inc.

A first main observation is that, for any different choice of the basis in the state space, there is a different set of matrices F, G, and H corresponding to the same input–output behavior. Indeed, a change of basis amounts to defining a new state vector, say $\tilde{x}(t)$, related to the original state vector $x(t)$ as

$$\tilde{x}(t) = Tx(t), \quad \det\ T \neq 0.$$

It is easy to see that, in the new representation, the system is described by

$$\tilde{x}(t+1) = \tilde{F}\tilde{x}(t) + \tilde{G}u(t),$$
$$y(t) = \tilde{H}\tilde{x}(t) + \tilde{K}u(t),$$

with

$$\tilde{F} = TFT^{-1}, \quad \tilde{G} = TG, \quad \tilde{H} = HT^{-1}, \quad \tilde{K} = K.$$

Thus, one obtains a different quadruple of matrices, apparently a "new" system. However, the transfer function does not change (see Appendix A).

There is a second reason of multiplicity in the state representation: The order is not uniquely defined. This point is discussed in the following elementary

Example 7.1 In the system of order 2

$$x_1(t+1) = 0.5x_1(t) + 2u(t), \tag{7.2a}$$

$$x_2(t+1) = 0.7x_1(t) + u(t), \tag{7.2b}$$

$$y(t) = x_1(t), \tag{7.2c}$$

input $u(\cdot)$ has effect on both $x_2(\cdot)$ and $x_1(\cdot)$; however, $y(\cdot)$ depends only upon $x_1(\cdot)$, so that the system

$$x_1(t+1) = 0.5x_1(t) + 2u(t), \tag{7.3a}$$

$$y(t) = x_1(t), \tag{7.3b}$$

has identical external behavior of the previous one, even if its order is only 1. This is also confirmed by the fact that the transfer functions of the above systems coincide:

$$G(z) = \frac{2}{z - 0.5}.$$

The reason of order multiplicity pointed out in this example can be removed by introducing the notion of *minimal realization*.

Definition 7.1 Given a transfer function $G(z)$, a system described by the triple (F, G, H) having $G(z)$ as transfer function is said to be a *realization* of $G(z)$.

A realization is said to be minimal if there does not exist any other realization with lower order. The number of state variables of a minimal realization is named *Smith–McMillan degree* of the system.

Example 7.2 For the system considered in the previous example, with the transfer function

$$G(z) = \frac{2}{z - 0.5},$$

representation (7.3) is minimal, whereas (7.2) is not.

In view of the above discussion, we reformulate the original state identification problem as follows:

State space identification problem
Given the measurements of input $u(\cdot)$ and output $y(\cdot)$ over a certain time interval, whereas state $x(\cdot)$ is not measurable, find

- order n of a minimal realization,
- matrices F, G, and H of a minimal realization.

A fundamental result concerning the minimal realization is provided by the following statement (see Appendix A).

Proposition 7.1 (Minimal realization) *A realization is minimal if and only if the corresponding system is reachable and observable.*

Hence, finding a minimal realization is equivalent to finding a reachable and observable realization.

7.2 Hankel Matrix

We will now see how we can proceed to find a minimal realization of a system from the measurement of its impulse response. To this purpose, we introduce the following.

Definition 7.2 (Hankel matrix) Given system (7.1), the $k \times k$ matrix

$$\mathcal{H}^k = \begin{bmatrix} HG & HFG & \cdots & HF^{k-1}G \\ HFG & HF^2G & \cdots & HF^kG \\ \vdots & & & \\ HF^{k-1}G & HF^kG & \cdots & HF^{2k-2}G \end{bmatrix}$$

is called the *Hankel matrix* of order k.

This matrix can be constructed from experimental data. Indeed, the response of system (7.1) to input

$$u(t) = \begin{cases} 1, & t = 0, \\ 0, & t > 0, \end{cases}$$

for an initial state $x(0) = 0$, is given by

$$HF^{t-1}G,$$

as it is easy to verify. The input is an impulse; hence, this is the impulse response, denoted as $h(\cdot)$

$$h(t) = HF^{t-1}G.$$

We can therefore reformulate the previous definition as follows.

Definition 7.3 (Experimental Hankel matrix)   For a given system with impulse response $h(\cdot)$, the $k \times k$ matrix

$$\mathcal{H}_e^k = \begin{bmatrix} h(1) & h(2) & \cdots & h(k) \\ h(2) & h(3) & \cdots & h(k+1) \\ \vdots & & & \\ h(k) & h(k+1) & \cdots & h(2k-1) \end{bmatrix}$$

is named *experimental Hankel matrix* of order k.

Note that to construct this $k \times k$ matrix, we need $2k - 1$ samples of the impulse response.

From the Hankel matrix, we can determine the system matrices as described in the subsequent sections.

7.3 Order Determination

It is straightforward to see that $\mathcal{H}^k$ can be factorized as

$$\mathcal{H}^k = \begin{bmatrix} H \\ HF \\ \vdots \\ HF^{k-1} \end{bmatrix} [G \quad FG \quad \cdots \quad F^{k-1}G].$$

In other words, the Hankel matrix of order k is the product of the observability matrix $\mathcal{O}^k$ times the reachability matrix $\mathcal{R}^k$:

$$\mathcal{H}^k = \mathcal{O}^k \mathcal{R}^k. \tag{7.4}$$

Suppose that the *Smith–McMillan degree* of the system is n. Furthermore, suppose that the dimension k of the Hankel matrix is such that $k \geq n$. In view of Proposition 7.1, a minimal realization must satisfy the reachability condition and the observability condition. Therefore, the reachability matrix and the observability matrix must have full rank, namely

$$\text{rank}[\mathcal{O}^k] = n, \quad \forall k \geq n; \qquad \text{rank}[\mathcal{R}^k] = n, \quad \forall k \geq n.$$

On the other hand, expression (7.4) shows that the Hankel matrix is the product of two matrices. We can then resort to a celebrated matrix analysis result on the rank of the product of two matrices, the so-called *Sylvester inequality*, see Appendix B. By applying such result to (7.4), one can conclude that

$$\text{rank}[\mathcal{H}^k] = n, \quad \forall k \geq n.$$

This fact provides a fundamental clue to identify the order n of a minimal realization: By constructing the *experimental Hankel matrix* from data, order n can be deduced by determining the rank of that matrix. Of course, this statement is correct under the condition that the dimension k of the Hankel matrix is large enough to guarantee that $k \geq n$.

7.4 Determination of Matrices *G* and *H*

Once the appropriate order n has been found by a rank analysis of the experimental $k \times k$ Hankel matrix, one can work out the $n \times n$ submatrix, $\mathcal{H}_e^n$, by restricting to the first n rows and the first n columns. In turn, such restricted Hankel matrix can be factorized as the product of two matrices, both of dimension $n \times n$, $\mathcal{O}_e^n$ and $\mathcal{R}_e^n$ as

$$\mathcal{H}_e^n = \mathcal{O}_e^n \mathcal{R}_e^n.$$

This factorization is not unique; indeed, if Q is any $n \times n$ matrix such that $QQ' = I$, we can write

$$\mathcal{H}_e^n = \mathcal{O}_e^n QQ' \mathcal{R}_e^n = \tilde{\mathcal{O}}_e^n \tilde{\mathcal{R}}_e^n,$$

where $\tilde{\mathcal{O}}_e^n = \mathcal{O}_e^n Q$ and $Q' \tilde{\mathcal{R}}_e^n$. This multiplicity of factors of the Hankel matrix corresponds to the multiplicity of minimal realizations: All such representations share the same order, but they exhibit different system matrices corresponding to all possible changes of basis in the state space.

Once a factorization has been chosen, we can proceed to "destructure" $\mathcal{O}_e^n$ and $\mathcal{R}_e^n$ into their rows and columns as

$$
\mathcal{H}_e^n = \mathcal{O}_e^n \mathcal{R}_e^n = \begin{bmatrix} H_e \\ H_e F_e \\ \vdots \\ H_e F_e^{n-1} \end{bmatrix} [G_e \quad F_e G_e \quad \cdots \quad F_e^{n-1} G_e].
$$

Therefore, the first row H_e of the first factor $\mathcal{O}_e^n$ provides an estimate of the output matrix H of the system, and the first column G_e of the second factor $\mathcal{R}_e^n$ provides an estimate of the input matrix G.

7.5 Determination of Matrix F

From the factorization of the $n \times n$ matrix $\mathcal{H}^n$, one obtains the observability matrix:

$$
\mathcal{O}_e^n = \begin{bmatrix} H_e \\ H_e F_e \\ \vdots \\ H_e F_e^{n-1} \end{bmatrix}.
$$

This $n \times n$ matrix is invertible since n has been chosen as the rank of the Hankel matrix, a choice guaranteeing that the pair F_e, H_e is observable.

Consider now the Hankel matrix of order $n + 1$:

$$
\mathcal{H}_e^{n+1} = \begin{bmatrix} H_e \\ H_e F_e \\ \vdots \\ H_e F_e^{n-1} \\ H_e F_e^n \end{bmatrix} [G_e \quad F_e G_e \quad \cdots \quad F_e^{n-1} G \quad F_e^n G_e].
$$

Its factorization as the product of an $(n + 1) \times (n)$ matrix times an $(n) \times (n + 1)$ matrix provides as first factor

$$
\begin{bmatrix} H_e \\ H_e F_e \\ \vdots \\ H_e F_e^{n-1} \\ H_e F_e^n \end{bmatrix}.
$$

Here, if we omit the first row, we obtain the $n \times n$ matrix:

$$\mathcal{O}_e^{n\uparrow} = \begin{bmatrix} H_e F_e \\ \vdots \\ \vdots \\ H_e F_e^{n-1} \\ H_e F_e^n \end{bmatrix}.$$

Comparing $\mathcal{O}_e^{n\uparrow}$ with $\mathcal{O}_e^n$, it is apparent that the following relation, known as *shift invariance property*, holds true

$$\mathcal{O}_e^{n\uparrow} = \mathcal{O}_e^n F_e.$$

Therefore, an estimate F_e of F can be found as

$$F_e = (\mathcal{O}_e^n)^{-1} \mathcal{O}_e^{n\uparrow}.$$

7.6 Mid Summary: An Ideal Procedure

At this point, we are well advised to make a concise summary of the procedure seen above. After constructing the $k \times k$ Hankel matrix from the experimental impulse response, we have to examine its rank. If the data were generated by a finite dimension dynamic system with minimal representation of order n, then the rank of the experimental Hankel matrix is constant $\forall k \geq n$. So, the minimal order can be found by a search of the rank of such matrix. Having determined n, one can focus on the $n \times n$ Hankel matrix. By its factorization, one can work out the observability matrix and the reachability matrix. The first row of the observability matrix provides H and the first column of reachability matrix G. As for F, it can be computed via the shift invariance property.

7.7 Order Determination with SVD

In the above procedure, there is a critical point. To be precise, Achille's heel is the rank determination. Indeed, there are many examples showing that even a small perturbation in an entry of a matrix may result in a displacement of the eigenvalues and in a jump in the rank. This issue, properly discussed in many books such as Wilkinson, J.H. (1988), Chapter 2 (Perturbation Theory), is illustrated in the following examples.

Example 7.3 Consider the $n \times n$ matrix

$$A = \begin{bmatrix} s & 1 & 0 & 0 & \cdots & 0 & 0 \\ 0 & s & 1 & 0 & \cdots & 0 & 0 \\ \vdots & \vdots & \vdots & \vdots & \cdots & \vdots & \vdots \\ 0 & 0 & 0 & 0 & \cdots & s & 1 \\ \epsilon & 0 & 0 & 0 & \cdots & 0 & s \end{bmatrix}.$$

If $\epsilon = 0$, the matrix is upper triangular so that the eigenvalues are the entries along the diagonal. Hence, there are n eigenvalues, all coincident and equal to s. Note that for $\epsilon = 0$, the above matrix is a unique Jordan block (see Appendix B).

If $\epsilon \neq 0$, then the eigenvalues split into n distinct eigenvalues of the form $s + \epsilon^{1/n}\alpha_i$, where α_i is one of the n square roots of 1.

Example 7.4 This example, drawn from the book by Strang (2016), shows that a small perturbation in the entry of a matrix may lead to a rank variation. When $\epsilon = 0$, the eigenvalues of the 4×4 matrix

$$A = \begin{bmatrix} 0 & 1 & 0 & 0 \\ 0 & 0 & 2 & 0 \\ 0 & 0 & 0 & 3 \\ \epsilon & 0 & 0 & 0 \end{bmatrix}$$

are all null. As for the rank, there are three independent columns so that rank$[A] = 3$.

Consider now the same matrix with $\epsilon = \frac{1}{60\ 000}$. The eigenvalues become $\pm\frac{1}{10}$ and $\pm\frac{i}{10}$, so that rank$[A] = 4$.

How to overcome the problems in the determination of the order for black box identification has been the subject of many contributions. Two landmark papers are the following: Zeiger and McEwen (1974) and Kung (1978).

The basic idea is to resort to a numerically reliable algorithm to assess the rank of the Hankel matrix, notwithstanding the noise affecting data. To this purpose, instead of focusing on the eigenvalues, one focuses on the so-called *singular values*, which will be introduced shortly after this paragraph. Indeed, it can be shown that the rank of a matrix coincides not only with the number of non-null eigenvalues but also with the number of non-null singular values, and the singular values are much less sensitive to perturbations in the Hankel matrix.

To define the singular values, we introduce the concept of *singular value decomposition* (SVD). We limit our attention to the Hankel matrix; for more details on the SVD, see Appendix B.

Proposition 7.2 (SVD theorem applied to the Hankel matrix) *It is possible to factorize the n × n Hankel matrix $\mathcal{H}_e^n$ as follows:*

$$\mathcal{H}_e^n = U\,S\,V', \tag{7.5}$$

where U, S, V are n × n matrices such that

- *U is orthogonal – namely $UU' = I$,*
- *V is orthogonal – namely $VV' = I$,*
- *S is a diagonal with diagonal entries that are* real *and* non-negative. *Moreover, such entries, denoted as $\delta_1, \delta_2, \delta_3, \ldots, \delta_k$, can be set in decreasing order, i.e. $\delta_1 \geq \delta_2 \geq \delta_3 \cdots \geq \delta_k \geq 0$.*

The entries on the diagonal of S are named *singular values*. The columns u_i and v_i of U and V are called *left singular vectors* and *right singular vectors*, respectively. Note that the left singular vectors (right singular vectors) are independent of each other due to the orthogonality property (of U and V, respectively). Finally, observe that, while there is a multiplicity of matrices U and V, which can be associated to factorization (7.5), the singular values are unique.

Remark 7.1 (Spreading a matrix into a sum of rank 1 matrices) The SVD formula (7.5) can be equivalently rewritten as

$$\mathcal{H}_e^n = \sum{}_1^k \delta_i u_i v_i'. \tag{7.6}$$

Each matrix $u_i v_i'$ appearing in such sum has rank 1. Moreover, each matrix $u_i v_i'$ is independent of another matrix $u_j v_j', j \neq i$, in the sense that $\alpha u_i v_i' + \beta u_j v_j' = 0$ if and only if $\alpha = 0$ and $\beta = 0$. Therefore, from expression (7.6), we learn that the rank of the matrix is equal to the number of non-null singular values.

7.8 Reliable Identification of a State Space Model

We are now in a position to provide a sound procedure for the estimation of a state space model. First, we reconsider the problem of finding the appropriate order. As previously seen, the issue is "simply" to find the rank of a matrix. This is a clean and nice statement; however, even small errors in data make rank determination hard. To safely face the problem, we have to move from the notion of *rank* to that of *numerical rank*, as provided by the analysis of the singular values. To be precise, with the data one constructs a $k \times k$ Hankel matrix $\mathcal{H}_e^k$, with k large enough so that one can safely believe that k is larger than the dimension of the minimal realization, $k > n$. Then perform the SVD of $\mathcal{H}_e^k$, as in (7.5), and organize the singular values in decreasing order

$$\delta_1 \geq \delta_2 \geq \cdots \geq \delta_r \geq \delta_{r+1} \geq \cdots \geq \delta_k \geq 0.$$

Normally, the number of non-null singular values is much higher than n. This is due to noisy and rounding effects, imprecisions, and so on. The SVD approximation procedure originates from the idea that some of the smaller singular values may be negligible. Assume for instance that the Hankel matrix has 10 singular values, given by

$$10, 5, 1, 10^{-1}, 10^{-2}, 10^{-8}, 10^{-9}, 10^{-10}, 0, 0.$$

We see that the *rank* of the matrix is 8 as there are 8 non-null singular values. However, if the original data has only six digits of accuracy, any singular value $\delta_i < 10^{-6}$ should be considered zero. Hence, we would consider the *numerical rank* to be 5. Correspondingly, we have a partition of the 10 singular values into two families: the *dominant singular values* (the first five in the example) and the remaining ones, the *secondary singular values*. On the other hand, considering the gap between the first three singular values and the remaining ones, another possibility to consider as appropriate a model of order 3 only. The basic idea is that dominant singular values capture the dynamics in data, the other ones can be seen as effect of noise.

The problem of course is to decide "how small is small," that is, at what point to consider a singular value negligible. There are situations where there is not a clear gap in the sequence of singular values, so the numerical rank depends entirely upon the choice of the zero tolerance threshold.

Once the dominant singular values of the Hankel matrix have been chosen, the order n of the system is identified as the number of such singular values. Correspondingly, in the SVD factorization of matrix $\mathscr{H}_e^k = USV'$, we can partition matrix S into four blocks as follows:

$$S = \begin{bmatrix} \hat{S} & 0 \\ 0 & \tilde{S} \end{bmatrix},$$

where $\hat{S}$ and $\tilde{S}$ are the submatrices of S with the dominant singular values and the secondary singular values, respectively. To proceed further, in the SVD, we force to zero the secondary singular values, thus replacing matrix S with

$$S^* = \begin{bmatrix} \hat{S} & 0 \\ 0 & 0 \end{bmatrix}.$$

Correspondingly, we partition U and V and define matrices

(a) $\hat{U}$ given by the first n columns of U,
(b) $\hat{V}$ given by the first n columns of V.

Thus, as suggested in Figure 7.1, we can replace

$$\mathscr{H}_e^k = US^*V'$$

with

$$\hat{\mathscr{H}}_e^k = \hat{U}\hat{S}\hat{V}'.$$

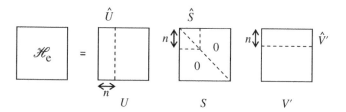

Figure 7.1 Approximating the Hankel matrix.

Matrix $\hat{\mathcal{H}}_e^k$ has the same dimension of the original Hankel matrix; also, it keeps the Hankel structure in the sense that it has identical entries over counterdiagonals. The main point is that $\hat{\mathcal{H}}_e^k$ has rigorously rank n.

The elements of $\hat{\mathcal{H}}_e^k$ can be seen as the samples of the true impulse response of an nth-order realization, uncorrupted by noise.

Then, we can resort to the identification procedure of the previous section to identify triple (F, G, H) by defining

$$\hat{S}^{1/2} = \text{diag } \{\delta_1^{1/2}, \delta_2^{1/2}, \dots, \delta_n^{1/2}\},$$
$$\hat{\mathcal{O}}^n = \hat{U}\hat{S}^{1/2},$$
$$\hat{\mathcal{R}}^n = \hat{S}^{1/2}\hat{V},$$

we have the decomposition

$$\mathcal{H}_e^{k*} = \hat{\mathcal{O}}^n \hat{\mathcal{R}}^n.$$

The first row of $\hat{\mathcal{O}}^n$ supplies H, and the first row of $\hat{\mathcal{R}}^n$ matrix G. F is obtained by the *shift invariance property*.

Overall, the method presented herein is named *Kung's subspace identification method*. Another term widely used is the *4SID* method, the acronym coming from the initials of *state space system subspace identification*.

Example 7.5 Consider the ARX system

$$y(t) = 0.2y(t-1) + 0.35y(t-2) + u(t-1) - 5u(t-2) + \eta(t),$$

where $\eta(t)$ is a white noise with zero mean value and variance $1/25$. The impulse response is represented in Figure 7.2. With the samples of the impulse response, the 30×30 Hankel matrix was constructed. The first 10 singular values are represented in Figure 7.3.

The analysis of the singular values of such matrix clearly indicates that $n = 2$ is the appropriate order choice.

By Kung's method, we obtain

$$F = \begin{bmatrix} 0.7849 & 0.3225 \\ 0.3007 & -0.5477 \end{bmatrix}, \quad G = \begin{bmatrix} 1.4113 \\ -1.7662 \end{bmatrix}, \quad H = [-1.4515 \ -1.7496],$$

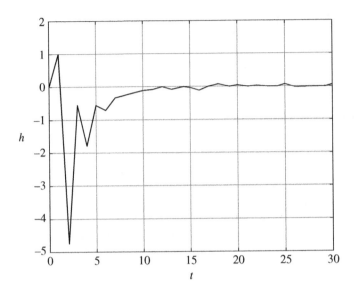

Figure 7.2 Impulse response of the system of Example 7.5.

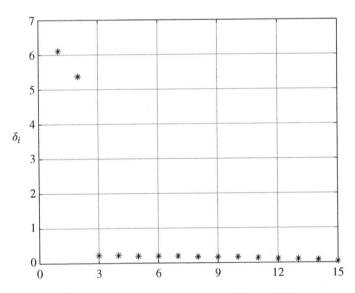

Figure 7.3 Singular values of the Hankel matrix of Example 7.5.

the transfer function of which is

$$\hat{G}(z) = 1.0417 \frac{z - 4.912}{(z - 0.7076)(z + 0.4705)}.$$

Comparing with the ARX data generation mechanism transfer function

$$G(z) = \frac{z - 5}{(z - 0.7)(z + 0.5)},$$

one can assess the accuracy in poles, zeros, and gain identification.

8

Predictive Control

8.1 Introduction

A control problem arises whenever we want to impose a behavior on a given system that is as close as possible to the behavior desired, by acting on quantities that can be manipulated. For example, in order to keep the temperature in a room as close as possible to a reference value (set point), we cannot act directly on the temperature, rather we modulate the amount of heat injected into the room. For an interesting and entertaining introduction, see Albertos and Mareels (2010).

In this chapter, we will introduce simple digital control methods based on black box models. For simplicity, we will make reference to single-input single-output systems, with $u(t)$ denoting the control variable and $y(t)$ the controlled variable. The desired behavior of $y(t)$ is denoted by $y°(t)$ and is called *reference signal* or *set point*.

The techniques we present are known as *predictive control methods*. The basic idea which they rely on is to exploit the model of the system to be controlled in order to predict its response to a given input and choose the input in such a way that the predicted output be as close as possible to the desired output.

The simplest among such techniques is *minimum variance* (MV) control, introduced in Section 8.2. Its main feature is that the design of the controller can be performed by means of elementary computations. On the other hand, MV does not allow the appropriate modulation of the effort required to achieve the control objective. The second technique we present (Section 8.3) is named *generalized minimum variance* (GMV) control; it resembles the MV approach, but it allows achieving a trade-off between the performance and the effort to achieve it.

These methods open the road to the so-called *model-based predictive control* (MPC), dealt with in Section 8.4. With MPC, the design can be carried over by taking also into account the constraints that must be respected in a practical design control problem.

Model Identification and Data Analysis, First Edition. Sergio Bittanti.
© 2019 John Wiley & Sons, Inc. Published 2019 by John Wiley & Sons, Inc.

In MV, GMV, and MPC methods, the control synthesis is performed assuming that the model of the system to be regulated has been identified and is available to the designer. A last question we address is the following one: Is it possible to design a controller directly from data? Should this be possible, then we could establish a bridge connecting control science and data analysis, so conciliating two fields that have developed separately for decades. In Section 8.5, we introduce a data-driven controller synthesis technique, the *virtual reference feedback tuning* (VRFT). This method can be seen as an iconic paradigm of the results achievable by the cross-fertilization between these (apparently) far fields, control science and data analysis.

8.2 Minimum Variance Control

In classical control theory, it is often assumed that the set point has some special features, for example, it is a step function, a ramp function, and so on. In other words, the design problem is tackled by making reference to some standard signals. In the approach we present now, it is assumed that the reference signal is a stationary process.

A preliminary important observation is in order. It is obvious that, given the system to be controlled, the characteristics of the overall control system depend upon the adopted design method. Hence, prior to the completion of the design, nothing about such characteristics can be anticipated. However, a fact we know from the very beginning is that the overall control system obtained after connecting the controller to the given system must be stable. Only under such condition, we can avoid any disturbance acting on the plant from producing permanent effects on the output. In conclusion, we expect that the control system will be stable.

On the other hand, the inputs of the control system will be reference signal $y°(t)$ plus disturbances acting on the plants. Let's assume that such disturbances are stationary processes as well. Then, the overall system will be stable and subject to stationary inputs and, we expect $y(\cdot)$ and $y(\cdot) - y°(\cdot)$ to be stationary processes too. Hence, the objective that $y(\cdot)$ be as close as possible to $y°(\cdot)$ can be stated by requiring that the variance of the error $y(\cdot) - y°(\cdot)$ be minimal. This motivates the introduction of the performance index

$$J = E[(y(t) - y°(t))^2]. \tag{8.1}$$

Note that, thanks to stationarity, J does not depend upon time.

For simplicity, we will initially focus on the *regulation problem*, when the signal $y°(t)$ is a constant, $y°(t) = \bar{y}°$, so that

$$J = E[(y(t) - \bar{y}°)^2].$$

J being independent of time, we can write this expression in equivalent form as $J = E[(y(t+1) - \bar{y}°)^2]$ or $J = E[(y(t+2) - \bar{y}°)^2]$ and so on.

The problem of designing the controller so as to minimize J is known as the MV control problem.

Example 8.1 Consider the system

$$\mathcal{S} : y(t+1) = ay(t) + b_1 u(t) + b_2 u(t-1) + \eta(t+1), \quad \eta(\cdot) \sim \text{WN}(0, \lambda^2).$$

We know that the optimal one-step-ahead predictor is

$$\hat{y}(t+1|t) = ay(t) + b_1 u(t) + b_2 u(t-1), \tag{8.2}$$

so that we can split $y(t+1)$ as

$$y(t+1) = \hat{y}(t+1|t) + \eta(t+1). \tag{8.3}$$

Now plug expression (8.3) in the loss function

$$J = E[(y(t+1) - \bar{y}^\circ)^2]. \tag{8.4}$$

We have

$$J = E[(\hat{y}(t+1|t) + \eta(t+1) - \bar{y}^\circ)^2]$$
$$= E[(\hat{y}(t+1|t) - \bar{y}^\circ)^2] + E[\eta(t+1)^2] + 2E[\eta(t+1)(\hat{y}(t+1|t) - \bar{y}^\circ)].$$

From (8.2), we see that $\hat{y}(t+1|t)$ depends upon $u(\cdot)$ and $y(\cdot)$ up to time t at most. Since we expect a control law of the type

$$u(t) = f(u(t-1), u(t-2), \dots, y(t), y(t-1), \dots, y^\circ(t), y^\circ(t-1), \dots),$$

$u(\cdot)$ and $y(\cdot)$ up to t will depend upon noise $\eta(\cdot)$ up to t at most. For this reason, $\hat{y}(t+1|t) - \bar{y}^\circ$ and $\eta(t+1)$ are uncorrelated to each other, so that, recalling that $\eta(\cdot)$ is zero mean,

$$E[\eta(t+1)(\hat{y}(t+1|t) - \bar{y}^\circ)] = 0.$$

Hence,

$$J = E[(\hat{y}(t+1|t) - \bar{y}^\circ)^2] + \lambda^2.$$

Since λ^2 cannot be modified with a control action, the minimum of J is achieved by imposing

$$\hat{y}(t+1|t) = \bar{y}^\circ.$$

Substituting into (8.2), we have

$$u(t) = \frac{1}{b_1}(-ay(t) - b_2 u(t-1) + \bar{y}^\circ)$$

as control law. As we see, this law generates $u(t)$ as a linear combination of the previous input sample $u(t-1)$, the feedback signal $y(t)$, and the set point $\bar{y}^\circ$. We observe that the coefficients of such combination are well defined provided that $b_1 \neq 0$. If, on the opposite, $b_1 = 0$, then the input–output delay would be greater than 1 and the loss function should be suitably reformulated, as is discussed in the sequel.

We now tackle the MV control problem (with a constant set point) in the general ARMAX case:

$$\mathcal{S} : A(z)y(t) = B(z)u(t-k) + C(z)\eta(t), \quad \eta(\cdot) \sim \mathrm{WN}(0, \lambda^2), \qquad (8.5)$$

where polynomials $A(z)$, $B(z)$, and $C(z)$ have the usual meaning and k is the input–output delay ($k \geq 1$). We assume that the leading parameter of $B(z)$, b_1, is non-null, so that k is the actual input–output delay.

We write the performance index in the form

$$J = E[(y(t+k) - \bar{y}^\circ)^2].$$

Here, we split $y(t+k)$ into the sum of its optimal k-steps-ahead prediction and the corresponding prediction error:

$$y(t+k) = \hat{y}(t+k|t) + \epsilon(t+k).$$

The predictor being optimal, $\epsilon(t+k)$ and $\hat{y}(t+k|t)$ are uncorrelated to each other. Therefore, the minimization of J is obtained by imposing that

$$\hat{y}(t+k|t) = \bar{y}^\circ.$$

This is the basic principle of predictive control:

> *The prediction of the output equals the set point*

8.2.1 Determination of the MV Control Law

According to the above principle, the control law is derived by finding first the optimal predictor $\hat{y}(t+k|t)$. For example, we perform the long division of $C(z)$ by $A(z)$ for k steps, so obtaining

$$C(z) = A(z)E(z) + z^{-k}\tilde{F}(z), \qquad (8.6)$$

with

$$E(z) = e_0 + e_1 z^{-1} + \cdots + e_{k-1} z^{-k+1}.$$

Note that, $A(z)$ and $C(z)$ being monic, the leading coefficient of $E(z)$ is unitary, $e_0 = 1$. As already seen in Section 3.8, equation (8.6) is called *Diophantine equation*.

Multiply now both members of (8.5) by $E(z)$. We have

$$A(z)E(z)y(t+k) = B(z)E(z)u(t) + C(z)E(z)e(t+k).$$

Adding at both members $C(z)y(t+k)$, it follows that

$$C(z)y(t+k) = B(z)E(z)u(t) + (C(z) - A(z)E(z))y(t+k) + C(z)E(z)e(t+k).$$

By dividing both members by $C(z)$, and taking into account the Diophantine equation (8.6), $y(t + k)$ can be given the expression

$$y(t + k) = \frac{\tilde{F}(z)}{C(z)}y(t) + \frac{B(z)E(z)}{C(z)}u(t) + E(z)e(t + k). \tag{8.7}$$

On the right-hand side of this expression, three terms appear:

(i) The first term is the output of a system with transfer function $\tilde{F}(z)/C(z)$ fed by input $y(t)$. Therefore, it depends upon $y(\cdot)$ up to time t at most.

(ii) Analogously, the second term depends upon $u(\cdot)$ up to time t at most.

(iii) Recalling the expression of polynomial $E(z)$, the third term is given by

$$E(z)e(t + k) = e_0\eta(t + k) + e_1\eta(t + k - 1) + \cdots + e_{k-1}\eta(t + 1).$$

This is a linear combination of snapshots of $\eta(\cdot)$ at the time instants $t + 1, t + 2, \ldots, t + k$. So, the third term on the right-hand side of (8.7) concerns the future (with respect to time t).

Hence, in (8.7), we see a split between the past (from which the first two terms depend) and the future (the third term, which is unpredictable from the past). Consequently, the optimal predictor must satisfy the condition

$$\hat{\mathcal{S}} : \quad C(z)\hat{y}(t + k|t) = \tilde{F}(z)y(t) + B(z)E(z)u(t).$$

Once the predictor has been determined, the passage to the controller is straightforward: According to the principle of predictive control, $\hat{y}(t + k|t)$ must be replaced by the set point $\bar{y}^\circ$. In this way, we obtain

$$C(z)\bar{y}^\circ = \tilde{F}(z)y(t) + B(z)E(z)u(t), \tag{8.8}$$

from which, solving for $u(t)$, the control law follows

$$\mathcal{C} : u(t) = \frac{C(z)}{B(z)E(z)}\bar{y}^\circ + \frac{\tilde{F}(z)}{B(z)E(z)}y(t). \tag{8.9}$$

Remark 8.1 (Time-varying reference signal) When the reference signal is not a constant, the natural choice for the performance index is

$$J = E[(y(t + k) - y^\circ(t + k))^2]. \tag{8.10}$$

Notice in passing that, thanks to the assumed stationarity of signals, expression (8.10) is equivalent to (8.1).

Assume first that the value taken by the reference signal at time $t + k$ is known at time t. This is not such a remote possibility since in certain cases the future value of the reference signal is decided in advance. Then we impose that

$$\hat{y}(t + k \mid t) = y^\circ(t + k)$$

and the control law is obtained from equation

$$C(z)y^\circ(t + k) = \tilde{F}(z)y(t) + B(z)E(z)u(t). \tag{8.11}$$

in place of (8.8).

If the future values of y^o are not known, then one can consider building a predictive model of the reference signal and impose that the predictor of y coincides with that of y^o:

$$\hat{y}(t + k \mid t) = \hat{y}^o(t + k \mid t).$$

Often it may be difficult to find predictor $\hat{y}^o(t + k \mid t)$; then, a common choice is to replace it with the last available value, namely to impose simply:

$$\hat{y}(t + k \mid t) = y^o(t).$$

In such a case the control law is obtained from:

$$c(z)y^o(t) = \tilde{F}(z)y(t) + B(z)E(z)u(t). \tag{8.12}$$

so that the controller is governed by the law

$$\mathcal{C} : u(t) = \frac{c(z)}{B(z)E(z)}y^o(t) - \frac{\tilde{F}(z)}{B(z)E(z)}y(t) \tag{8.13}$$

Interestingly enough, such law can be straightforwardly derived by considering

$$J = E[(y(t + k) - y^o(t))^2] \tag{8.14}$$

as loss function from the very begining in place of (8.10).

In the literature, expressions of the type (8.14) are often encountered.

Remark 8.2 (Input–output delay)   Consider again expression (8.8). To find explicitly the control law, one has to solve this equation with respect to $u(t)$. In this regard, note that $u(t)$ is here multiplied by the product of the leading coefficients of polynomials $B(z)$ and $E(z)$. Since the leading coefficient of $E(z)$ is 1, the coefficient multiplying $u(t)$ in (8.8) is b_1. Therefore, to solve (8.8) with respect to $u(t)$, b_1 must be different from 0, namely the input–output delay must be precisely k.

Remark 8.3 (Role of polynomial $A(z)$)   In formula (8.13), we don't see $A(z)$, so, at a first sight, the control law seems independent from such polynomial. Of course this is false; indeed, $A(z)$ enters Diophantine equation (8.6).

Remark 8.4 (Controller polynomials computation)   Once the plant polynomials $A(z)$, $B(z)$, and $C(z)$ are given, the determination of the controller requires to find $E(z)$ and $\tilde{F}(z)$. $E(z)$ and $\tilde{F}(z)$ are obtained by solving the Diophantine equation (8.6), namely by performing a long division (for k steps) between two plant polynomials, $C(z)$ and $A(z)$. Then one has to compute the product between $B(z)$ and $E(z)$. All these are elementary operations. Thus, the control design based on the MV approach requires very simple computations.

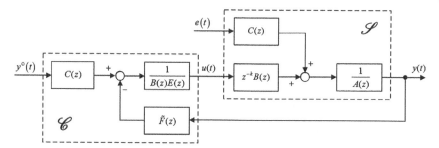

Figure 8.1 Minimum variance control system.

8.2.2 Analysis of the MV Control System

To study the overall MV control system, we focus on three main points: its structure, its stability, and the relationship relating the reference signal to the output.

8.2.2.1 Structure

From (8.5) and (8.13), the block diagram of the overall feedback system can be drawn, as depicted in Figure 8.1.

As we can see,

(i) reference $y^\circ(t)$ is transferred to the system after a filtering action through polynomial $C(z)$, leading to the filtered signal $y_F^\circ(t) = C(z)y^\circ(t)$;

(ii) the output $y(t)$ is fed back after filtering through polynomial $\tilde{F}(z)$, providing the filtered output signal $y_F(t) = \tilde{F}(z)y(t)$;

(iii) the difference $y_F^\circ(t) - y_F(t)$ between the above filtered signals determines the steering action through transfer function $1/B(z)E(z)$: $u(t) = (1/B(z)E(z))(y_F^\circ(t) - y_F(t))$.

8.2.2.2 Stability

To investigate the stability of such a control system, we first observe that the scheme of Figure 8.1 is composed of three subsystems: an inner feedback loop plus two outer blocks. One outer block is the reference signal filter, namely the block fed by $y^\circ(t)$ to generate $y_F^\circ(t)$; the other is the disturbance filter, namely the block where white noise $\eta(t)$ is filtered to form the disturbance $C(z)\eta(t)$ acting on the ARMAX model.

A system with such structure is stable if and only if the inner loop is stable and the two outer blocks are stable.

Now, the outer blocks have the same transfer function, $C(z)$: this is a moving average and therefore the outer blocks are guaranteed stable.

Passing to the loop, in the following remark, we make a preliminary observation on feedback systems.

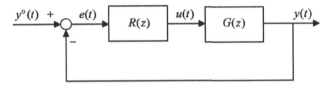

Figure 8.2 A typical feedback control system.

Remark 8.5 (Characteristic polynomial of a feedback loop) Consider the typical feedback loop depicted in Figure 8.2.

It easy to see that the overall transfer function (from $y^°(t)$ to $y(t)$) is given by

$$F(z) = \frac{L(z)}{1 + L(z)}, \tag{8.15}$$

where

$$L(z) = R(z)G(z)$$

is the so-called *open-loop transfer function*. Introducing the numerator and denominator of $L(z)$, i.e. by writing

$$L(z) = \frac{N(z)}{D(z)},$$

we obtain

$$F(z) = \frac{N(z)}{D(z) + N(z)}.$$

Letting

$$\Delta(z) = D(z) + N(z),$$

we conclude that the poles of $F(z)$ are the solutions of

$$\Delta(z) = D(z) + N(z) = 0. \tag{8.16}$$

$\Delta(z)$ is the *characteristic polynomial* and (8.16) the *characteristic equation* of the feedback system. The solutions of $\Delta(z) = 0$ are the *closed-loop poles*. By contrast, the solutions of $D(z) = 0$ are the *open-loop poles*.

Observe that, since the degree of $D(z)$ is greater than or equal to the degree of $N(z)$, the degree of $\Delta(z)$ coincides with that of $D(z)$. Therefore, if no cancellation occurs between $N(z)$ and $D(z)$, the number of closed-loop poles coincides with the number of open-loop poles.

We also observe that, if

$$G(z) = \frac{N_G(z)}{D_G(z)}, \quad R(z) = \frac{N_R(z)}{D_R(z)},$$

then Eq. (8.16) becomes

$$\Delta(z) = D_G(z)D_R(z) + N_G(z)N_R(z) = 0.$$

This expression can be extended to feedback system with various blocks in series along the loop: the open-loop transfer function $L(z)$ is given by the product of all transfer functions of the various blocks encountered in the loop, and, if the feedback is negative, *the characteristic polynomial* $\Delta(z)$ *is the sum of the products of all denominators of the transfer functions in the loop plus the products of all numerators of the same transfer functions.*

We now return to the study of MV control systems. From Figure 8.1, we see that the *open-loop transfer function* $L(z)$ is given by

$$L(z) = z^{-k} \frac{B(z)\tilde{F}(z)}{A(z)B(z)E(z)}.$$

Moreover, we see that the feedback is negative (along the loop there is only one $-$ sign). Therefore, the characteristic polynomial $\Delta(z)$ of the loop is given by

$$\Delta(z) = B(z)(A(z)E(z) + z^{-k}\tilde{F}(z)).$$

Considering the Diophantine equation (8.6), we come to the conclusion that the characteristic polynomial is

$$\Delta(z) = B(z)C(z).$$

Hence, the closed-loop poles are given by the singularities of $C(z)$ and the singularities of $B(z)$.

We expect $C(z)$ to be stable. Indeed, in the ARMAX model, $C(z)\eta(t)$ is a residual, which is assumed to be stationary, so that, in view of the spectral factorization theorem, $C(z)$ can be taken as a stable filter. So, we conclude that the MV control system is stable if and only if the singularities of polynomial $B(z)$ lie inside the unit disk. In this regard, we observe that the transfer function of system (8.5) from $u(t)$ to $y(t)$ is $z^{-k}B(z)/A(z)$; this means that the singularities of $B(z)$ are the zeros of the process to be controlled. Since a discrete time process with all zeros inside the unit disk is called *minimum phase*, we can say that the stability condition of the MV control system is that the plant is minimum phase.

Closed-Loop Relationship $y°(t) \rightarrow y(t)$

As for the closed-loop transfer function $S(z)$ from the reference signal $y°(t)$ to the output $y(t)$, we leave its easy computation to our readers:

$$S(z) = C(z)\frac{z^{-k}B(z)}{A(z)B(z)E(z) + z^{-k}B(z)\tilde{F}(z)}.$$

On the other hand, as already seen,

$$A(z)B(z)E(z) + z^{-k}B(z)\bar{F}(z) = \Delta(z) = B(z)C(z),$$

so that

$$S(z) = C(z)\frac{z^{-k}B(z)}{B(z)C(z)} = z^{-k}. \tag{8.17}$$

Note that, in this computation, polynomial $B(z)$ has been canceled.

We come to the conclusion that the overall control system behaves as a pure delay of k steps. This means that, except for such delay, a step variation in $y°(t)$ is replicated without any distortions at the output. Such a nice result does not come for free: As a rule, a brilliant performance in the input–output behavior is accompanied by overdemanding transients of other variables of the loop, in particular the $u(\cdot)$. Control signals of large intensity may be disruptive, also to the violation of the feasible physical constraints of the equipment.

In conclusion, the MV control concept is an interesting principle, leading to a simple design method. However, it has basic limitations: It cannot be used for non-minimum phase plants, and it does not allow the possibility of keeping under control the other variables in the control system.

More flexibility in the design is provided by the method that is presented in the subsequent section.

8.3 Generalized Minimum Variance Control

In loss function (8.13), the control variable does not appear, implicitly meaning that, in our design concept, the features of $u(t)$ do not matter. We now modify such performance criterion by introducing the GMV control with performance index:

$$J = E[(y_G(t + k) - y°(t))^2], \tag{8.18}$$

where the generalized output $y_G(\cdot)$ is defined as

$$y_G(t + k) = P(z)y(t + k) + Q(z)u(t), \tag{8.19}$$

see Figure 8.3.

Above, the symbol $P(z)y(t + k)$ denotes the stationary process at the output of a dynamical system with transfer function $P(z)$ and input $y(t + k)$. Analogously, $Q(z)u(t)$ is the process generated by $u(t)$ through $Q(z)$.

In this new control concept, $P(z)$ and $Q(z)$ are the design knobs. By their suitable tuning, one strives to achieve an acceptable accuracy in control without overloading other functionalities.

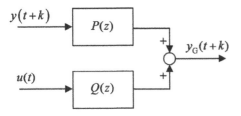

Figure 8.3 Generalized output signal.

We will now determine the GMV control law, again under the assumption that the system is described by (8.5). For example, define the *generalized output*

$$\tilde{y}(t + k) = P(z)y(t + k).$$ (8.20)

Introducing the *generalized reference signal*

$$\tilde{y}^\circ(t) = y^\circ(t) - Q(z)u(t),$$ (8.21)

we have

$$J = E[(\tilde{y}(t + k) - \tilde{y}^\circ(t))^2].$$ (8.22)

By writing $P(z)$ and $Q(z)$ as

$$P(z) = \frac{P_N(z)}{P_D(z)}, \quad Q(z) = \frac{Q_N(z)}{Q_D(z)},$$

we have

$$\tilde{y}(t) = P(z)y(t) = \frac{P_N(z)}{P_D(z)}y(t)$$

$$= \frac{P_N(z)}{P_D(z)}\left(\frac{B(z)}{A(z)}u(t - k) + \frac{C(z)}{A(z)}e(t)\right).$$

Therefore, $\tilde{y}(t)$ satisfies the ARMAX equation:

$$\tilde{A}(z)\tilde{y}(t) = \tilde{B}(z)u(t - k) + \tilde{C}(z)e(t),$$ (8.23)

where

$$\tilde{A}(z) = P_D(z)A(z),$$
$$\tilde{B}(z) = P_N(z)B(z),$$
$$\tilde{C}(z) = P_N(z)C(z).$$

Observe that (8.22) together with (8.23) defines an MV control problem of the type studied in the Section 8.2. We can therefore use the previously obtained results. To this end, we introduce the new Diophantine equation

$$\tilde{C}(z) = \tilde{A}(z)E(z) + z^{-k}\tilde{F}(z),$$ (8.24)

where $E(z)$ is again composed by k terms, $E(z) = e_0 + e_1 z^{-1} + \cdots + e_{k-1}z^{-k+1}$. Taking into account the definitions of $\tilde{A}(z)$, $\tilde{B}(z)$, and $\tilde{C}(z)$, the above equation can be written by means of the original polynomials as

$$P_N(z)C(z) = P_D(z)A(z)E(z) + z^{-k}\tilde{F}(z).$$ (8.25)

The control equation (8.12) becomes

$$\tilde{C}(z)\tilde{y}^\circ(t) = \tilde{F}(z)\tilde{y}(t) + \tilde{B}(z)E(z)u(t).$$

Thus, in view of definitions (8.20) and (8.21), we have

$$P_N(z)C(z)\left(y^0(t) - \frac{Q_N(z)}{Q_D(z)}u(t)\right) =$$
$$\tilde{F}(z)\frac{P_N(z)}{P_D(z)}y(t) + P_N(z)B(z)E(z)u(t).$$

From this expression, we can derive the GMV control law:

$$\mathscr{C} : \quad G(z)u(t) + F(z)y(t) - H(z)y^0(t) = 0, \tag{8.26a}$$

$$G(z) = P_D(z)[B(z)Q_D(z)E(z) + C(z)Q_N(z)], \tag{8.26b}$$

$$F(z) = \tilde{F}(z)Q_D(z), \tag{8.26c}$$

$$H(z) = C(z)P_D(z)Q_D(z). \tag{8.26d}$$

In analogy with MV, the controller is obtained by performing the long division (8.25) for k steps. This leads to polynomials $E(z)$ and $\tilde{F}(z)$. Given transfer functions $P(z)$ and $Q(z)$, one can then determine the three controller polynomials $F(z)$, $G(z)$, and $H(z)$ via (8.26b)–(8.26d).

The overall feedback control system is represented in Figure 8.4.

We will now see some typical choices for $P(z)$ and $Q(z)$.

8.3.1 Model Reference Control

Model reference control is a GMV control system obtained when, in (8.18), $Q(z)$ is set to zero, so that

$$J = E[(P(z)y(t+k) - y^0(t))^2].$$

In this way, the control signal does not enter the performance index. By letting $Q_N(z) = 0$ and $Q_D(z) = 1$, from the GMV controller expressions, we derive the

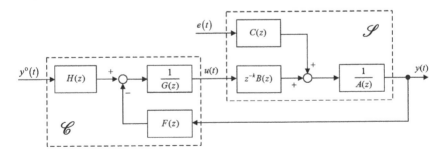

Figure 8.4 Generalized minimum variance control system.

MR controller:

$$\mathcal{S} : \quad G(z)u(t) + F(z)y(t) - H(z)y°(t) = 0, \tag{8.27a}$$

$$G(z) = P_D(z)B(z)E(z), \tag{8.27b}$$

$$F(z) = \tilde{F}(z), \tag{8.27c}$$

$$H(z) = C(z)P_D(z). \tag{8.27d}$$

The characteristic polynomial is now

$$\Delta(z) = B(z)C(z)P_N(z).$$

Hence, for stability, the singularities of $B(z)$ must be inside the unit disk. So, the MR approach can be used for minimum phase systems only (same as MV).

From the expression of $\Delta(z)$, we also see that the singularities of $P_N(z)$ (namely the zeros of $P(z)$) must also lie inside the unit disk. As for $C(z)$, recalling the discussion on the stability of the MV control system, we can assume they have modulus lower than 1.

An easy computation shows that the transfer function $S(z)$ from $y°(t)$ to $y(t)$ is given by

$$S(z) = H(z)\frac{z^{-k}B(z)}{\Delta(z)} = z^{-k}[P(z)]^{-1}.$$

Hence, apart from the k steps delay, the overall transfer function is the inverse of $P(z)$. This is why

$$M(z) = [P(z)]^{-1}$$

is named *model reference*.

The tuning of $P(z)$ is based on this fact: once the controller is designed, we know that the closed-loop behavior will be determined by transfer function $M(z)$. We will therefore impose that (i) the zeros of $P(z)$ be inside the unit disk (to avoid closed-loop instability), (ii) the poles of $P(z)$ be inside the unit disk (to guarantee that the generalized output $\tilde{y}(t)$ defined in (8.20) be stationary), and (iii) the gain of $P(z)$ be equal to 1: $P(1) = 1$. This last condition is dictated by the fact that, in discrete time, the steady-state output against a constant input is given by the product of the input times the transfer function evaluated at $z = 1$. Therefore, by imposing that the gain of $P(z)$ be equal to 1, we guarantee that, when the reference signal is a constant, $y°(t) = \bar{y}°$, the mean value of the output $y(t)$ coincides with the reference signal, see Appendix A.

A typical choice is

$$P(z) = (1 + \alpha)^{-n}(1 + \alpha z^{-1})^n.$$

In this way, the model reference has n coincident poles in $z = -\alpha$ and unit gain. By choosing integer n and tuning α, various types of step responses can be obtained. For example, if we set $n = 1$, the 90% and 99% durations of the transients of the model reference step responses are provided in the following table.

α	$t_{resp\ 90\%}$	$t_{resp\ 99\%}$
0.2	2	3
0.4	3	6
0.5	4	7
0.6	5	10
0.7	7	13
0.8	11	21
0.85	15	29
0.9	22	44
0.95	45	90
0.99	230	459

In general, we expect that a more rapid closed-loop response can be achieved at the price of larger efforts in other variables in the feedback loop, in particular an increased intensity in the transients of $u(\cdot)$. So, a rationale that is often adopted is to consider various reference models and choose an appropriate one by a trial-and-error iterative procedure carried out on a simulator of the plant until an acceptable trade-off among the various control requirements is achieved.

8.3.2 Penalized Control Design

As we have seen, the MV and MR approaches cannot be used for non-minimum phase plants. We now introduce a GMV technique, named *penalized control* (PC), which can be used for those systems too.

To be precise in the GMV performance index, we impose from the beginning that $P(z) = 1$:

$$J = E[(y(t+k) + Q(z)u(t) - y^0(t))^2],\tag{8.28}$$

so that the design knob is now $Q(z)$.

If we set $P_N(z) = 1$ and $P_D(z) = 1$, Eq. (8.25) becomes

$$C(z) = A(z)E(z) + z^{-k}\tilde{F}(z),$$

the same Diophantine equation we used in MV. Moreover, from (8.26), we obtain

$$\mathcal{S}\ :\ G(z)u(t) + F(z)y(t) - H(z)y^0(t) = 0,\tag{8.29a}$$

$$G(z) = B(z)Q_D(z)E(z) + C(z)Q_N(z),\tag{8.29b}$$

$$F(z) = \tilde{F}(z)Q_D(z), \tag{8.29c}$$

$$H(z) = C(z)Q_D(z). \tag{8.29d}$$

Correspondingly, the characteristic polynomial $\Delta(z)$ from $y^\circ(t)$ to $y(t)$ is given by

$$\Delta(z) = C(z)[B(z)Q_D(z) + A(z)Q_N(z)], \tag{8.30}$$

and the closed-loop transfer function is

$$S(z) = \frac{z^{-k}}{1 + Q(z)\dfrac{A(z)}{B(z)}}. \tag{8.31}$$

From (8.30), we see that the poles of the closed-loop system are the singularities of $C(z)$ and the solutions of

$$B(z)Q_D(z) + A(z)Q_N(z) = 0. \tag{8.32}$$

The roots of (8.32) can be moved in the complex plane by choosing $Q_N(z)$ and $Q_D(z)$. In particular, we see that the zeros of $B(z)$ are no more closed-loop poles (unless $Q_N(z) = 0$). Thus, contrary to MV and MR, the PC technique can be applied to non-minimum phase systems.

We will now examine a couple of typical choices for $Q(z)$.

8.3.2.1 Choice A for $Q(z)$

A possibility is to take for $Q(z)$ a constant

$$Q(z) = \beta,$$

so that (8.28) becomes

$$J = E[(y(t + k) + \beta u(t) - y^\circ(t))^2]. \tag{8.33}$$

Thus, parameter β is the weight of the control variable $u(t)$ in the performance index. Obviously, for $\beta = 0$, we go back to the MV approach (compare (8.33) with (8.14)).

Equation (8.32) then becomes

$$B(z) + \beta A(z) = 0 \tag{8.34}$$

or, equivalently,

$$\beta^{-1}B(z) + A(z) = 0. \tag{8.35}$$

From (8.34), we see that for $\beta = 0$ the solutions of (8.32) are the singularities of $B(z)$ (this was expected since, for $Q(z) = 0$, the PC approach becomes the MV approach). From (8.35), we see that for $\beta \to \infty$ such solutions tend to the singularities of $A(z)$. This entails that, if the system to be controlled is stable, then the poles of the closed-loop system can be pushed inside the stability region for a sufficiently high value of β.

Remark 8.6 To avoid any ambiguities, when we refer to the solutions of (8.34), we mean the roots of the corresponding equation written in positive powers of z. For instance, if $A(z)$ and $B(z)$ are given as $A(z) = 1 - a_1 z^{-1} - a_2 z^{-1}$ and $B(z) = b_0 + b_1 z^{-1} + b_2 z^{-2}$, expression (8.34) must be written as

$$b_0 z^2 + b_1 z + b_2 + \beta(z^2 - a_1 z - a_2) = 0.$$

The corresponding root locus will be constituted by two branches, moving from the two solutions of $b_0 z^2 + b_1 z + b_2 = 0$ (the roots of $B(z)$) to the two solutions of $z^2 - a_1 z - a_2 = 0$ (the roots of $A(z)$) as β goes from 0 to ∞.

For the sake of completeness, let's also consider the case when $A(z)$ and $B(z)$ have different degrees. For example, assume that $A(z) = 1 - a_1 z^{-1}$ and $B(z) = b_0 + b_1 z^{-1} + b_2 z^{-2}$. Then the root locus equation (8.34) is

$$(b_0 z^2 + b_1 z + b_2) + \beta(z^2 - a_1 z) = 0,$$

which can be equivalently written as

$$(z^2 - a_1 z) + \beta^{-1}(b_0 z^2 + b_1 z + b_2) = 0.$$

In particular, we see that for $\beta = 0$, the equation is $b_0 z^2 + b_1 z + b_2 = 0$ while, for $\beta \to \infty$, $z^2 - a_1 z = 0$. Hence, the locus is constituted by two branches departing $(\beta = 0)$ from the two roots of $B(z)$; for $\beta \to \infty$, one of the two lines ends in the root of $A(z)$ $(z = a_1)$, the other one tends to the origin of the complex plane.

If, vice versa $A(z) = 1 - a_1 z^{-1} - a_2 z^{-2}$ and $B(z) = b_0 + b_1 z^{-1}$, then the root locus equation (8.34) becomes

$$(b_0 z^2 + b_1 z) + \beta(z^2 - a_1 z - a_2) = 0$$

or equivalently

$$(z^2 - a_1 z - a_2) + \beta^{-1}(b_0 z^2 + b_1 z) = 0.$$

Thus, for $\beta = 0$, we have $b_0 z^2 + b_1 z = 0$, and for $\beta \to \infty$, $z^2 - a_1 z - a_2 z = 0$. Therefore, one of the two branches of the locus departs from the root of $B(z)$ $(z = -b_1/b_0)$, the other one from the origin. The branches end in the two roots of $A(z)$.

Summarizing, with the PC approach based on the choice $Q(z) = \beta$, parameter β can be tuned by studying the root locus of the closed-loop poles, so as to achieve a nice positioning of such poles.

Passing to the study of the transfer function (8.31), when $Q(z) = \beta$, we have

$$S(z) = \frac{z^{-k}}{1 + \beta \dfrac{A(z)}{B(z)}}. \tag{8.36}$$

In particular,

$$S(1) = \frac{1}{1 + \dfrac{\beta}{\mu}},$$

where $\mu = B(1)/A(1)$ is the gain of the system to be controlled. Therefore, $S(1) \neq 1$ (unless $\beta = 0$). Thus, we expect a bias in the output $y(t)$, the mean value of which will not coincide with $\bar{y}^0$, when the reference signal is constant.

8.3.2.2 Choice B for Q(z)

Another frequent choice for $Q(z)$ is

$$Q(z) = \beta(1 - z^{-1}).$$

Correspondingly, Eq. (8.28) takes the expression

$$J = E\left[(y(t+k) + \beta(1 - z^{-1})u(t) - y^0(t))^2\right]. \tag{8.37}$$

The advantage of such choice is that $Q(1) = 0$, $\forall \beta$, so that the closed-loop gain $S(1)$ is unitary for any value of the tuning parameter β (see (8.31)).

As for the closed-loop poles, observe that Eq. (8.32) becomes

$$B(z) + \beta A(z)(1 - z^{-1}).$$

Therefore, the branches of the corresponding root locus start from the roots of $B(z)$ (for $\beta = 0$) to end into the roots of $A(z)(1 - z^{-1})$ (for $\beta \to \infty$). Hence, one of such branches will end in $z = 1$, the remaining ones will end in the roots of $A(z)$.

Remark 8.7 From (8.28), we can conclude that the best we can hope for the minimization of the performance index is that

$$y(t+k) - y^0(t) = Q(z)u(t),$$

namely

$$u(t) = Q(z)^{-1}\delta(t),$$

where

$$\delta(t) = y(t+k) - y^0(t)$$

is the discrepancy between the reference and the output signals.

This provides an interesting interpretation of $Q(z)$: Its inverse can be seen as the transfer function of a controller acting on error $\delta(t)$ to decide $u(t)$. Since $1/(1 - z^{-1})$ is the discrete-time integrator, choosing $Q(z) = \beta(1 - z^{-1})$ is like inserting an integrator in the feedback loop, a classical remedy against biases in the output response.

For MV and GMV, the papers by Clarke et al. (1987a,b), are cornerstone contributions. The possibility of equipping MV control techniques with tracking identification methods (Section 5.8) has been studied in an extensive literature, giving rise to the field of adaptive identification and control, see, e.g. Goodwin and Sin (1984), Bitmead et al. (1990), and Landau et al. (1998). A version of MV methods for nonlinear systems is discussed in Bittanti and Piroddi (1994, 1997).

Remark 8.8 (Self-Tuning controller)   Minimum variance controllers have been worked out under the tacit assumption that the model of the system to be governed is given and known to the designer. In the case that the model is not available, it can be estimated from data by an identification procedure. A possibility is to resort to a recursive identification method so as to perform the model estimation in real time. This leads to the realm of *self-organizing* (or *adaptive*) *control* systems. In this framework, a typical situation arises when the controller is a Minimum Variance one, where the unknown model parameters are replaced by their estimates obtained by the Recursive Least Squares algorithm applied to the measurements of the system's input and output. This is the so-called *self tuning controller*, analysed in many textbooks and papers, see for example (Åström, Wittemark, 1989). A tutorial on self tuners can be found in the paper *Least Squares Based Self Tuning Control Systems* in (Bittanti, Picci, 1996).

8.4   Model-Based Predictive Control

The methods seen in the previous sections of this chapter have been further elaborated giving rise to a control design approach named MPC, very successful in the industrial applications of our days.

The model of the plant is used to predict the output signal over an extended horizon in response to a certain input sequence. A simple formulation of the problem refers to the following performance index:

$$J_{\mathrm{MPC}} = \left[\sum_{k}^{p} (\hat{y}(t+k) - \bar{y}^{\circ})^2 + q\, u(t+k-1)^2 \right], \tag{8.38}$$

where

- $\bar{y}^{\circ}$ is the reference signal (here assumed to be a constant),
- $\hat{y}$ is the predicted output,
- p is the prediction horizon,
- q is a parameter to be tuned by the designer.

Suppose that p and q have been chosen. Then, for an assigned sequence of inputs $u(t), u(t+1), \ldots$, by means of the model, it is possible to determine the samples of the predicted output. In this way, the value of J_{MPC} can be computed. With the same p and q, by changing the input sequence, the

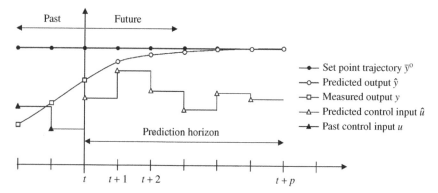

Past   Future

— Set point trajectory $\bar{y}^0$
— Predicted output $\hat{y}$
— Measured output y
— Predicted control input $\hat{u}$
— Past control input u

Prediction horizon

$t \quad t+1 \quad t+2 \qquad\qquad t+p$

Figure 8.5 Model predictive control essentials.

corresponding predicted output sequence can be found, and then the new value of J_{MPC} computed. Proceeding this way, one can solve a number of open-loop problems with the objective of minimizing J_{MPC}. In such optimization, one can also take into account constraints of various type (such as $u_{\min} \leq u(t+k-1) \leq u_{\max}, \forall k$). Once the optimal input sequence has been found, only the first sample $u(t)$ is applied to the system.

The procedure is iteratively repeated by moving the horizon one step ahead each time, i.e. from the interval from $t+1$ to $t+p$ to the interval from $t+2$ to $t+p+1$, and then from $t+3$ to $t+p+2$, and so on. At each step, only the first computed sample of $u(\cdot)$ is applied to the system, see Figure 8.5.

This is a peculiar design strategy (named *receding horizon control*), where the feedback action is iteratively computed by solving a sequence of open-loop problems.

The described rationale is often compared to the expense planning in a family. Imagine a family that is making a budget for the following month. Once they do the shopping for the first week, then the next week they will analyze the new economic situation by taking into account the expenses budgeted for the month that begin in that week (expenses that might be changed following unplanned spending in the previous week). By studying again the costs over a month, they will decide the expenses of the next week.

MPC is treated in a number of books such as Rawlings et al. (2017), Bemporad and Morari (2017), Magni and Scattolini (2014), Camacho and Bordons Alba (2007), Maciejowski (2002), and Allgöwer and Zheng (2000).

8.5 Data-Driven Control Synthesis

Up to now, the design of a control system has been performed on the basis of the preliminary knowledge of the model. In this section, we see an alternative

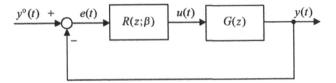

Figure 8.6 A feedback system with parametrized controller.

principle, thanks to which the controller can be designed from input–output sequences of data collected on the system to be controlled. This leads to the *data-driven control synthesis* methods.

Among them, we focus here on the VRFT technique. To introduce it, consider the classical feedback control scheme of Figure 8.2, where reference signal (set point) $y°(t)$ is compared with the output $y(t)$ to form error $e(t)$; the error feeds the controller, whose task is to decide the control signal $u(t)$ to govern the system.

For simplicity, we adopt a fully deterministic point of view and denote by $G(z)$ the transfer function of the system. As for the controller, we assume it has a given structure depending on a set of parameters β: hence, its transfer function is denoted by $R(z; \beta)$ and the scheme of Figure 8.2 becomes that of Figure 8.6.

Example 8.2 (P and PI controllers) Controller

$$R(z; \beta) = \beta \tag{8.39}$$

imposes that

$$u(t) = \beta e(t). \tag{8.40}$$

The value of the control action being proportional to that of the error, (8.40) is named *proportional controller* or simply *P controller.*
Another possibility is to assume

$$R(z; \beta) = \frac{\beta}{1 - z^{-1}}. \tag{8.41}$$

Then

$$u(t) = \frac{\beta}{1 - z^{-1}} e(t).$$

Therefore,

$$u(t) = u(t - 1) + \beta e(t),$$

so that if $u(0) = 0$ and $e(t) = \bar{e}, \forall t$, then $u(t) = t\beta\bar{e}$. If $\beta = 1$, then the output is the (discrete-time) integral of the input. This is why (8.41) is named *integral controller* or *I controller.*

One can combine the two controllers above and consider the *proportional and integral controller* or *PI controller*:

$$R(z; \beta) = \beta_1 + \frac{\beta_2}{1 - z^{-1}} \tag{8.42}$$

characterized by the two parameters β_1 and β_2.

The controller tuning can be performed according to various criteria. One of them is the *model reference* approach. We already encountered this principle in Section 8.3.1 as a special case of *GMV control*. To be precise, assume that a reference model, with transfer function $M(z)$, is given; the objective of the control design is that the transfer function of the overall system should be as close as possible to the reference model. Now, according to (8.15), the closed-loop transfer function (from $y^\circ(t)$ to $y(t)$) is given by

$$\frac{G(z)R(z; \beta)}{1 + G(z)R(z; \beta)}. \tag{8.43}$$

Therefore, the control objective is to choose β so as to minimize the discrepancy between the transfer functions of the control system and the reference model; this can be stated as the minimization with respect to β of

$$J_{\text{model}}(\beta) = \left\| \frac{G(z)R(z; \beta)}{1 + G(z)R(z; \beta)} - M(z) \right\|_2^2. \tag{8.44}$$

When the model $G(z)$ of the system is unknown, solving such problem is not an easy task. To get around the obstacle, we take a completely different point of view, a data-based approach, and reformulate the problem as follows. Suppose we have recorded the signals at the input and output of the system to be controlled over a certain interval of time. So, two set of data, one with the input sequence $u(\cdot)$ and the other one with the output sequence $y(\cdot)$, are available. N denotes the number of snapshots. We exploit the information contained in such data to design the controller, according to the following considerations. Focus on the output sequence, and let $\bar{y}^\circ(t)$ be an input of the reference model such that

$$M(z)\bar{y}^\circ(t) = y(t).$$

Since $M(z)$ is the desired behavior of the closed-loop system once the design has been properly fulfilled, output $y(t)$ can be interpreted as the desired output of the control system when $M(z)$ is fed with $\bar{y}^\circ(t)$. Therefore, $\bar{y}^\circ(\cdot)$ is named *virtual reference*. Correspondingly, one can determine the *virtual error* $e(t) = \bar{y}^\circ(t) - y(t)$ and construct another sequence of data, sequence $e(\cdot)$. This new set adds to the previously introduced ones, $u(\cdot)$ and $y(\cdot)$.

And here is the spark igniting the idea of the new method: True, $G(z)$ is unknown; however, we know that, in the ideal control system (the one resulting in a closed loop with transfer function $M(z)$), the controller generates $u(\cdot)$

when fed by $e(\cdot)$. This is a very useful piece of information, because it tell us which data sequences are at the input and at the output of the best controller one can hope for, the controller guaranteeing that the overall system behaves as the reference model. Hence, the best controller can be determined by an identification procedure that captures the dynamical relationship between sequence $e(\cdot)$ (as input) and sequence $u(\cdot)$ (as output), by minimizing the data-based cost index:

$$J_{\text{data}}(\beta) = \frac{1}{N} \sum_{t}^{N} (u(t) - R(z; \beta)e(t))^2. \tag{8.45}$$

If $R(z; \beta)$ is linear in β, a simple application of least squares enables finding the controller from experimental observations with an explicit formula. Otherwise, the minimization of $J_{\text{data}}(\beta)$ can be performed with a Newton-like technique similar to that seen in Section 5.4.

This method of control design is known as VRFT (Virtual Reference Feedback Tuning).

Remark 8.9 (Model reference set-up versus data-driven methods) Among the many contributions on data-driven techniques in control science, we mention the article (Campi et al. 2002). In that paper, it is also shown that, under suitable conditions, the data-based cost index (8.45) is nearly optimal for the model reference criterion (8.44), namely it shares the same minimizer.

We conclude this chapter with this food for thought. From their birth, the two worlds of data analysis and control developed independently of one another. As witnessed by data-driven control methods, one of the great results of modern scientific progress is that they can instead interact fruitfully; indeed, using data analysis, it is possible to treat problems of control in a simple manner. Thus, the field of influence of data science is further expanded, encompassing the world of decision and control. The VRFT method is a paradigmatic example of this happy union, a union that is sure to bear more fruit in the years to come. For a vision on this new approach, we refer to the book (Campi and Garatti, 2018).

9

Kalman Filtering and Prediction

9.1 Introduction

In the first part of this book, we have treated the prediction problem for stationary processes as the problem of estimating the future value of a signal $y(\cdot)$ from the observation of its past snapshots. In other words, given the observations $y(t), y(t-1), y(t-2), \ldots$, of $y(\cdot)$, one is willing to find $y(t+r)$ where $r \geq 1$. We now address a more general problem arising when the unknown variable $x(t)$ may be different from the observed variable $y(t)$. The estimation of $x(t)$ is performed by taking advantage of the information on it hidden in $y(\cdot)$ through the relationship linking $x(t)$ to $y(t)$, as expressed by a mathematical model. Of course, in such a framework, the problem can take a variety of facets. In particular, we can deal with the issue of estimating $x(t)$, given the observations $y(t), y(t-1), y(t-2), \ldots$, or that of estimating $x(t+r)$, given the observations $y(t), y(t-1), y(t-2), \ldots$ The first is called *filtering problem*, while the second is named *prediction* if $r > 0$ and *smoothing* if $r < 0$.

Uncountable are the fields where these problems arise, from aerospace to geophysics, from biology to process control, from economy to natural sciences, to name a few of them.

The theory we present, based on the state space representation, is known as *Kalman filtering*, since it originates from the contributions of R.E. Kalman (1930–2016).

Remark 9.1 (Kalman contribution) Born in Hungary in 1930, Rudolf E. Kalman (Figure 9.1) was taken by his father to America in 1943. He studied Electrical Engineering at MIT (close to Boston) and Columbia University (New York), receiving his PhD in 1957. He worked first at IBM and then spent some years at the Research Institute of Advanced Studies–RIAS, in Baltimore. In RIAS, an institute associated with an aerospace company, Kalman, employed as *Research Mathematician*, found a welcoming environment inspiring a fruitful activity. It was during these years that he worked out some of his milestone contributions. In 1960, in the *Transactions of ASME, Series D, Journal of Basic*

Model Identification and Data Analysis, First Edition. Sergio Bittanti.
© 2019 John Wiley & Sons, Inc. Published 2019 by John Wiley & Sons, Inc.

Figure 9.1 R.E. Kalman, picture taken by Sergio Bittanti in Bologna, Italy, during the conferral of the Sigillum Magnum – 4 September 2002. The Sigillum Magnum is the highest recognition of the Alma Mater Studiorum, as the University of Bologna (founded 1088) is called.

Engineering, he published the article "A New Approach to Linear Filtering and Prediction Problems," marking the birth of the Kalman filter. Again in 1960, he participated in the *1st World Congress* of the *International Federation of Automatic Control (IFAC)* held in Moscow. There he presented the paper "The General Theory of Control Systems," a fundamental contribution to pose the control problem on solid mathematical grounds. In 1964, he became professor in Stanford, and then in 1971, he moved to the University of Florida in Gainesville, where he worked up to 1992. Starting in 1973 he was also associated with the ETH in Zürich, from where he retired in 1997. Kalman was one of the main protagonists of the sort of genetic mutation of engineering during the second half of the twentieth century. In our days, the systems and control discipline is among the fundamental ones in engineering. In April 2017, the *IEEE Control Systems Magazine* published a section of commentaries on Kalman legacy.

The author of this book had the privilege of meeting Rudy Kalman in a number of occasions. In 2015, he promoted the conferral of the *honoris causa* PhD degree in *Information Technology* at the Politecnico di Milano, Italy, to him. After the approval of the Italian Ministry of Education, the ceremony was planned for 12 September 2016. It so happens that 12 September is also the anniversary of his wedding that took place in 1959. But he passed away on 2 July 2016, so that the honorary certificate was posthumously given to his wife. For more details on this ceremony, the readers can visit the webpage of the author of this book.

9.2 Kalman Approach to Prediction and Filtering Problems

In its simplest formulation, the Kalman filtering problem can be stated as follows. Consider the dynamical system

$$x(t + 1) = Fx(t) + v_1(t), \tag{9.1a}$$

$$y(t) = Hx(t) + v_2(t), \tag{9.1b}$$

where state $x(t)$ has n entries,

$$x(t) = [x_1(t)\, x_2(t)\, \cdots\, x_n(t)]',$$

and the output $y(t)$ has p entries,

$$y(t) = [y_1(t)\, y_2(t)\, \cdots\, y_p(t)]'.$$

$v_1(t)$ and $v_2(t)$ are two random vectors (with n and p entries, respectively) representing the disturbances acting on the state and output equations, respectively. We describe them as white noises with zero mean values and given variance matrices, V_1 and V_2:

$$v_1(\cdot) \sim \mathrm{WN}(0, V_1),$$

$$v_2(\cdot) \sim \mathrm{WN}(0, V_2).$$

We also assume that they are uncorrelated to each other:

$$E[v_1(i)v_2(j)'] = 0 \quad \forall i, j,$$

$$E[v_k(i)x(1)'] = 0 \quad \forall i; \; k = 1, 2,$$

and uncorrelated with the initial condition $x(1)$. We also make the assumption that its mean value is known, null for simplicity,

$$E[x(1)] = 0.$$

As for the variance, $\mathrm{Var}[x(1)]$ is assumed to be known (as a symmetric and positive semi-definite matrix). We denote it by $P(1)$:

$$\mathrm{Var}[x(1)] = P(1).$$

Note that the hypothesis $E[x(1)] = 0$, together with condition $E[v_1(t)] = 0, \forall t$, implies that $E[x(t)] = 0, \forall t$. Then, being $E[v_2(t)] = 0, \forall t$, we have $E[y(t)] = 0, \forall t$. In this way, at any time point, $x(t)$ and $y(t)$ are (stationary or nonstationary) stochastic processes with zero mean value.

Overall the system is described by the $n \times n$ dynamic matrix F, the $p \times n$ output matrix H, the $n \times n$ variance matrix V_1 of $v_1(\cdot)$, the $p \times p$ variance matrix V_2 of $v_2(\cdot)$, and the variance matrix $P(1)$ of the initial state. All such matrices are assumed to be known.

In its general form, the problem we address is as follows:

Kalman problem Estimate $x(N + r)$, given the observations of $y(t)$ for all time points from $t = 1$ to time $t = N$.

Our path will be articulated in various steps as follows:

One-step-ahead prediction problem: $r = 1$
Estimate $x(N + 1)$, given the observations of $y(t)$ from $t = 1$ to time $t = N$.

Multi-steps-ahead prediction problem: $r > 1$
Estimate $x(N + r)$, given the observations of $y(t)$ from $t = 1$ to time $t = N$.
Positive integer r is called *prediction horizon*.

Filtering problem: $r = 0$
Estimate $x(N)$, given the observations of $y(t)$ from $t = 1$ to time $t = N$.

We first solve the one-step-ahead prediction problem, and then we pass to the remaining problems.

Remark 9.2 (Generalizations of the Kalman problem) Throughout this book, we always approach a new problem in its simplest formulation and then generalize it. This rationale is adopted here as well: we solve Kalman problem in the above formulation; then we make extensions. To be precise, we anticipate that we will later consider the state estimation problem for the system:

$$x(t + 1) = Fx(t) + Gu(t) + v_1(t),$$
$$y(t) = Hx(t) + v_2(t),$$

where $u(\cdot)$ is an exogenous variable. $v_1(\cdot)$ and $v_2(\cdot)$ are again zero-mean white noises but possibly correlated to each other:

$$E[v_1(i)v_2(j)'] = \begin{cases} V_{12} & \forall i = j, \\ 0, & \forall i \neq j. \end{cases}$$

Moreover, the expected value of the initial state may be non-null: $E[x(1)] = x(1)_m$.

Further generalizations concern (i) time-varying systems (i.e. matrices F, G, H, V_1, V_2, and V_{12} may be function of time), (ii) systems with colored (non-white) disturbances, and (iii) nonlinear systems.

9.3 The Bayes Estimation Problem

By inspecting state equation (9.1a) and output equation (9.1b), it is apparent that, since disturbances $v_1(t)$ and $v_2(t)$ are assumed to be white noises, state $x(t)$ and output $y(t)$ have to be treated as random vectors. Therefore, in its essence, filtering is a problem where both the data $(y(N), y(N - 1), \ldots, y(1))$ and the unknown $(x(N + r))$ are random vectors. Estimating a random variable from the observation of another random variable is known as Bayes problem, from the name of Thomas Bayes (1702–1761), a clergyman whose fundamental contribution was published posthumous in 1763. So we first study the following:

Bayes problem
Estimate a random variable or random vector θ, given the observation of a random variable or a random vector d.

Again we will pose it in the simplest formulation and subsequently generalize it.

9.3.1 Bayes Problem – Scalar Case

Assume that θ and d are scalar random variables with zero mean value. Moreover,

i) θ has variance $\text{Var}[\theta] = E[\theta^2] = \lambda_{\theta\theta}$.
ii) d has variance $\text{Var}[d] = E[d^2] = \lambda_{dd}$.
iii) The covariance θ-d is given by $E[\theta d] = \lambda_{\theta d}$; obviously, the covariance d-θ is given by $E[d\theta] = \lambda_{d\theta} = \lambda_{\theta d}$.

We can therefore say that

$$E\begin{bmatrix} \theta \\ d \end{bmatrix} = \begin{bmatrix} 0 \\ 0 \end{bmatrix},$$

$$\text{Var}\begin{bmatrix} \theta \\ d \end{bmatrix} = \begin{bmatrix} \lambda_{\theta\theta} & \lambda_{\theta d} \\ \lambda_{d\theta} & \lambda_{dd} \end{bmatrix}.$$

These two expressions concerning the mean value and the variance can be jointly written as

$$\begin{bmatrix} \theta \\ d \end{bmatrix} \sim \left(\begin{bmatrix} 0 \\ 0 \end{bmatrix}, \begin{bmatrix} \lambda_{\theta\theta} & \lambda_{\theta d} \\ \lambda_{d\theta} & \lambda_{dd} \end{bmatrix} \right).$$

Any rule to find θ from d is named estimator and denoted as $\hat{\theta}$.

Among all possible estimation rules, a simple one is the *linear estimator*:

$$\hat{\theta} = \hat{\theta}(\alpha, \beta) = \alpha d + \beta, \tag{9.2}$$

where parameters α and β are real numbers. For each pair of parameters, we can assess the estimation accuracy by evaluating

$$J(\alpha, \beta) = E[(\hat{\theta}(\alpha, \beta) - \theta)^2]. \tag{9.3}$$

As best linear estimator, we take the one minimizing $J(\alpha, \beta)$. Substituting (9.2) into (9.3), we have

$$\begin{aligned} J(\alpha, \beta) &= E[(\alpha d + \beta - \theta)^2] \\ &= E[(\alpha^2 d^2 + \beta^2 + \theta^2 - 2\alpha d\theta + 2\alpha\beta d - 2\beta\theta)] \\ &= \alpha^2 \lambda_{dd} + \beta^2 + \lambda_{\theta\theta} - 2\alpha\lambda_{\theta d}. \end{aligned}$$

Here, we see that $J(\alpha, \beta)$ depends on β only through the additive term β^2. Hence, the optimal value of parameter β is $\beta = 0$. With such choice,

$$J(\alpha, 0) = \alpha^2 \lambda_{dd} + \lambda_{\theta\theta} - 2\alpha\lambda_{\theta d}.$$

This is a quadratic function in α whose minimum is achieved for $\alpha = \lambda_{\theta d}/\lambda_{dd}$. In conclusion, the optimal linear estimator is

$$\hat{\theta} = \frac{\lambda_{\theta d}}{\lambda_{dd}} d. \tag{9.4}$$

This expression in known as *Bayes formula*. Often, in place of $\hat{\theta}$, the optimal estimate is written as

$$E[\theta|d] = \frac{\lambda_{\theta d}}{\lambda_{dd}} d.$$

The symbol $E[\theta|d)]$ has the advantage of clearly pointing out the data set d on which the estimate of θ is based. The corresponding estimation error is given by

$$E[(\hat{\theta} - \theta)^2] = J\left(\frac{\lambda_{\theta d}}{\lambda_{dd}} d, 0\right) = E\left[\left(\frac{\lambda_{\theta d}}{\lambda_{dd}} d - \theta\right)^2\right]$$

$$= \frac{\lambda_{\theta d}^2}{\lambda_{dd}^2} \lambda_{dd} + \lambda_{\theta\theta} - 2\frac{\lambda_{\theta d}}{\lambda_{dd}} \lambda_{\theta d}$$

$$= \lambda_{\theta\theta} - \frac{\lambda_{\theta d}^2}{\lambda_{dd}}. \tag{9.5}$$

We can also write

$$E[(\hat{\theta} - \theta)^2] = \lambda_{\theta\theta}(1 - \rho^2), \tag{9.6}$$

where

$$\rho = \frac{\lambda_{\theta d}}{\sqrt{\lambda_{dd}}\sqrt{\lambda_{\theta\theta}}}$$

is the covariance coefficient (see Section 1.4.1) between random variables θ and d.

Expression (9.6) is worthy commenting. When we don't have any data, namely *a priori*, what we can say about unknown θ is that it is a random variable with zero mean value and uncertainty characterized by its standard deviation $\lambda_{\theta\theta}^{1/2}$. For instance, if θ were Gaussian, then the probability that it takes value in the interval $-2\sqrt{\lambda_{\theta\theta}}$ and $+2\sqrt{\lambda_{\theta\theta}}$ around its mean value is about 95%. When we have at our disposal the observation of variable d, the information conveyed by such data point enables reducing uncertainty; indeed, the *a posteriori* variance is $\lambda_{\theta\theta}(1 - \rho^2)$, from which we see that the standard deviation reduces to $\sqrt{\lambda_{\theta\theta}(1 - \rho^2)}$. So the higher the correlation between θ and d (i.e. the higher ρ in its feasibility interval $[0, 1]$), the better is the estimation accuracy.

9.3.2 Bayes Problem – Vector Case

We pass now to the problem of estimating a random vector θ, given the observation of a random vector d. Note that θ and d may have any dimensions, not necessarily coincident.

Assume that

$$\begin{bmatrix} \theta \\ d \end{bmatrix} \sim \left(\begin{bmatrix} 0 \\ 0 \end{bmatrix}, \begin{bmatrix} \Lambda_{\theta\theta} & \Lambda_{\theta d} \\ \Lambda_{d\theta} & \Lambda_{dd} \end{bmatrix} \right),$$

where

$$\Lambda_{\theta\theta} = E[\theta\theta'], \quad \Lambda_{\theta d} = E[\theta d'],$$
$$\Lambda_{d\theta} = E[d\theta'] = E[\theta d']' = \Lambda_{\theta d}', \quad \Lambda_{dd} = E[dd']$$

are the various variance and cross-variance matrices.

The scalar Bayes formula (9.4) can be easily generalized to the vector case as follows:

$$\hat{\theta} = \Lambda_{\theta d}\Lambda_{dd}^{-1}d. \tag{9.7}$$

The expression of the variance of the estimation error (Eq. (9.5) in the scalar case) becomes

$$E[(\hat{\theta} - \theta)(\hat{\theta} - \theta)'] = \Lambda_{\theta\theta} - \Lambda_{\theta d}\Lambda_{dd}^{-1}\Lambda_{d\theta}. \tag{9.8}$$

Remark 9.3 (Bayes formula in case of non-null mean values)   The hypothesis of null mean value of the random variables or random vectors can be removed. If

$$E[\theta] = \theta_m, \quad E[d] = d_m,$$

it is easy to reduce the new situation to the previous one by defining the unbiased random variables

$$\tilde{\theta} = \theta - \theta_m,$$
$$\tilde{d} = d - d_m.$$

Since $\tilde{\theta}$ and $\tilde{d}$ have zero mean value, the standard Bayes formula can be applied to them. Then, by replacing $\tilde{\theta}$ with $\theta - \theta_m$ and $\tilde{d}$ with $d - d_m$, it is straightforward to see that the (vector) Bayes formula is

$$\hat{\theta} = \theta_m + \Lambda_{\theta d}\Lambda_{dd}^{-1}(d - d_m),$$

while the formula for the variance of the estimation error, Eq. (9.8), remains unchanged.

9.3.3 Recursive Bayes Formula – Scalar Case

All expressions previously seen for the Bayes estimate are *batch* formulas. This means that all data, from the first one to the last one, are required to perform the computations needed to determine the estimate $\hat{\theta}$. We now derive a recursive version of the Bayes formula, enabling the computation by updating the estimate once a new snapshot is available.

For the sake of simplicity, let's consider a data vector d constituted by two scalar snapshots, $d = [d(1)\ d(2)]'$; moreover, let the unknown θ be scalar. All variables have zero mean value and their covariance matrix is given, as summarized below:

$$\begin{bmatrix} \theta \\ d(1) \\ d(2) \end{bmatrix} \sim \left(\begin{bmatrix} 0 \\ 0 \\ 0 \end{bmatrix}, \begin{bmatrix} \lambda_{\theta\theta} & \lambda_{\theta 1} & \lambda_{\theta 2} \\ \lambda_{1\theta} & \lambda_{11} & \lambda_{12} \\ \lambda_{2\theta} & \lambda_{21} & \lambda_{22} \end{bmatrix} \right).$$

Here, the meaning of the symbols is self-evident, for example $\lambda_{\theta 1} = E[\theta d(1)]$, $\lambda_{12} = E[d(1)d(2)]$, and so on.

While the Bayes optimal estimate given the single snapshot $d(1)$ is given by

$$E[\theta | d(1)] = \frac{\lambda_{\theta 1}}{\lambda_{11}} d(1),$$

in view of (9.7) the optimal estimate given $d(1)$ and $d(2)$ is

$$E[\theta | d(1), d(2)] = [\lambda_{\theta 1}\ \ \lambda_{\theta 2}] \begin{bmatrix} \lambda_{11} & \lambda_{12} \\ \lambda_{21} & \lambda_{22} \end{bmatrix}^{-1} \begin{bmatrix} d(1) \\ d(2) \end{bmatrix}, \quad \lambda_{21} = \lambda_{12}.$$

By computing the inverse matrix, we obtain

$$E[\theta | d(1), d(2)] = \frac{1}{\lambda_{11}\lambda^2}(-\lambda_{\theta 1}\lambda_{21} + \lambda_{\theta 2}\lambda_{11})d(2)$$

$$+ \frac{1}{\lambda_{11}\lambda^2}(\lambda_{\theta 1}\lambda_{22} - \lambda_{\theta 2}\lambda_{12})d(1), \tag{9.9}$$

where

$$\lambda^2 = \lambda_{22} - \frac{\lambda_{12}^2}{\lambda_{11}}. \tag{9.10}$$

In expression (9.9), add and subtract the term $(\lambda_{\theta 1}/\lambda_{11})d(1)$:

$$E[\theta | d(1), d(2)] = \frac{1}{\lambda_{11}\lambda^2}(-\lambda_{\theta 1}\lambda_{21} + \lambda_{\theta 2}\lambda_{11})d(2)$$

$$+ \frac{1}{\lambda_{11}\lambda^2}(\lambda_{\theta 1}\lambda_{22} - \lambda_{\theta 2}\lambda_{12})d(1) - \frac{\lambda_{\theta 1}}{\lambda_{11}}d(1) + \frac{\lambda_{\theta 1}}{\lambda_{11}}d(1).$$

In this sum, there are four terms; for the first and second ones, we bring coefficient $1/\lambda_{11}$ inside the corresponding parentheses; moreover, we group together the third and the second ones, so that

$$E[\theta | d(1), d(2)] = \frac{1}{\lambda^2}\left(\lambda_{\theta 2} - \lambda_{\theta 1}\frac{\lambda_{12}}{\lambda_{11}}\right)d(2)$$

$$+ \frac{1}{\lambda^2}\left(\lambda_{\theta 1}\frac{\lambda_{22}}{\lambda_{11}} - \lambda_{\theta 2}\frac{\lambda_{12}}{\lambda_{11}} - \frac{\lambda_{\theta 1}}{\lambda_{11}}\lambda^2\right)d(1) + \frac{\lambda_{\theta 1}}{\lambda_{11}}d(1).$$

Focus now on coefficient

$$\left(\lambda_{\theta 1} \frac{\lambda_{22}}{\lambda_{11}} - \lambda_{\theta 2} \frac{\lambda_{12}}{\lambda_{11}} - \frac{\lambda_{\theta 1}}{\lambda_{11}} \lambda^2 \right).$$

By replacing λ^2 with expression (9.10), we have

$$\left(\lambda_{\theta 1} \frac{\lambda_{22}}{\lambda_{11}} - \lambda_{\theta 2} \frac{\lambda_{12}}{\lambda_{11}} - \frac{\lambda_{\theta 1}}{\lambda_{11}} \lambda^2 \right) = \frac{\lambda_{12}}{\lambda_{11}} \left(-\lambda_{\theta 2} + \lambda_{\theta 1} \frac{\lambda_{12}}{\lambda_{11}} \right).$$

In conclusion,

$$E[\theta | d(1), d(2)] = \frac{\lambda_{\theta 1}}{\lambda_{11}} d(1)$$
$$+ \frac{1}{\lambda^2} \left(\lambda_{\theta 2} - \lambda_{\theta 1} \frac{\lambda_{12}}{\lambda_{11}} \right) \left[d(2) - \frac{\lambda_{12}}{\lambda_{11}} d(1) \right]. \tag{9.11}$$

In this formula, we see how the optimal estimate with two snapshots $d(1)$ and $d(2)$ can be formed starting from the optimal estimate based on $d(1)$ only, namely $(\lambda_{\theta 1}/\lambda_{11})d(1)$, by adding a contribution that takes into account $d(2)$. So, this is a truly recursive expression of the Bayes (scalar) formula.

9.3.4 Innovation

In formula (9.11), we see the appearance of a most important new character, deserving a definition by itself.

Definition 9.1 (Innovation) Given two random variables $d(1)$ and $d(2)$, the random variable

$$e = d(2) - E[d(2) | d(1)] = d(2) - \frac{\lambda_{12}}{\lambda_{11}} d(1)$$

is named innovation of $d(2)$ with respect to $d(1)$.

Its main properties are

a) mean value

$$E[e] = 0$$

b) variance

$$\lambda_{ee} = \text{Var}[e] = E \left[\left(d(2) - \frac{\lambda_{12}}{\lambda_{11}} d(1) \right)^2 \right]$$
$$= \lambda_{22} + \frac{\lambda_{12}^2}{\lambda_{11}^2} \lambda_{11} - 2 \frac{\lambda_{12}^2}{\lambda_{11}} = \lambda^2$$

c) correlation with θ

$$\lambda_{\theta e} = E\left[\theta\left(d(2) - \frac{\lambda_{12}}{\lambda_{11}}d(1)\right)\right] = \lambda_{\theta 2} - \lambda_{\theta 1}\frac{\lambda_{12}}{\lambda_{11}}$$

d) correlation with $d(1)$

$$\lambda_{1e} = E\left[d(1)\left(d(2) - \frac{\lambda_{12}}{\lambda_{11}}d(1)\right)\right] = \lambda_{12} - \frac{\lambda_{12}}{\lambda_{11}}\lambda_{11} = 0.$$

In view of the definition of innovation, and thanks to properties (b) and (c), expression (9.11) can be rewritten as follows:

$$E[\theta|d(1), d(2)] = \frac{\lambda_{\theta 1}}{\lambda_{11}}d(1) + \frac{\lambda_{\theta e}}{\lambda_{ee}}e. \tag{9.12}$$

On the other hand, $(\lambda_{\theta 1}/\lambda_{11})d(1)$ is the Bayes formula supplying the optimal estimate of θ given $d(1)$, while $(\lambda_{\theta e}/\lambda_{ee})e$ is again Bayes formula applied to the estimation of θ given e. So, we can write:

$$E[\theta|d(1), d(2)] = E[\theta|d(1)] + E[\theta|e]. \tag{9.13}$$

Expression (9.13) shows that the optimal estimate of θ from data $d(1)$ and $d(2)$ does not coincide with the sum of the optimal estimates of θ given $d(1)$ only and $d(2)$ only. The only case when this is true is when $d(1)$ and $d(2)$ are uncorrelated, so that $E[d(2)|d(1)] = 0$, and $e = d(2)$:

$$E[\theta|d(1), d(2)] = E[\theta|d(1)] + E[\theta|d(2)] \quad \text{false in general.}$$

To better understand the meaning of innovation, consider this ideal experiment: instead of having $d(1)$ and $d(2)$ as data, suppose having $d(1)$ and e. What would then be the optimal estimate of θ? To find the reply, we have to compute $E[\theta|d(1), e]$. For example, we resort again to the Bayes formula:

$$E[\theta|d(1), e] = [\lambda_{\theta 1} \lambda_{\theta e}]\begin{bmatrix} \lambda_{11} & \lambda_{1e} \\ \lambda_{e1} & \lambda_{ee} \end{bmatrix}^{-1}\begin{bmatrix} d(1) \\ e \end{bmatrix}.$$

Since $d(1)$ and e are uncorrelated, i.e. $\lambda_{e1} = 0$, it is easy to see that

$$E[\theta|d(1), e] = E[\theta|d(1)] + E[\theta|e]. \tag{9.14}$$

By comparing (9.13) with (9.14), we come to the conclusion that the estimate of θ by means of $d(1)$ and $d(2)$ coincides with the estimate obtained with $d(1)$ and e. In other words, to estimate θ, once $d(1)$ is given, the information content of $d(2)$ is the same as the information content of e.

This conclusion is not surprising. Indeed, from the very definition of innovation, we can write $d(2)$ as

$$d(2) = E[d(2)|d(1)] + e. \tag{9.15}$$

Here, $d(2)$ is decomposed into the sum of two terms, the first of which is its optimal prediction given $d(1)$, $E[d(2)|d(1)]$. Thus, the second term (the innovation) is the "part" of $d(2)$ that cannot be predicted knowing $d(1)$. Note that, in view of the innovation properties, the terms in decomposition (9.15) are uncorrelated to each other.

We can summarize all these considerations by saying that the innovation contains all and only the *fresh information* conveyed by $d(2)$ with respect to $d(1)$. This is why, once $d(1)$ is given, knowing $d(2)$ or e is the same.

9.3.5 Recursive Bayes Formula – Vector Case

The recursive Bayes expressions can be generalized to the vector case. For example, assume that θ, $d(1)$, and $d(2)$ are random vectors (of any dimensions) such that

$$\begin{bmatrix} \theta \\ d(1) \\ d(2) \end{bmatrix} \sim G\left(\begin{bmatrix} 0 \\ 0 \\ 0 \end{bmatrix}, \begin{bmatrix} \Lambda_{\theta\theta} & \Lambda_{\theta 1} & \Lambda_{\theta 2} \\ \Lambda_{1\theta} & \Lambda_{11} & \Lambda_{12} \\ \Lambda_{2\theta} & \Lambda_{21} & \Lambda_{22} \end{bmatrix} \right), \tag{9.16}$$

where $\Lambda_{\theta 1} = \Lambda'_{1\theta}$, $\Lambda_{\theta 2} = \Lambda'_{2\theta}$, and $\Lambda_{12} = \Lambda'_{21}$. The definition of innovation of $d(2)$ with respect to $d(1)$ is again

$$e = d(2) - E[d(2)|d(1)]. \tag{9.17}$$

Note that e is now a vector of the same dimensions as $d(2)$.

In view of (9.7), we can also write

$$e = d(2) - \Lambda_{21}\Lambda_{11}^{-1}d(1). \tag{9.18}$$

The recursive Bayes formula is

$$\begin{aligned} E[\theta|d(1), d(2)] &= \Lambda_{\theta 1}\Lambda_{11}^{-1}d(1) + \Lambda_{\theta e}\Lambda_{ee}^{-1}e \\ &= E[\theta|d(1)] + E[\theta|e]. \end{aligned} \tag{9.19}$$

Remark 9.4 (Recursive Bayes formula in the case of non-null mean values)
Unknown and data may have non-null mean value so that:

$$\begin{bmatrix} \theta \\ d(1) \\ d(2) \end{bmatrix} \sim \left(\begin{bmatrix} \theta_m \\ d(1)_m \\ d(2)_m \end{bmatrix}, \begin{bmatrix} \Lambda_{\theta\theta} & \Lambda_{\theta 1} & \Lambda_{\theta 2} \\ \Lambda_{1\theta} & \Lambda_{11} & \Lambda_{12} \\ \Lambda_{2\theta} & \Lambda_{21} & \Lambda_{22} \end{bmatrix} \right).$$

Even in these conditions, we adopt as definition of innovation expression (9.17). To compute the mean value of e, observe that

$$E[d(2)] = d(2)_m.$$

Moreover, being

$$E[d(2)|d(1)] = d(2)_m + \Lambda_{21}\Lambda_{11}^{-1}(d(1) - d(1)_m),$$

we have

$$E[E[d(2)|d(1)]] = d(2)_m + \Lambda_{21}\Lambda_{11}^{-1}E[d(1) - d(1)_m] = d(2)_m.$$

Therefore,

$$E[e] = d(2)_m - d(2)_m = 0.$$

Hence, whatever the mean values be, the mean value of innovation is always null. To find the optimal estimator, we again reduce the problem to the null mean value case by de-biasing, i.e. we pose

$$\tilde{\theta} = \theta - \theta_m,$$
$$\tilde{d}(i) = d(i) - d(i)_m, \quad i = 1, 2.$$

Then we can apply the standard recursive Bayes formula:

$$E[\tilde{\theta}|\tilde{d}(1), \tilde{d}(2)] = \Lambda_{\theta 1}\Lambda_{11}^{-1}\tilde{d}(1) + \Lambda_{\theta e}\Lambda_{ee}^{-1}\tilde{e}, \tag{9.20}$$

where $\tilde{e}$ is the innovation of $\tilde{d}(2)$ given $\tilde{d}(1)$. It is easy to see that $\tilde{e} = e$. Then, recalling the de-biasing definitions, we have

$$E[\theta|d(1), d(2)] = \theta_m + \Lambda_{\theta 1}\Lambda_{11}^{-1}(d(1) - d(1)_m) + \Lambda_{\theta e}\Lambda_{ee}^{-1}e. \tag{9.21}$$

The variance of the estimation error is given by (9.8).

Finally, the variance of the innovation is again

$$\Lambda_{ee} = \Lambda_{22} - \Lambda_{21}\Lambda_{11}^{-1}\Lambda_{12} = \Lambda_{22} - \Lambda_{12}'\Lambda_{11}^{-1}\Lambda_{12}. \tag{9.22}$$

9.3.6 Geometric Interpretation of Bayes Estimation

The Bayes problem can be given a very useful geometric interpretation. First, we give the interpretation of formulas (9.4) and (9.5). Then we pass to the recursive Bayes formula.

9.3.6.1 Geometric Interpretation of the Bayes Batch Formula

In order to set the Bayes problem into a geometric framework, we need – so to say – to "transform" random variables into vectors. To this end, denoting by $\mathbb{G}$ the set of all scalar random variables with zero mean value, we construct a vector space $\mathscr{G}$ over it in a very natural way, by defining the sum of two elements of $\mathbb{G}$ and the product of an element of $\mathbb{G}$ by a real number as follows. Given $g_1 \in \mathbb{G}$ and $g_2 \in \mathbb{G}$, their sum is given by the usual notion of sum of random variables (i.e. for each outcome s of the underlying random experiment, the sum $g_1 + g_2$ is the random variable taking the values $g_1(s) + g_2(s)$ for all s). In this way, the sum is another random variable belonging to $\mathbb{G}$. As for the product of an element $g \in \mathbb{G}$ by a real number, say μ, it is the random variable taking the values $\mu g(s)$ for all s, again a variable belonging to $\mathbb{G}$.

With such operations, the set $\mathbb{G}$ forms a vector space over the real numbers, which is denoted by $\mathscr{G}$, as already said.

We can define the scalar product over $\mathcal{G}$ by letting

$$(g_1, g_2) = E[g_1 g_2].$$

Thus,

$$(g_1, g_2) = (g_2, g_1),$$
$$(g, g) = \text{Var } g \geq 0,$$
$$(g, g) = 0 \Leftrightarrow g \sim (0, 0),$$
$$(\mu_1 g_1 + \mu_2 g_2, g_3) = \mu_1(g_1, g_3) + \mu_2(g_2, g_3).$$

Hence, the defined operator is indeed a scalar product.

The norm of a vector in the space so introduced can be defined as

$$\|g\| = \sqrt{(g, g)}.$$

In conclusion, the random variables can be seen as vectors in a space. The "length" of a vector ($\|g\|$) is the standard deviation of that variable.

We are now in a position to define the angle α between two vectors g_1 and g_2. In full analogy with the definition of angle between two vectors in the plane, we pose

$$\cos \alpha = \frac{(g_1, g_2)}{\|g_1\| \|g_2\|}.$$

Note that

$$\frac{(g_1, g_2)}{\|g_1\| \|g_2\|} = \frac{E[g_1 g_2]}{(\text{Var } g_1)^{1/2} (\text{Var } g_2)^{1/2}}.$$

Therefore, the cosine of the angle between vectors g_1 and g_2 is just the correlation coefficient ρ of the two random variables g_1 and g_2. Consequently, two vectors in $\mathcal{G}$ are *orthogonal* when the corresponding random variables are *uncorrelated*. The two vectors are aligned when $\rho = \pm 1$. To be precise, they have the same direction if $\rho = 1$, and opposite direction if $\rho = -1$.

We now come to the geometrical interpretation of the Bayes formula. For example, consider two random variables g_1 and g_2 and denote by λ_{11} the variance of g_1 and by λ_{12} the cross-variance $g_1 - g_2$. The Bayes formula for the estimation of g_2 given g_1 is

$$E[g_2 | g_1] = \frac{\lambda_{12}}{\lambda_{11}} g_1 = \frac{E[g_1 g_2]}{\text{Var } g_1} g_1 = \frac{(g_1, g_2)}{\|g_1\|^2} g_1.$$

Here, multiplying and dividing by the norm of g_2, we have

$$E[g_2 | g_1] = \frac{(g_1, g_2)}{\|g_1\| \|g_2\|} \frac{g_1}{\|g_1\|} \|g_2\|$$

$$= \|g_2\| \cos \alpha \frac{g_1}{\|g_1\|}.$$

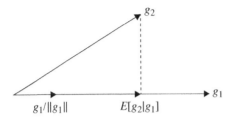

Figure 9.2 Geometric interpretation of the Bayes formula.

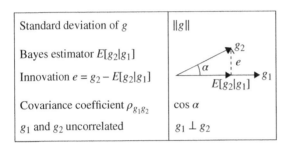

Figure 9.3 Probability and geometry – table of correspondences.

Standard deviation of g	$\|g\|$
Bayes estimator $E[g_2\|g_1]$	
Innovation $e = g_2 - E[g_2\|g_1]$	
Covariance coefficient $\rho_{g_1 g_2}$	$\cos \alpha$
g_1 and g_2 uncorrelated	$g_1 \perp g_2$

Since $g_1/\|g_1\|$ is a unit norm vector with the same direction of g_1, this expression says that the Bayes estimator of random variable g_2 given the random variable g_1 is nothing but the *projection* of vector g_2 on vector g_1 (Figure 9.2).

The variance of the estimation error of g_2 given g_1 is the squared norm of vector $g_2 - E[g_2/g_1]$. This is nothing but the area of a square having $g_2 - E[g_2/g_1]$ as side. Such area can be simply determined via the Pythagorean theorem applied to the right triangle of Figure 9.2. Hence,

$$\text{Var} \left[g_2 - E[g_2|g_1]\right] = \|g_2\|^2 - \|E[g_2|g_1]\|^2 = \lambda_{22} - \lambda_{12}^2/\lambda_{11},$$

where λ_{22} is the variance of g_2. This is the well-known expression of the *a posteriori* variance (9.5).

In Figure 9.3, the correspondence between probabilistic concepts and geometric interpretations is outlined.

9.3.6.2 Geometric Interpretation of the Recursive Bayes Formula

We now deduce the recursive Bayes formula by purely geometric considerations (Figure 9.4). Consider three random variables, θ, $d(1)$, and $d(2)$, and the associated three vectors in $\mathcal{G}$. The plane (subspace) $\mathcal{H}[d(1), d(2)]$ is the set of all linear combinations of vectors $d(1)$ and $d(2)$. The optimal estimate of $d(2)$ given $d(1)$ is the projection of $d(2)$ on $d(1)$. Consequently, the innovation of $d(2)$ with respect to $d(1)$ is the vector e given by the difference of $d(2)$ minus the projection of $d(2)$ on $d(1)$. Obviously, e lies in the plane $\mathcal{H}[d(1), d(2)]$ and is orthogonal to $d(1)$. As for the optimal estimate of θ given $d(1)$, $E[\theta|d(1)]$, it is given by projection of θ on $d(1)$, while the optimal estimate of θ from innovation e, $E[\theta|e]$, is the projection of θ on the vector e previously constructed. Note that, since e is orthogonal to $d(1)$, the vectors $E[\theta|d(1)]$ and $E[\theta|e]$ *are*

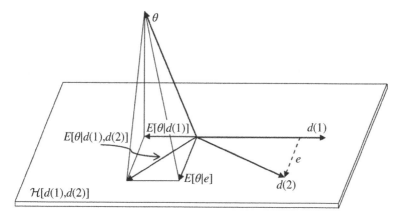

Figure 9.4 Geometric interpretation of the recursive Bayes formula.

orthogonal to each other. This is why the projection of vector θ on the plane of $d(1)$ and $d(2)$, $E[\theta|d(1), d(2)]$, coincides with the sum of $E[\theta|d(1)]$ and $E[\theta|e]$. This leads to the recursive Bayes formula (9.13). Interestingly enough, if instead we sum $E[\theta|d(1)]$ and $E[\theta|d(2)]$, we do not obtain $E[\theta|d(1), d(2)]$, as a moment reflection on the geometrical view of Bayes formula as in Figure 9.4 reveals.

Remark 9.5 (Geometric interpretation–vector case)   The above-explained geometrical view can be extended to random vectors (in place of random variables).

Given a random vector d_1, we shall denote by $\mathcal{H}[d_1]$ the set of all random variables obtained by all linear combinations of the various entries of d_1. This set forms a subspace in $\mathcal{G}$. Suppose now to have another random vector $d(2)$. The optimal estimate $E[d_2|d_1]$ is the vector (of the same dimensions of d_2) whose ith entry is the vector of $\mathcal{G}$ obtained by projecting the ith entry of d_2 on $\mathcal{H}[d_1]$.

The difference of d_2 and its projection on $\mathcal{H}[d_1]$ is orthogonal to $\mathcal{H}[d_1]$, in the sense that all elements of d_2 minus their projections on $\mathcal{H}[g_2]$ are random variables that, as vectors in $\mathcal{G}$, are orthogonal to the vectors of $\mathcal{H}[d_1]$.

Note that, in case of random vectors, the same word "vector" is used with two meanings. First, it means a set of random variables; second, it can denote an element of a vector space. So, the entries of a random vector are also vectors of the space $\mathcal{G}$.

9.4   One-step-ahead Kalman Predictor

We can now address the problem of predicting the state of dynamical system ((9.1a) and (9.1b)). Our aim is to find a recursive expression for the optimal estimator of $x(N + 1)$ given the data $y(t)$ up to time N: $\hat{x}(N + 1|N)$.

In such a problem, a main role is played by the notion of innovation, a notion from which we start our study.

9.4.1 The Innovation in the State Prediction Problem

The set of available data over the interval of time $[1, N]$ can be grouped into a unique vector:

$$y^N = [y(N)' \; y(N-1)' \; \cdots \; y(1)']'.$$

In the space $\mathscr{G}$ of random variables, the various entries of y^N generate a subspace, which we denote as $\mathscr{H}[y^N]$. We name it the *subspace of the past*.

The innovation conveyed by snapshot $y(N+1)$ with respect to past data (y^N) is defined as

$$e(N+1) = y(N+1) - E[y(N+1)|y^N]. \tag{9.23}$$

As seen in the previous sections, $E[y(N+1)|y^N]$ is the projection of $y(N+1)$ on $\mathscr{H}[y^N]$. Therefore, the innovation is orthogonal to subspace $\mathscr{H}[y^N]$, as schematically depicted in Figure 9.5. We can say that the innovation is orthogonal to the past.

9.4.2 The State Prediction Error

Another player of main importance in Kalman filtering is the state prediction error, denoted as $v(N+1)$.

Definition 9.2 The state prediction error (at time $N+1$) is

$$v(N+1) = x(N+1) - E[x(N+1)|y^N)].$$

While the output prediction error, i.e. the innovation e, has p components, as many as the system output, the state prediction error has n components, as many as the system state. As e, v has also zero mean value.

The variance matrix of the sate prediction error is denoted by $P(N)$:

$$P(N) = E[v(N)v(N)']. \tag{9.24}$$

As we will see, this matrix plays a main role in Kalman theory.

$e(t)$ and $v(t)$ are related to each other. Indeed, in view of (9.1b),

$$\begin{aligned} e(N) &= y(N) - \hat{y}(N|N-1) = Hx(N) + v_2(N) - H\hat{x}(N|N-1) \\ &= H(x(N) - \hat{x}(N|N-1)) + v_2(N) \\ &= Hv(N) + v_2(N). \end{aligned} \tag{9.25}$$

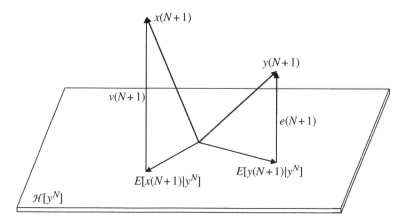

Figure 9.5 Innovation $e(N + 1)$ and state prediction error $v(N + 1)$ for a sequence of data, $y(1), y(2), \ldots, y(N), y(N + 1)$.

In the same way that $E[y(N + 1|N)]$ is the projection of $y(N + 1)$ on $\mathcal{H}[y^N]$, $E[x(N + 1|N)]$ is the projection of $x(N + 1)$ on $\mathcal{H}[y^N]$. Consequently, both e and $v(N + 1)$ are orthogonal to the past, as represented in Figure 9.5.

9.4.3 Optimal One-Step-Ahead Prediction of the Output

The optimal one-step-ahead of the output is

$$\hat{y}(N + 1|N) = E[y(N + 1)|y^N].$$

In view of Eq. (9.1b), we have

$$\hat{y}(N + 1|N) = E[Hx(N + 1) + v_2(N + 1)|y^N]$$
$$= HE[x(N + 1)|y^N] + E[v_2(N + 1)|y^N].$$

Here, the second term is null. Indeed, from (9.1a), it is apparent that $x(N)$ depends upon $v_1(\cdot)$ up to time $N - 1$, besides initial condition $x(1)$:

$$x(N) = f(v_1^{N-1}, x(1)). \tag{9.26}$$

Hence, from (9.1b),

$$y(N) = f(v_1^{N-1}, x(1), v_2(N)), \tag{9.27}$$

so that

$$y^N = f(v_1^{N-1}, x(1), v_2^N). \tag{9.28}$$

Obviously, in (9.26)–(9.28), $f(\cdot)$ is a generic function.

Now, $v_2(\cdot)$ being a white noise, $v_2(N+1)$ is uncorrelated with its past, v_2^N. Moreover, thanks to the assumption we have made that $v_2(\cdot)$ is independent, $v_1(\cdot)$ is uncorrelated with $v_2(\cdot)$ and with initial condition $x(1)$. Therefore, $v_2(N+1)$ is uncorrelated with $y^N = f(v_1^{N-1}, x(1), v_2^N)$. In other words, the information conveyed by y^N on $v_2(N+1)$ is null. Hence, $E[v_2(N+1)|y^N]$ coincides with $E[v_2(N+1)] = 0$.

In conclusion, the optimal prediction of the output can be straightforwardly computed from the optimal prediction of the state

$$\hat{y}(N+1|N) = H\hat{x}(N+1|N). \tag{9.29}$$

9.4.4 Optimal One-Step-Ahead Prediction of the State

To find the optimal predictor of the state, we resort again to the Bayes formula:

$$\hat{x}(N+1|N) = E[x(N+1)|y^N].$$

To obtain its recursive expression, we consider the set y^N as the union of the set y^{N-1} and last snapshot $y(N)$:

$$\hat{x}(N+1|N) = E[x(N+1)|y^{N-1}, y(N)].$$

With the recursive Bayes formula, we have

$$\hat{x}(N+1|N) = E[x(N+1)|y^{N-1}] + E[x(N+1)|e(N)], \tag{9.30}$$

where $e(N)$ is the innovation of $y(N)$ with respect to y^{N-1}.

We now compute the two terms at the right-hand side of this formula.

(a) *Computation of $E[x(N+1)|y^{N-1}]$*:
   By means of (9.1a), this term can be written as

$$E[x(N+1)|y^{N-1}] = E[Fx(N) + v_1(N)|y^{N-1}]$$
$$= FE[x(N)|y^{N-1}] + E[v_1(N)|y^{N-1}].$$

Now, $v_1(N)$ is uncorrelated with y^{N-1}; therefore, $E[v_1(N)|y^{N-1}] = E[v_1(N)] = 0$. Hence,

$$E[x(N+1)|y^{N-1}] = F\hat{x}(N|N-1). \tag{9.31}$$

(b) *Computation of $E[x(N+1)|e(N)]$*:
   By the batch Bayes formula, we have

$$E[x(N+1)|e(N)] = \Lambda_{x(N+1)e(N)}\Lambda_{e(N)e(N)}^{-1}e(N). \tag{9.32}$$

We now address the problem to compute the two matrices $\Lambda_{x(N+1)e(N)}$ and $\Lambda_{e(N)e(N)}$.

(b1) *Computation of $\Lambda_{x(N+1)e(N)}$*:
   By definition

$$\Lambda_{x(N+1)e(N)} = E[x(N+1)e(N)'].$$

Recalling the relationship between the innovation and the state prediction error, Eq. (9.25), we have

$$
\begin{aligned}
\Lambda_{x(N+1)e(N)} \\
&= E[(Fx(N) + v_1(N))\{H(x(N) - \hat{x}(N|N-1)) + v_2(N)\}'] \\
&= FE[x(N)(x(N) - \hat{x}(N|N-1))']H' \\
&\quad + FE[x(N)v_2(N)'] \\
&\quad + E[v_1(N)\{H(x(N) - \hat{x}(N|N-1)) + v_2(N)\}'] \quad\quad (9.33)
\end{aligned}
$$

Here, at the right-hand side, we see three terms.

Consider first the second one, $FE[x(N)v_2(N)']$. Since $v_2(\cdot)$ is by assumption uncorrelated with $v_1(\cdot)$ and with the initial condition, random vectors $x(N)$ and $v_2(N)$ are uncorrelated. Hence, the expected value of their product is the product of the expected values and therefore is null $(E[x(N)v_2(N)'] = 0)$.

As for the last term, $E[v_1(N)\{H(x(N) - \hat{x}(N|N-1)) + v_2(N)\}']$, it is composed by the sum of the following three elements:

$$
E[v_1(N)x(N)']H',
$$

$$
- E[v_1(N)\hat{x}(N|N-1)']H',
$$

$$
E[v_1(N)v_2(N)'].
$$

Equation (9.26) shows that $x(N)$ depends upon $v_1(\cdot)$ up to time $N-1$ only. $v_1(\cdot)$ being a white noise, $v_1(N)$ is uncorrelated with its past, v_1^{N-1}. Since the initial condition $x(1)$ is uncorrelated with disturbances, we can conclude that $v_1(N)$ and $x(N)$ are uncorrelated, so that $E[v_1(N)x(N)']H' = 0$.

Analogous considerations hold for $E[v_1(N)\hat{x}(N|N-1)']H'$. Indeed, $\hat{x}(N|N-1)$ is a predictor; hence, it is a function of $y(\cdot)$ up to $N-1$. On the other hand, from Eq. (9.28), it appears that y^{N-1} depends upon the initial condition $x(1)$, from $v_1(\cdot)$ up to time $N-2$ and upon $v_2(\cdot)$ up to time $N-1$. All these three quantities are uncorrelated with $v_1(N)$, so that $E[v_1(N)\hat{x}(N|N-1)']H' = 0$.

Finally, it is apparent that $E[v_1(N)v_2(N)'] = 0$.

In conclusion, from (9.33), we have

$$
\Lambda_{x(N+1)e(N)} = FE[x(N)(x(N) - \hat{x}(N|N-1))']F'.
$$

By adding and subtracting the same quantity, we replace here $x(N)$ with $x(N) - \hat{x}(N|N-1) + \hat{x}(N|N-1)$ to obtain

$$
\begin{aligned}
FE&[x(N)(x(N) - \hat{x}(N|N-1))']H' \\
&= FE[(x(N) - \hat{x}(N|N-1))(x(N) - \hat{x}(N|N-1))']H' \\
&\quad + FE[\hat{x}(N|N-1)(x(N) - \hat{x}(N|N-1))']H'.
\end{aligned}
$$

Recalling that the state prediction error is denoted as $v(N)$, this expression can also be written as

$$FE[x(N)(x(N) - x(\hat{N}|N-1))']H'$$
$$= FE[v(N)v(N)']H' + FE[\hat{x}(N|N-1)v(N)']H'. \qquad (9.34)$$

The state prediction error $v(N)$ is orthogonal to the subspace of the past $\mathcal{H}[y^{N-1}]$. On the contrary, $\hat{x}(N|N-1)$ is a linear combination of data, $y(N-1), y(N-2), \dots, y(1)$, and therefore lies on that subspace. So, they are orthogonal, and $E[\hat{x}(N|N-1)v(N)'] = 0$.

In conclusion, from (9.34), it follows that

$$\Lambda_{x(N+1)e(N)} = FP(N)H', \qquad (9.35)$$

where $P(N)$ is the variance matrix of $v(N)$; see (9.24).

(b2) *Computation of* $\Lambda_{e(N)e(N)}$:

From Eq. (9.25), we have

$$\Lambda_{e(N)e(N)} = E[e(N)e(N)'] = HE[v(N)v(N)']H' + V_2$$
$$+ HE[v(N)v_2(N)'] + E[v_2(N)v(N)']H'.$$

We now let our readers show that the last two terms of this expression are null, so that the variance matrix of the innovation is

$$\Lambda_{e(N)e(N)} = HP(N)H' + V_2. \qquad (9.36)$$

From (9.30)–(9.32), (9.35), and (9.36), we obtain the recursive expression of the one-step-ahead optimal predictor:

$$\hat{x}(N+1|N) = F\hat{x}(N|N-1) + K(N)e(N), \qquad (9.37)$$

where

$$K(N) = FP(N)H'[HP(N)H' + V_2]^{-1}. \qquad (9.38)$$

9.4.5 Riccati Equation

In the previous expressions, the variance matrix of the state estimation error $v(N)$ appears, denoted as $P(N)$. We will now see how to compute recursively such matrix. We start by first deriving the recursive expression of $v(\cdot)$. From the definition of the state error,

$$v(N+1) = x(N+1) - \hat{x}(N+1|N),$$

by means of (9.1a) and (9.37), we obtain

$$v(N+1) = Fv(N) + v_1(N) - K(N)e(N),$$

where $K(N)$ is given by (9.38). Resorting to Eq. (9.25) for innovation $e(N)$, we have the recursion for $v(\cdot)$:

$$v(N+1) = [F - K(N)H]v(N) + V_1(N) - K(N)V_2(N). \qquad (9.39)$$

From (9.39), we compute the recursion for $P(\cdot)$

$$P(N+1) = \text{Var}\,[v(N+1)] = E[v(N+1)v(N+1)']$$

$$= E[(F - K(N)H)v(N)v(N)'(F - K(N)H)'] + E[V_1(N)V_1(N)']$$
$$+ E[K(N)V_2(N)V_2(N)'K(N)'] + E[(F - K(N)H)v(N)V_1(N)']$$
$$- E[(F - K(N)H)v(N)V_2(N)']K(N)' - E[V_1(N)V_2(N)']K(N)'$$
$$+ E[V_1(N)v(N)'(F - K(N)H)']$$
$$- E[K(N)V_2(N)v(N)'(F - K(N)H)'].$$

We let our patient readers show that the last five terms at the right-hand side of this equation are null. Therefore,

$$P(N+1) = (F - K(N)H)P(N)(F - K(N)H)' + V_1 + K(N)V_2K(N)'. \tag{9.40}$$

By expanding the product $(F - K(N)H)P(N)(F - K(N)H)$, we have

$$P(N+1) = FP(N)F' - K(N)HP(N)F' - FP(N)H'K(N)'$$
$$+ K(N)HP(N)H'K(N)' + V_1 + K(N)V_2K(N)'. \tag{9.41}$$

Focus now the sum of the fourth and sixth terms:

$$K(N)HP(N)H'K(N)' + K(N)V_2K(N)' = K(N)(HP(N)H' + V_2)K(N)'.$$

By means of (9.38), this quantity can be rewritten as

$$K(N)HP(N)H'K(N)' + K(N)V_2K(N)' = FP(N)H'K(N)'.$$

Therefore, the fourth and sixth terms at the right-hand side of (9.41) cancel out with the third one:

$$P(N+1) = FP(N)F' - K(N)HP(N)F' + V_1. \tag{9.42}$$

In view of (9.38),

$$P(N+1) = FP(N)F' + V_1 - FP(N)H'(HP(N)H' + V_2)^{-1}HP(N)F'. \tag{9.43}$$

Now, by exploiting Eq. (9.38), we can also write

$$P(N+1) = FP(N)F' + V_1 - K(N)(HP(N)H' + V_2)K(N)'. \tag{9.44}$$

Equation (9.43) or equivalently (9.40) or again (9.44) is named *difference Riccati equation* (DRE) or simply *Riccati equation*.

Remark 9.6 (Count Riccati story) Born in Venice in 1676, Count Riccati (Figure 9.6) studied at the University of Padua, where he enrolled in 1694 as a student of law. During the studies, his natural curiosity led him to attend

Figure 9.6 Count Jacopo Francesco Riccati.

a course on astronomy. The teacher, Stefano Degli Angeli, was fond of Isaac Newton's *Philosophiae Naturalis Principia*, which he passed on to the young pupil around 1695. This was the event that raised his interest for mathematics.

Riccati was a self-taught gentleman, whose profound knowledge came from readings. After getting married in 1696, he settled in Castelfranco Veneto, a nice town in the countryside of Venice, where he spent most of his mature life by taking care of his large family and estate and also the town's administration. His main contribution deals with the first-order nonlinear differential equation, which is now named after him. The paper, written in Latin, was published in the *Acta Eruditorum Lipsiae* in 1724. Believe it or not, an appendix to that paper was published in the same Journal in 1723. In the decades of the second part of the past century, the generalization of the scalar equation to a matrix differential equation, as well as the analogous difference nonlinear equation, also took his name. After the death of his wife, Count Riccati moved from Castelfranco Veneto to the nearby town of Treviso in 1749. There, he passed away in 1754. More on the Riccati life and equation can be found in the booklet *Count Riccati and the early days of the Riccati equation*, available at the website of the author.

A workshop fully devoted to the Riccati equation took place in Como (Italy) in 1989, following which the volume *The Riccati Equation* was published by Springer in 1991. Various sessions have been organized in many congresses. In particular, during the *International Symposium on the Mathematical Theory of Networks and Systems* (MTNS) held in Padua (Italy) in 1998, the so-called *Riccati Pilgrimage* was organized with an excursion of the attendees to Castelfranco Veneto. A recent session on the Riccati equation was held in summer 2017 at the 20th IFAC World Congress of Toulouse (France).

The city of Castelfranco Veneto has undertaken several initiatives over the years to honor the memory of its illustrious citizen. In 1990, the city organized the conference *The Riccati's and the Culture of the Castelfranco Region in 18-th Century Europe*, leading to a book published by Leo S. Olschki Editore. Subsequently, in 2000, the UNESCO international year of mathematics, the local *Popular University* promoted the conference *The Mathematicians of Castelfranco Veneto*. The proceedings of such conference were published 17 years later, a fact celebrated with a second conference entitled *The Mathematicians of Castelfranco Veneto* in 2017. All the events in Castelfranco Veneto took place in the beautiful theater named *Teatro Accademico* designed by Riccati's sons.

9.4.6   Initialization

According to what we have seen, the formation of the prediction requires the step-by-step updating of the main equation (9.37) as well as the updating of the auxiliary matrix $P(N)$ provided by the Riccati equation (9.43). Hence, we have to specify the appropriate initial conditions for these equations.

The value of $\hat{x}(1|0)$ can be easily found by noting that $E[x(1)|y^0]$ is the estimate of $x(1)$ given no data, as the first snapshot is at time 1. In absence of information, we take the *a priori* estimate, i.e.

$$\hat{x}(1|0) = E[x(1)].$$

As for matrix $P(\cdot)$, we recall that, by definition,

$$P(1) = E[(x(1) - \hat{x}[1|0])(x(1) - \hat{x}[1|0])'].$$

However, $\hat{x}(1|0) = E[x(1)]$; hence,

$$P(1) = E[(x(1) - E[x(1)])(x(1) - E[x(1)]')] = \text{Var}[x(1)].$$

Remark 9.7 Until now we have assumed that the mean value of the initial state is null, $E[x(1)] = 0$. This assumption can be generalized to the case of any mean, $E[x(1)] = x(1)_m$. The above equations still hold true provided that we set as initial condition for the state equation of the predictor

$$\hat{x}(1|0) = E[x(1)] = x(1)_m.$$

9.4.7 One-step-ahead Optimal Predictor Summary

Hypotheses

The data generation mechanism is described by the system

$$\mathcal{S} : \begin{cases} x(t+1) = Fx(t) + v_1(t), \\ y(t) = Hx(t) + v_2(t), \end{cases}$$

where

a) the initial condition for $x(1)$ is characterized in mean and variance as

$$x(1) \sim (x(1)_m, P_1);$$

b) the disturbances are described as

$$v_1(t) \sim \text{WN}(0, V_1), \quad v_2(t) \sim \text{WN}(0, V_2);$$

they are uncorrelated to each other and uncorrelated with $x(1)$; obviously, as all variance matrices, V_1 and V_2 are symmetric and positive semi-definite; we assume V_2 to be definite positive

c) $F, H, V_1, V_2, x(1)_m$, and $P(1)$ are known.

Problem

Find $\hat{x}(N+1|N) = E[x(N+1)|y^N]$ in recursive form.

Kalman Predictor

The optimal one-step-ahead predictor $\hat{\mathcal{S}}_1$ is described by the updating equation of the state and by an algebraic equation providing the optimal prediction of the output given that of the state:

$$\hat{\mathcal{S}}_1 : \begin{cases} \hat{x}(N+1|N) &= F\hat{x}(N|N-1) + K(N)e(N), \\ \hat{y}(N+1|N) &= H\hat{x}(N+1|N). \end{cases}$$

As initial condition, we take

$$\hat{x}(1/0) = x(1)_m.$$

Matrix $K(N)$, named *predictor gain*, is

$$K(N) = FP(N)H'(HP(N)H' + V_2)^{-1}, \tag{9.45}$$

where $P(N)$ is the variance of the state prediction error, obtained by solving the DRE

$$P(N+1) = FP(N)F' + V_1 - FP(N)H'(HP(N)H' + V_2)^{-1}HP(N)F',$$

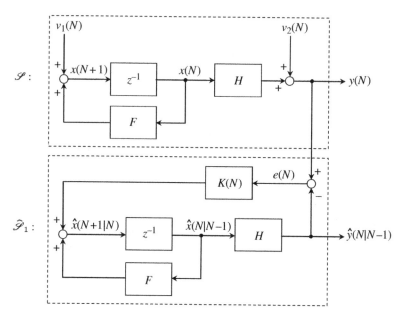

Figure 9.7 Kalman predictor block scheme.

with the initial condition

$$P(1) = P_1.$$

The block schemes of the data generation mechanism $\mathscr{S}$ and of the corresponding one-step-ahead predictor $\hat{\mathscr{S}}_1$ are given in Figure 9.7.

Various comments on the optimal predictor are in order.

- The solution of the Riccati equation is the variance matrix of the state prediction error. As such, it must be a symmetric positive semi-definite matrix. This is confirmed by Eq. (9.40), from which we see that, if $P(N)$ is symmetric and positive semi-definite, $P(N + 1)$ must enjoy the same property. As a consequence, if the initial condition of the Riccati equation is symmetric positive semi-definite (and P_1 must certainly satisfy such requirements being the variance matrix of $x(1)$), $P(N)$ will also keep such property at each N.

- In the Riccati equation, we see the inverse of matrix $HP(N)H' + V_2$. Here, we see the sum of variance V_2 of the output disturbance plus $HP(N)H'$. This last matrix is positive semi-definite since $P(N)$ is positive semi definite, as noted above. Therefore, if we assume that variance v_2 is positive definite, then $HP(N)H' + V_2$ is positive definite too, and hence invertible.

- As seen in the derivation of the optimal predictor, $HP(N)H' + V_2$ is the variance of the innovation $e(N)$. $HP(N)H'$ being positive semi-definite, we can say that $\text{Var}[e] \geq V_2$, where $\geq$ is to be interpreted in the matrix sense (namely $\text{Var}[e] - V_2$ is positive semi-definite). This means that the variance of the innovation cannot be smaller than that of the output disturbance v_2.

- We now analyze the structure of the optimal predictor. In absence of observations, no information would be conveyed by the innovation, so that the prediction would be based on the *a priori information* only:

$$\hat{x}(N+1|N) = F\hat{x}(N|N-1),$$
$$\hat{y}(N+1|N) = H\hat{x}(N+1|N).$$

We say that the predictor operates in *open loop*.

The availability of fresh information via innovation $e(N)$ leads to a modification of these equations, with the introduction of the additional term $K(N)e(N)$:

$$\hat{x}(N+1|N) = F\hat{x}(N|N-1) + K(N)e(N),$$
$$\hat{y}(N+1|N) = H\hat{x}(N+1|N).$$

Thanks to this additional term, the information hidden in the sequence of data is suitably taken into account to wisely modify the *a priori estimate*.

Note that, since $e(N) = y(N) - \hat{y}(N|N-1)$, the output of the predictor (i.e. $\hat{y}(N+1|N)$) depends upon its previous value $\hat{y}(N|N-1)$. In other words, the predictor is a feedback system with signal $\hat{y}(\cdot+1|\cdot)$ flowing in the loop. Hence, we say that it works in *closed loop*.

Finally, we observe that the gain $K(N)$ plays a main role in determining the modification of the open loop prediction $F\hat{x}(N|N-1)$ due to the additional term $K(N)e(N) = K(N)(y(N) - \hat{y}(N|N-1))$.

Example 9.1 (Predictor with null output matrix)   The output matrix of the system

$$x(t+1) = Fx(t) + v_1(t)$$
$$y(t) = v_2(t)$$

is null, $H = 0$. Therefore, the gain $K(N) = FP(N)H'(HP(N)H' + V_2)^{-1}$ is also null, so that

$$\hat{x}(N+1|N) = F\hat{x}(N|N-1).$$

The optimal predictor makes no use of data y. This is exactly what we could expect since, in such a situation, being $y(t) = v_2(t)$, the measured signal $y(t)$ is just a noise uncorrelated with the disturbance acting on the state equation, and therefore, it does not contain any information of the unknown (the system state). The estimate is only based on the assumed initial state; to be precise, if $\hat{x}(1|0) = P_1$, then $\hat{x}(N+1|N) = F^{N-1}P(1)$.

Example 9.2 (Predictor with null dynamic matrix)   In the system

$$x(t+1) = v_1(t)$$
$$y(t) = Hx(t) + v_2(t),$$

matrix $F = 0$. Therefore, the predictor is simply

$$\hat{x}(N + 1|N) = 0.$$

In other words, the optimal prediction is the trivial one, i.e. the mean value. This fact is what we could expect from the very beginning. Indeed, the state equation says that the state is white noise, $x(t + 1) = v_1(t)$, uncorrelated with the disturbance corrupting the output. Hence, as in the previous example, $y(t)$ does not contain any information on the state. Moreover, the initial state knowledge is of no use as $F = 0$. The only state estimation making sense is that of the mean value.

Example 9.3 (Estimating a constant) Consider now a scalar-state and scalar-output system described by

$$x(t + 1) = x(t)$$
$$y(t) = x(t) + v(t),$$

where $v(t) \sim WN(0, 1)$ and $x(1) \sim (x(1)_m, 1)$.

Since the state equation is not subject to any disturbance, $V_1 = 0$, so that the Riccati equation is

$$P(N + 1) = P(N)/(1 + P(N)).$$

Note that unknown $P(N)$ is scalar in this example. Since the initial condition is

$$P(1) = \text{Var}[x(1)] = 1,$$

the solution is

$$P(N) = 1/N$$

and the corresponding gain

$$K(N) = P(N)/(1 + P(N)) = 1/(1 + N).$$

The optimal predictor is

$$\hat{x}(N + 1|N) = \hat{x}(N|N - 1) + e(N)/(1 + N),$$

where

$$e(N) = y(N) - \hat{x}(N|N - 1)$$

and the initial condition

$$\hat{x}(1|0) = x(1)_m.$$

The solution of this equation is

$$\hat{x}(N|N - 1) = \frac{x(1)_m + y(1) + y(2) + \cdots + y(N - 1)}{N}.$$

Thus, the predictor performs a mean over all data plus the mean value of the initial state (as a sort of further snapshot). This is no surprise since the state equation ($x(t + 1) = x(t)$) says that the state is a constant. Moreover, the output equation ($y(t) = x(t) + v(t)$) tells us that such constant is measured up to an additive white noise. In other words, we are here in the classical situation where an unknown constant is to be estimated given its noisy measurements.

9.4.8 Generalizations

The derivations above are open to wide generalizations. First, one can consider correlated disturbances v_1 and v_2. Second, it is possible to include an exogenous input. Finally, the theory can be extended to time-varying systems. We state now a general result.

9.4.8.1 System

$$\mathcal{S} : \begin{cases} x(t + 1) = F(t)x(t) + G(t)u(t) + v_1(t) \\ y(t) = H(t)x(t) + v_2(t) \end{cases}$$

where disturbances $v_1(\cdot)$ and $v_2(\cdot)$ have the following characteristics:

$$E[v_1(t)] = 0, \ \forall t; \quad E[v_2(t)] = 0, \ \forall t,$$

$$E[v_1(t_1)v_1(t_2)'] = \begin{cases} V_1(t), & t_2 = t_1 = t, \\ 0, & \forall t_2 \neq t_1, \end{cases}$$

$$E[v_2(t_1)v_2(t_2)'] = \begin{cases} V_2(t), & t_2 = t_1 = t, \\ 0, & \forall t_2 \neq t_1, \end{cases}$$

$$E[v_1(t_1)v_2(t_2)'] = \begin{cases} V_{12}(t), & t_2 = t_1 = t, \\ 0, & \forall t_2 \neq t_1. \end{cases}$$

We also assume that the variance matrix $V_2(t)$ is non-singular for each t (or, equivalently, that it is positive definite for each t).

As for initial condition $x(1)$, it is characterized by conditions

$$E[x(1)] = x(1)_m,$$

$$E[(x(1) - x_1)(x(1) - x_1)'] = P_1.$$

The disturbances and the initial condition are uncorrelated

$$E[v_i(t)(x(1) - x(1)_m)'] = 0, \quad \forall i = 1, 2.$$

9.4.8.2 Predictor

$$\hat{\mathcal{S}}_1 : \begin{cases} \hat{x}(N + 1|N) = F(N)\hat{x}(N|N - 1) + G(N)u(N) + K(N)e(N) \\ \hat{y}(N + 1|N) = H(N)\hat{x}(N + 1|N) \\ e(N) = y(N) - \hat{y}(N|N - 1) \end{cases}$$

with gain $K(N)$ given by

$$K(N) = [F(N)P(N)H(N)' + V_{12}(N)][H(N)P(N)H(N)' + V_2(N)]^{-1}.$$
(9.46)

Here, $P(N)$ is the solution of the Riccati equation

$$P(N+1) = F(N)P(N)F(N)' + V_1(N)$$
$$- K(N)\{H(N)P(N)H(N)' + V_2(N)\}K(N)'.$$
(9.47)

The initial conditions are

$$\hat{x}(1|0) = x(1)_m, \quad P(1) = P_1.$$

Remark 9.8 Compare expression (9.46) for the prediction gain in presence of correlation between the state and output disturbance with expression (9.38) in absence of such correlation. As expected, if $V_{12}(t) = 0$, these two expressions coincide.

As for the Riccati equation, in Section 9.4.5, we have seen that there are various ways in which such equation can be written, in particular Eqs. (9.43) and (9.44). Note that, due to the cross-term V_{12}, (9.47) does not hold any more when the two disturbances are correlated. Expression (9.44) (replicated in (9.47) for the time-varying case) provides the valid Riccati equation.

9.5 Multistep Optimal Predictor

We now pass to the problem of finding the r-steps ahead predictor. For simplicity, we consider the case when the system is time invariant and not subject to exogenous variables, i.e.

$$x(t+1) = Fx(t) + v_1(t)$$
$$y(t) = Hx(t) + v_2(t),$$

with $v_1(\cdot)$ and $v_2(\cdot)$ zero-mean white noises, uncorrelated with the initial condition $x(1)$, and uncorrelated to each other, namely, $E[v_i(t_1)v_j(t_2)'] = 0, \forall t_2 \neq t_1$.

The task of the r-steps-ahead predictor, with $r > 1$, is to estimate $x(N+r)$ using data up to time point N, i.e. from the observation of $y(N), y(N-1), \ldots$ The optimal predictor $\hat{x}(N+r|N)$ is given by

$$\hat{x}(N+r|N) = E[x(N+r)|y^N].$$

The state at $t+r$ can be written as $x(N+r) = Fx(N+r-1) + v_1(N+r-1)$. Moreover, the data $y(\cdot)$ up to N depend upon $v_1(\cdot)$ up to tine $N-1$ and upon $v_2(\cdot)$ up to N. Therefore,

$$\hat{x}(N+r|N) = F\hat{x}(N+r-1|N).$$

Iterating, we obtain

$$\hat{x}(N + r|N) = F^{r-1}\hat{x}(N + 1|N). \tag{9.48}$$

Passing to the output, since for $r > 1$, $v_2(N + r)$ is uncorrelated with the data up to N, we have

$$\hat{x}(N + r|N) = H\hat{x}(N + r|N). \tag{9.49}$$

In conclusion, the r-steps-ahead optimal predictor is easily derived from the one-step-ahead optimal predictor by means of (9.48) and (9.49).

The variance of the two-steps-ahead predictor can be easily determined. Since

$$\hat{x}(N + 2|N) = F\hat{x}(N + 1|N),$$

we have

$$x(N + 2) - \hat{x}(N + 2|N) = F(x(N + 1) - \hat{x}(N + 1|N)) + v_1(N).$$

Denoting by $P^{(2)}(N)$ the variance of $x(N + 2) - \hat{x}(N + 2|N)$, an easy computation leads to the link between the one-step-ahead and the two-steps-ahead variance matrices:

$$P^{(2)}(N) = FP(N)F' + V_1. \tag{9.50}$$

As we will later investigate in Section 9.7, under suitable assumptions matrix $P(N)$ will converge to a limit matrix $\bar{P}$, and therefore, $P^{(2)}(N)$ will also converge to a limit matrix, say $\bar{P}^{(2)}$, given by

$$\bar{P}^{(2)} = F\bar{P}F' + V_1. \tag{9.51}$$

Since $P(N)$ is a positive semi-definite solution of (9.47), $\bar{P}$ is a positive semi-definite solution of the algebraic equation

$$\bar{P} = F\bar{P}F' + V_1 - \bar{K}\{H\bar{P}H' + V_2\}\bar{K}'. \tag{9.52}$$

Can we compare $\bar{P}^{(2)}$ with $\bar{P}$? Intuitively, we expect that predicting a signal two steps in advance is more demanding than forming a single-step-ahead prediction, so that $\bar{P}^{(2)}$ should be greater (in the positive semi-definite sense, see Appendix B) than $\bar{P}$. To check if this fact holds true, consider again Eq. (9.52). In view of (9.50), such equation can be rewritten as

$$\bar{P} = \bar{P}^{(2)} - \bar{K}\{H\bar{P}H' + V_2\}\bar{K}'. \tag{9.53}$$

Since matrix $\bar{K}\{H\bar{P}H' + V_2\}\bar{K}'$ is positive semi-definite, we see that inequality

$$\bar{P}^{(2)} \geq \bar{P}$$

holds true indeed.

9.6 Optimal Filter

We now pass to the problem of estimating the state at time N, given the data up to time N, namely $y(N)$, $y(N - 1), \ldots, y(1)$.

To find $\hat{x}(N|N)$, we resort again to the recursive Bayes formula:

$$\hat{x}(N|N) = E[x(n)|y^N] = E[x(N)|y^{N-1}, y(N)]$$
$$= E[x(N)|y^{N-1}] + E[x(N)|e(N)]$$
$$= \hat{x}(N|N - 1) + \Lambda_{x(N)e(N)} \Lambda_{e(N)e(N)}^{-1} e(N).$$

The innovation variance, $\Lambda_{e(N)e(N)}$, was previously computed (see Eq. (9.36)). As for matrix $\Lambda_{x(N)e(N)}$, the derivation can be performed in full analogy with the computation of $\Lambda_{x(N+1)e(N)}$ (see Eqs. (9.33)–(9.35)):

$$\Lambda_{x(N)e(N)} = E[\{x(N)\}\{H(x(N) - \hat{x}(N|N - 1)) + v_2(N)\}']$$
$$= E[\{x(N) - \hat{x}(N|N - 1) + \hat{x}(N|N - 1)\}$$
$$\times \{H(x(N) - \hat{x}(N|N - 1) + v_2(N)\}']$$
$$= E[\{x(N) - \hat{x}(N|N - 1)\}\{x(N) - \hat{x}(N|N - 1)\}']H'$$
$$+ E[\{x(N) - \hat{x}(N|N - 1)\}v_2(N)']$$
$$+ E[\{\hat{x}(N|N - 1)\}\{x(N) - \hat{x}(N|N - 1)\}']H'$$
$$+ E[\{\hat{x}(N|N - 1)\}v_2(N)'].$$

It is easy to see that

$$E[\{x(N) - \hat{x}(N|N - 1)\}v_2(N)'] = 0,$$
$$E[\{\hat{x}(N|N - 1)\}\{x(N) - \hat{x}(N|N - 1)\}']H' = 0,$$
$$E[\{\hat{x}(N|N - 1)\}v_2(N)' = 0].$$

Hence,

$$\Lambda_{x(N)e(N)} = P(N)H',$$

where $P(N)$ is again the variance matrix of the one-step-ahead prediction error $P(N) = E[(x(N) - \hat{x}(N|N - 1))(x(N) - \hat{x}(N|N - 1))'].$

Summing up, the optimal filter is obtained from the one-step-ahead optimal predictor as follows:

$$\hat{x}(N + 1|N) = \hat{x}(N|N - 1) + K(N)e(N), \tag{9.54a}$$
$$\hat{x}(N|N) = \hat{x}(N|N - 1) + K^{(0)}(N)e(N), \tag{9.54b}$$

with

$$K^{(0)}(N) = P(N)H'(HP(N)H' + V_2)^{-1}, \quad K(N) = FK^{(0)}(N). \tag{9.55}$$

In the previous sections, we already encountered the *predictor gain* $K(N)$; we have now introduced a new gain, the *filter gain* $K^{(0)}(N)$.

The expression for $\hat{x}(N|N-1)$ can be elaborated as follows:

$$\hat{x}(N|N-1) = E[Fx(N-1) + v_1(N-1)|y^{N-1}]$$
$$= F\hat{x}(N-1|N-1) + E[v_1(N-1)|y^{N-1}].$$

If $v_1(N)$ and $v_2(N)$ are uncorrelated, then $E[v_1(N-1)|y^{N-1}] = 0$, so that

$$\hat{x}(N|N-1) = F\hat{x}(N-1|N-1). \tag{9.56}$$

From (9.54) and (9.56), we conclude that, when the disturbances are uncorrelated,

$$\hat{x}(N|N) = F\hat{x}(N-1|N-1) + K^{(0)}(N)e(N). \tag{9.57}$$

Remark 9.9 (Optimal filter alternative expression) When matrix F is invertible and disturbances $v_1(\cdot)$ and $v_2(\cdot)$ are uncorrelated, from (9.56) we see that the optimal filter can also be computed as

$$\hat{x}(N|N) = F^{-1}\hat{x}(N+1|N).$$

Remark 9.10 (Filtering error variance) The variance of the filtering error

$$P^{(0)}(N) = \text{Var}\,[x(N) - \hat{x}(N|N)] \tag{9.58}$$

can be determined from that of the prediction error $P(N)$. Indeed, a simple computation leads to the conclusion that

$$P^{(0)}(N) = P(N) - P(N)H'(HP(N)H' + V_2)^{-1}HP(N). \tag{9.59}$$

Note that $P(N)H'(HP(N)H' + V_2)^{-1}HP(N)$ is positive semi-definite, so that this expression entails that $P^{(0)}(N) \leq P(N)$. This is indeed the result that everyone of good sense would expect since, in the determination of predictor $\hat{x}(N+1|N)$, the available data are $\{y(1), y(2), \ldots, y(N)\}$, whereas in the determination of filter $\hat{x}(N+1|N+1)$, the available data are $\{y(1), y(2), \ldots, y(N), y(N+1)\}$. The availability of the further snapshot $y(N+1)$ conveys more information on the unknown state, so that the variance of the filtering error is smaller.

9.7 Steady-State Predictor

Is the estimate of the unknown state provided by Kalman theory accurate or loose? To reply to this main question, we make reference to the variance of the prediction error: the smaller the variance the better the estimate. This bring us to the study of matrix $P(N)$, in particular its behavior when the number of data becomes very large, ideally infinity.

For this study, we focus on the case when the system is given by the usual time-invariant linear equations

$$\mathscr{S} : \begin{cases} x(t+1) = Fx(t) + v_1(t) \\ y(t) = Hx(t) + v_2(t), \end{cases}$$

where $v_1(t)$ and $v_2(t)$ are uncorrelated to each other, with constant variance matrices V_1 and V_2.

As seen in Section 9.4, the optimal predictor is a linear system. However, our readers certainly took notice of a surprising fact: even though the system is time-invariant, the predictor is *time varying*; indeed gain $K(N)$, as a function of the (time-varying) solution $P(N)$ of the Riccati equation, depends upon time.

Natural questions then arising are: Why the gain is not constant? Second, what happens with the passing of time? Does the gain converge when the number of snapshots tend to ∞?

If $K(N)$ were converging, say $\lim_{N \to \infty} K(N) = \bar{K}$, then we could consider using a suboptimal time-invariant estimator with gain given by $\bar{K}$ in place of the time-varying $K(N)$.

9.7.1 Gain Convergence

From expression (9.38) of the gain

$$K(N) = FP(N)H'(HP(N)H' + V_2)^{-1},$$

it appears that the time variability of $K(N)$ is due to the variability of the solution $P(N)$ of the Riccati equation. Therefore, if

$$\lim_{N \to \infty} P(N) = \bar{P},$$

then the gain is asymptotically convergent. We can then consider the *steady-state predictor*, associated with the *steady-state asymptotic gain*

$$\bar{K} = F\bar{P}H'(H\bar{P}H' + V_2)^{-1}.$$

Analogously one can introduce the *steady-state filter* Notice in passing that, if the convergence takes place, then the limit matrix $\bar{P}$ is positive semi-definite. Indeed, in the same way that the limit of a sequence of non-negative real numbers is non-negative, the limit of a sequence of positive semi-definite matrices is positive semi-definite, as it is easy to prove.

We also observe that if the convergence takes place, then P is a constant solution of the DRE. Indeed, if in Eq. (9.43), we set $P(N) = \bar{P}$, then we would obtain $P(N+1) = P(N) = \bar{P}$. Stated another way, $\bar{P}$ is an equilibrium point for the Riccati equation.

This suggests introducing the so-called *algebraic Riccati equation* (ARE) as

$$P = FPF' + V_1 - FPH'(HPH' + V_2)^{-1}HPF', \tag{9.60}$$

an equation obtained from the DRE by setting $P(N + 1) = P(N) = P$.

$\bar{P}$ is a positive semi-definite solution of the ARE. In turn, this means that $\bar{P}$ can be found by looking for the positive semi-definite solutions of an algebraic equation, the ARE, instead of solving a difference equation such as the DRE and then computing its limit as $N \rightarrow \infty$.

Note that the ARE is a *nonlinear* matrix equation. As such, it can have a multiplicity of solutions. The solutions that are not positive semi-definite have to be discarded, as in our estimation problem they cannot be seen as covariance matrices. Observe that the ARE may have a unique positive semi-definite solution or various positive semi-definite solutions. In the latter case, the "best" solution is that associated to the "minimal" $P(N)$ since it provides the more accurate estimate of the unknown state.

In practice, to find the steady-state optimal predictor, one looks for the positive semi-definite solution of the ARE (or the "best" among the positive semi-definite solutions of the ARE); with such matrix, one computes the optimal gain

$$\bar{K} = F\bar{P}H'(H\bar{P}H' + V_2)^{-1}.$$

The corresponding (time-invariant) predictor will be

$$\hat{\mathcal{S}}_\infty : \begin{cases} \hat{x}(N + 1|N) = F\hat{x}(N|N - 1) + \bar{K}e(N), \\ \hat{y}(N + 1|N) = H\hat{x}(N + 1|N), \\ e(N) = y(N) - \hat{y}(N|N - 1). \end{cases}$$

Remark 9.11 (Stability of the predictor) It is most important that the predictor be stable. Otherwise, any small disturbance would lead to the divergence of the estimate.

In this regard, focus on system $\hat{\mathcal{S}}_\infty$ above. This is a time-invariant system having $y(\cdot)$ as input and $\hat{x}(\cdot | \cdot -1)$ as state. The state equation of such system is:

$$\hat{x}(N + 1|N) = F\hat{x}(N|N - 1) + \bar{K}e(N)$$
$$= F\hat{x}(N|N - 1) + \bar{K}(y(N) - \hat{y}(N|N - 1))$$
$$= F\hat{x}(N|N - 1) + \bar{K}(y(N) - Hx(N|N - 1))$$
$$= (F - \bar{K}H)\hat{x}(N|N - 1) + \bar{K}y(N).$$

From this last equation, we see that the stability is determined by matrix

$$F - \bar{K}H = F - F\bar{P}H'(H\bar{P}H' + V_2)^{-1}H. \tag{9.61}$$

Establishing when all its eigenvalues are inside the unit disk is not an easy matter. Indeed, they depend upon the solution $\bar{P}$ of the ARE, and such solution is determined in a complex way from the matrices of the given system via a nonlinear equation. We will address this problem in the subsequent sections. For the moment, we limit ourselves to introduction of the following definition.

Definition 9.3 (Stabilizing solution)   A solution $\bar{P}$ of the ARE is said to be stabilizing if matrix $F - \bar{K}H$ is stable (i.e. all its eigenvalues are inside the unit disk).

Remark 9.12   Up to now, the convergence analysis of the solution of the Riccati equation has been mainly motivated from the search of a time-invariant suboptimal predictor. As a matter of fact, there is another motivation, even more important than the previous one. The solution $P(N)$ of the Riccati equation is the variance matrix of state prediction error. The convergence of such matrix therefore denotes the capacity of the predictor to supply an estimate with bounded errors. Should $P(N)$ be divergent, then the predicted state would drive away from the true one.

Example 9.4 (From a *naive* estimator to a Kalman estimator)   The scalar system

$$\begin{cases} x(t+1) = ax(t) \\ \quad y(t) = x(t), \end{cases}$$

is free of disturbances; hence, the only uncertainty concerns the initial state.

To find the state from the measurements of $y(t)$, we tentatively set up an elementary estimator mimicking the dynamics of the given system as follows:

$$\hat{x}(t+1) = a\hat{x}(t). \tag{9.62}$$

Now, the output equation, $y(t) = x(t)$, says that the measured output coincides with the unknown system state. Therefore, the fact that observed signal $y(t)$ coincides with the estimator output (i.e. $y(t) - \hat{x}(t) = 0$) would be an indication of a correct state estimate. Vice versa, the fact that $y(t) - a\hat{x}(t) \neq 0$ would be an indication of wrong estimate. So, we can try to improve the *naif* estimator (9.62) by taking into account such error with a suitable weight, say L, as follows:

$$\hat{x}(t+1) = a\hat{x}(t) + L(y(t) - \hat{x}(t)).$$

With such a modified estimation rule, the evolution of the state error $v(t) = x(t) - \hat{x}(t)$ is given by

$$v(t+1) = av(t) - L(y(t) - \hat{x}(t)) = (a - L)v(t).$$

From this simple recursion, it is apparent that the error tends to zero if $|a - L| < 1$. Therefore, the tuning of L must be done with constraint $a - 1 < L < a + 1$.

The natural question is then: Which is the best choice of L? In this regard, we start by observing that for $L = a$, we have $v(t+1) = 0$, namely the error is driven to zero in one step only. This is the so-called *dead-beat estimator*. On the other hand, such extraordinary performance has to be taken with caution as we treated an ideal case where no disturbances act on the system. If we pass to a more realistic situation with

$$y(t) = x(t) + v(t), \quad v(\cdot) \sim WN(0, 1),$$

the evolution of the estimation error is governed by

$$v(t + 1) = (a - L)v(t) - Lv(t).$$

Then

$$\text{Var } [v(t + 1)] = (a - L)^2 \text{Var } [v(t)] + L^2.$$

Thus, when $|a - L| < 1$, $v(t)$ converges to a stationary process with variance

$$\frac{L^2}{1 - (a - L)^2},$$

which is minimal when

$$\bar{L} = \frac{a^2 - 1}{a}.$$

This is the best choice of the gain against disturbances.

Note that the $\bar{L}$ is just the gain of the optimal steady-state Kalman estimator. Indeed, since $F = a, G = 0, H = 1, V_{12} = 0 \ V_1 = 0, V_2 = 1$, the Riccati equation is

$$P(t + 1) = \frac{a^2 P(t)}{1 + P(t)}.$$

At steady state, $\bar{P} = a^2 - 1$. Correspondingly, the Kalman gain is $\bar{K} = a\bar{P}/\bar{P} + 1 = [a^2 - 1]/a$, a value that coincides with $\bar{L}$.

9.7.2 Convergence of the Riccati Equation Solution

The Riccati equation is a nonlinear matrix equation. Finding the properties of its solutions is a complex task, considered by many scholars along various decades, especially in the second half of the twentieth century. We now outline the main results. In our analysis, we first deal with simple situations where it is possible to resort to intuitive reasoning. Later we prove a general convergence theorem.

9.7.2.1 Convergence Under Stability

Assume that the data generation mechanism

$$x(t + 1) = Fx(t) + v_1(t), \tag{9.63}$$

is stable.

We first study the sequence of the so generated random vectors $x(t)$, in mean (denoted as $x(t)_m$) and variance (denoted as $\Lambda(t) = \text{Var}[x(t)]$).

As for the mean value, by applying the expected value operator to both members of Eq. (9.63), we have

$$x(t + 1)_m = Fx(t)_m, \tag{9.64}$$

so that

$$x(t+1)_m = F^t x(1)_m.$$

If the system is stable, F^t tends to the zero matrix as t tends to ∞, so that, for any initial condition, $x(t+1)_m$ tends to zero as well: asymptotically, process $x(\cdot)$ is zero mean.

The variance matrix $\Lambda(t) = E[(x(t) - x(t)_m)(x(t) - x(t)_m)']$ can be worked out by observing that Eqs. (9.64) and (9.63) entail that

$$x(t+1) - x(t+1)_m = F(x(t) - x(t)_m) + v_1(t).$$

Hence,

$$E[(x(t+1) - x(t+1)_m)(x(t+1) - x(t+1)_m)']$$
$$= E[\{F(x(t) - x(t)_m) + v_1(t)\}\{F(x(t) - x(t)_m) + v_1(t)\}'].$$

The cross-term $E[F(x(t) - x(t)_m)v_1(t)']$ is null since $(x(t) - x(t)_m)$ depends on $v_1(\cdot)$ only up to $t-1$, and $v_1(t)$ is uncorrelated from its past being a white noise. Therefore,

$$E[(x(t+1) - x(t+1)_m)(x(t+1) - x(t+1)_m)']$$
$$= FE[(x(t) - x(t)_m)(x(t) - x(t)_m)']F' + V_1.$$

Hence

$$\Lambda(t+1) = F\Lambda(t)F' + V_1. \tag{9.65}$$

This is known as *discrete-time Lyapunov equation* (DLE).

Starting from an initial condition $E[(x(1) - x(1)_m)(x(1) - x(1)_m)'] = P_1$, the solution is

$$\Lambda(1) = P_1; \quad \Lambda(2) = FP_1F' + V_1; \quad \Lambda(3) = F^2P_1F'^2 + FV_1F' + V_1; \quad \ldots$$

and, in general

$$\Lambda(t) = (F^{t-1})P_1(F^{t-1})' + [V_1 + FV_1F' + \cdots + (F^{t-2})V_1(F^{t-2})'].$$

Here, we see the superposition of the effect of the initial condition P_1 on the state variance, $(F^{t-1})P_1(F^{t-1})'$, plus the effect of the disturbance, given by the sum under the square brackets. This is no surprise as the Lyapunov equation is linear (contrary to the Riccati equation). If the system is stable, matrix F^{t-1} tends to zero when t tends to infinity. Hence, the effect of the initial condition vanishes. Moreover, the infinite sum $[V_1 + FV_1F' + (F^2)V_1(F^2)' + \cdots]$ is convergent. In conclusion,

$$\lim_{t \to \infty} \Lambda(t) = \bar{\Lambda}.$$

The limit value $\bar{\Lambda}$ is an equilibrium point of the DLE (9.65) and therefore is a positive semi-definite solution of the *algebraic Lyapunov equation* (ALE):

$$\Lambda = F\Lambda F' + V_1. \tag{9.66}$$

We can summarize what we have seen by saying that, under the stability condition, system (9.63) generates a state sequence $x(\cdot)$ constituting a stationary process with zero mean value and constant variance matrix $\bar{\Lambda}$.

From this observation, we can propose a trivial predictor, providing as prediction the mean value of the process. Although its triviality, the variance of its prediction error is finite, as it coincides then with the process variance.

Since we can conceive a very elementary predictor with bounded error, we expect that the Kalman predictor, with its optimality, should have a limited prediction error too, lower than the error of the trivial one.

Always under the stability assumption, we can make the conjecture that the Kalman predictor must be stable. Indeed, in the current situation, $x(\cdot)$, and consequently $y(\cdot)$, is a stationary process. Now the predictor is a linear system having $y(\cdot)$ as input. If, by contradiction, the predictor is unstable, then, notwithstanding the stationarity of its input, its state would be divergent. So we would come to the conclusion according to which we would predict a stationary process with a nonstationary one, which would be complete nonsense.

As a matter of fact, one can prove the following proposition.

Proposition 9.1 (First convergence theorem for the DRE) *Suppose that the data generation mechanism is stable (all eigenvalues of matrix F inside the unit disk). Then,*

1) *For any (positive semi-definite) initial condition, the solution of the DRE asymptotically converges to*

$$\lim_{t \to \infty} P(t) = \bar{P}.$$

$\bar{P}$ is the same matrix whatever the initial condition $P(1) = P_1$ be (provided that $P_1 \geq 0$).
2) *$\bar{P}$ is the only positive semi-definite solution of the ARE equation (9.60).*
3) *The steady-state predictor is stable. This means that all eigenvalues of matrix*

$$F - F\bar{P}H'(H\bar{P}H' + V_2)^{-1}H,$$

see Eq. (9.61), are inside the unit disk.

9.7.2.2 Convergence Without Stability

The stability of the data generation mechanism is a sufficient condition for the convergence of the solution of the Riccati equation (and therefore of the gain of the Kalman predictor). However, it is not necessary, as shown in the next example.

Example 9.5 (Convergence without stability–a simple example) Consider the scalar system

$$x(t + 1) = \alpha x(t) + v_1(t)$$
$$y(t) = \gamma x(t) + v_2(t),$$

Figure 9.8 Graphical
determination of the
solutions of the
ARE – Example 9.5.

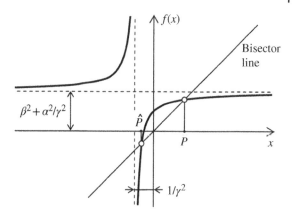

where α and γ are real parameters, $v_1(t)$ has variance β^2 and $v_2(t)$ unit variance. $v_1(t)$ and $v_2(t)$ are uncorrelated to each other and uncorrelated with the initial state of the system. We assume that $|\alpha| > 1$, i.e. the system is unstable. Therefore, the state is a diverging signal (its variance grows to ∞).

We study three cases: (i) $\beta \neq 0$, $\gamma \neq 0$; (ii) $\beta \neq 0$, $\gamma = 0$; (iii) $\beta = 0$, $\gamma \neq 0$.

i) $\beta \neq 0$ and $\gamma \neq 0$.

The DRE is then the scalar difference equation

$$P(t+1) = \beta^2 + \frac{\alpha^2 P(t)}{1 + \gamma^2 P(t)} \tag{9.67}$$

and the associated ARE is

$$P = \beta^2 + \frac{\alpha^2 P}{1 + \gamma^2 P}.$$

To study the solutions of Eq. (9.67), it is useful to resort to a graphical method. To this end, we introduce the function

$$f(x) = \beta^2 + \frac{\alpha^2 x}{1 + \gamma^2 x}.$$

$f(x)$ is just the second member of Eq. (9.67) with variable x in place of $P(t)$. Thus, if we evaluate this function at abscissa $x = P(t)$, we must obtain at the ordinate $y = P(t+1)$, as stated by Eq. (9.67). In such graphical vision, the condition $P(t+1) = P(t)$ corresponds to imposing the coincidence between abscissa and ordinate. In other words, the solutions of the ARE correspond to the points of intersection of $f(x)$ with the bisector line. In Figure 9.8, such intersections are denoted as $\bar{P}$ and $\hat{P}$. Between them, only $\bar{P}$ is acceptable since the $\hat{P} < 0$ and therefore cannot be a variance matrix. We now examine the evolution of the solution of the DRE starting from a given initial condition P_1 (with $P_1 \geq 0$). Again we resort to the graphical analysis as in Figure 9.9a. If at a given time point t, $P(t) < \bar{P}$, from the drawing we see that $P(t+1) > P(t)$, and then $P(t+2) > P(t+1)$ and so on: the

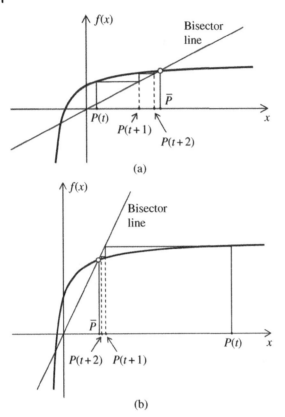

solutions evolve along a staircase converging toward $\bar{P}$. Thus, if $P_1 < \bar{P}$, the solution of the DRE is monotonically increasing and tends to $\bar{P}$. Vice versa, the graphical analysis shows that, if $P_1 > \bar{P}$, the staircase is monotonically decreasing and tends again to $\bar{P}$, see Figure 9.9b. Thus, all feasible solutions of the DRE (i.e. all solutions corresponding to non-negative initial conditions) tend to the same asymptotic value $\bar{P}$, the only non-negative solution of the ARE. We let our readers show that the corresponding steady-state predictor is stable (to this purpose, it must be shown that $|\alpha - \bar{K}\gamma| < 1$, where $\bar{K}$ is the steady-state Kalman gain). This means that $\bar{P}$ is the stabilizing solution of the Riccati equation (see Definition (9.3)).

ii) $\gamma = 0$.

When $\gamma = 0$, the Riccati equation becomes

$$P(t + 1) = \beta^2 + \alpha^2 P(t).$$

Figure 9.10 Graphical determination of
the solutions of the ARE – Example 9.5
with $\beta = 0$.

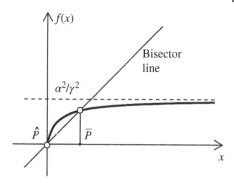

Being $|\alpha| > 1$, this is an unstable linear system. Hence, $P(\cdot)$ is divergent.
Being proportional to γ, the predictor gain is null, so that the predictor
operates in open loop and therefore it is unstable (as the system is).

iii) $\beta^2 = 0$ and $\gamma \neq 0$.

As shown in Figure 9.10, in this case, there are two feasible solutions of the
ARE, one positive and one null. From the same drawing we see that, if the
initial condition is positive, $P(t)$ converges to the positive solution of the
ARE. If instead the initial condition is null, then $P(t) = 0$ for each $t \geq 1$.
Note that only the non-null solution is stabilizing.

We now comment on the above results. First of all, we observe that there is a
fundamental difference between the case when $\gamma = 0$ and $\gamma \neq 0$. If $\gamma = 0$, then
$y(t) = v_2(t)$, namely the observations reduce to noise only, and as such they
do not bring any information on the unknown, the system state. Consequently,
the predictor works in open loop, trying to mimic the (unstable) behavior of the
system. Consequently, not only the variances of $x(t)$ and $\hat{x}(t|t-1)$ are divergent
but also the error $x(t) - \hat{x}(t|t-1)$ is divergent: the predictor is not able to track
the diverging system state.

If $\gamma \neq 0$, the state $x(t)$ is a diverging signal, and so is the output $y(t)$. Does
such measured output convey enough information to allow the tracking of the
unknown state? In this regard, we start considering the simple case when the
measurement is not corrupted by noise, namely $y(t) = \gamma x(t)$. Then, from such
formula we can find the state as $x(t) = y(t)/\gamma$. So, if the output equation is deterministic, the state can be found with no error from the output observation. In
the presence of noise, we have to consider the ARE, which has a unique finite
solution (to which all solutions of the DRE tend). Hence, the good news is that
the tracking is possible. Of course the tracking error will increase with the variance of corrupting noise $\mathrm{Var}[v_2(t)]$ (we invite our readers to show that $\bar{P}$ is
monotonically increasing with $\mathrm{Var}[v_2(t)]$). In conclusion, when $\gamma \neq 0$, we have a

stable predictor; being fed by the diverging signal $y(t)$, such predictor generates a diverging variance nonstationary signal $\hat{x}(t|t-1)$, able to track the unknown state $x(t)$.

The case when the system state in undisturbed ($\beta = 0$) is also worthy commenting. In that case, the optimal predictor is simply $\hat{x}(t+1|t) = a\hat{x}(t|t-1)$. If the initial state is known without any uncertainty, then the predictor and the data generation mechanism do coincide both in the dynamical law and in the initial condition. Then we have perfect prediction: $\hat{x}(t|t-1) = x(t)$. This implies that the estimation error $x(t) - \hat{x}(t|t-1) = 0$, so that its variance must be null, i.e. the solution of the DRE must be $P(t) = 0$ for each time point. In particular, $\bar{P} = 0$ solves the ARE.

If instead $P_1 \neq 0$, then, thanks to the information contained in data, the predictor is a closed-loop stable system, able to track the true state.

9.7.2.3 Observability

The study of convergence of the solution of the Riccati equation is a main topic in Kalman filtering: should the solution of the DRE be divergent, the uncertainty in state estimation would be greater and greater and the predictor would fail its goal.

In the simple case of the first-order system studied in Example 9.5, we have seen how important is the condition that $\gamma \neq 0$. In general, we have to find a condition expressing the capacity of the output to bring *enough information* on the state. This leads to the concept of *observability*.

Definition 9.4 (Observability) With reference to the deterministic system (with state of any dimension)

$$x(t+1) = Fx(t),$$
$$y(t) = Hx(t),$$

we say that the system (or the (F, H)) is observable if no pair of different initial states x_1 and x_2 exist such that the corresponding output signals coincide at each time point.

Note that

$$y(1) = Hx(1),$$
$$y(2) = HFx(1),$$
$$\vdots$$
$$y(N) = HF^{N-1}x(1),$$

so that we can write

$$y^N = \mathcal{O}^N x(1), \tag{9.68}$$

where

$$
\mathcal{O}^N =
\begin{bmatrix}
H \\
\cdots \\
HF \\
\cdots \\
HF^2 \\
\cdots \\
\vdots \\
\cdots \\
HF^{N-1}
\end{bmatrix}
$$

has dimension $pN \times n$. Note that this matrix has n columns, so that it cannot have a rank greater than n.

The pair (F, H) is observable if the set of linear equations (9.68) does not admit two different solutions. Since $x(1)$ is a vector with n elements, this happens when $\mathcal{O}^N$ has rank n.

Note that matrix $\mathcal{O}_N$ has a special structure, and as shown in Appendix A, its rank cannot increase for $N > n$. This observation leads to the following definition.

Definition 9.5 (Observability matrix)   The matrix

$$
\mathcal{O}^n =
\begin{bmatrix}
H \\
\cdots \\
HF \\
\cdots \\
HF^2 \\
\cdots \\
\vdots \\
\cdots \\
HF^{n-1}
\end{bmatrix},
$$

where n is the number of state variables, is named *observability matrix* of (F, H).

We can conclude our discussion on observability with a main result.

Proposition 9.2 (Observability test) (F, H) *is observable if and only if the observability matrix $\mathcal{O}^n$ has rank n.*

9.7.2.4 Reachability

The ARE may admit a multiplicity of positive semi-definite solutions. For instance, in the previous Example 9.5, there were *two* feasible solutions of the

ARE for $\gamma \neq 0$ and $\beta = 0$. On the opposite, when $\beta \neq 0$, the ARE admitted a unique positive semi-definite solution. What makes so different the two situations $\beta \neq 0$ and $\beta = 0$ is the fact that, in the latter case, the state of the system is *uncorrupted* by noise. If the initial state is not uncertain, then the evolution of the state can be perfectly predicted. Correspondingly, the solution of the DRE will be null at each time point. If instead there is an initial uncertainty, namely ($P_1 > 0$), then the state is uncertain, with an increasing variance since $|\alpha| > 1$. However, if ($\gamma \neq 0$), the output brings enough information and predictor fairly tracks the unknown state, so that $P(N)$ does not diverge. These considerations can be extended to multivariable systems: if there is a state variable (or a combination of state variables) free of disturbance, then the ARE equation admits more than one feasible solution.

The way in which the disturbance acts on the system state is determined by the structure of the variance matrix V_1 in relation to the structure of the dynamic matrix F. To be precise, consider any matrix G_v such that $G_v G_v' = V_1$. Such G_v can be seen as the "square root" of matrix V_1. For example, the column vector $[1 \quad 0]'$ is such a factor for the matrix 2×2

$$\begin{bmatrix} 1 & 0 \\ 0 & 0 \end{bmatrix}.$$

But also the matrix 2×2

$$\begin{bmatrix} 1 & 0 \\ 0 & 0 \end{bmatrix}$$

is a "square root" of the same matrix.

The introduction of G_v enables writing disturbance $v_1(\cdot)$ as $G_v \xi(\cdot)$, where $\xi(\cdot)$ is a random process such that

(a) as $v(\cdot)$, also $\xi(\cdot)$ is a white noise (namely $E[\xi(t_1)\xi(t_2)] = 0$ if $t_1 \neq t_2$);
(b) the variance matrix of $\xi(\cdot)$ is the identity I.

It is easy to check that, with such properties, $G_v \xi(\cdot)$ is a white noise with variance $G_v I G_v' = V_1$.

In conclusion, the state equation of the data generation mechanism can be written as

$$x(t+1) = Fx(t) + G_v \xi(t).$$

In such an expression, $\xi(\cdot)$ is a random vector with the identity as variance matrix. So, each entry of $\xi(t)$ is a white noise with unit variance, uncorrelated with the other entries. In principle, all of these disturbances could have the same effect on each of the state variables. Whether they actually have an effect or not depends on the structure of matrices F and G_v, as illustrated below.

Example 9.6 (A direction in the state space uncorrupted by noise)   Suppose that

$$V_1 = \begin{bmatrix} \alpha^2 & \alpha^2 \\ \alpha^2 & \alpha^2 \end{bmatrix},$$

while

$$F = \begin{bmatrix} a & 0 \\ 0 & a \end{bmatrix}.$$

A "square root" of V_1 is

$$G_v = \begin{bmatrix} \alpha \\ \alpha \end{bmatrix}.$$

Hence, the two state equations are

$$x_1(t + 1) = ax_1(t) + v_{11}(t)$$
$$x_2(t + 1) = ax_2(t) + v_{12}(t),$$

with $v_{11}(\cdot)$ and $v_{12}(\cdot)$ white noises, each of which with variance α^2, and with unit covariance coefficient.

The structure of these equations suggests a change of basis in the state space, by introducing new state variables, $\zeta_1(\cdot)$ and $\zeta_2(\cdot)$, defined as

$$\zeta_1(t) = x_1(t) + x_2(t)$$
$$\zeta_2(t) = x_1(t) - x_2(t).$$

The state equations in $\zeta_1(\cdot)$ and $\zeta_2(\cdot)$ are

$$\zeta_1(t + 1) = a\zeta_1(t) + w_{11}(t)$$
$$\zeta_2(t + 1) = a\zeta_2(t) + w_{12}(t),$$

where

$$w_{11}(t) = v_{11}(t) + v_{12}(t) \quad \text{and} \quad w_{12}(t) = v_{11}(t) - v_{12}(t).$$

It is straightforward to verify that $w_{12}(\cdot)$ has null variance. Therefore, we can simplify the system as

$$\zeta_1(t + 1) = a\zeta_1(t) + w_{11}(t)$$
$$\zeta_2(t + 1) = a\zeta_2(t).$$

Thus, there is a direction in the state space along which no disturbance acts, the ζ_2 axis in the $\zeta_1 - \zeta_2$ space, or, equivalently, the bisector of the second and fourth quadrant in the $x_1 - x_2$ space. Correspondingly, the variance of the estimation error of $\zeta_2(t) = x_1(t) - x_2(t)$ will be null at each time point if the initial condition on $x_1(\cdot) - x_2(\cdot)$ is null. If instead the variance of $x_1(1) - x_2(1)$ is non-null, the time evolution of the error will be determined by the equation governing $\zeta_2(t+1)$, namely $\zeta_2(t+1) = a\zeta_2(t)$. If $|a| < 1$, then the variance of $x_1(\cdot) - x_2(\cdot)$ will tend to 0 (implying that asymptotically $x_1(\cdot) - x_2(\cdot)$ can be estimated with no error). If instead $|a| > 1$, then the variance of $x_1(\cdot) - x_2(\cdot)$ will tend to some non-null value or will diverge (depending on the output equation).

We now re-elaborate the above observations from the point of view of the Riccati equation. To this purpose, we distinguish four cases:

1) $|a| < 1$ and $\mathrm{Var}[x_1(1) - x_2(1)] = 0$.
2) $|a| < 1$ and $\mathrm{Var}[x_1(1) - x_2(1)] \neq 0$.
3) $|a| > 1$ and $\mathrm{Var}[x_1(1) - x_2(1)] = 0$.
4) $|a| > 1$ and $\mathrm{Var}[x_1(1) - x_2(1)] \neq 0$.

The solution of the DRE is the variance of the state prediction error of the optimal predictor. As such, it must be not greater than the variance of the state. Now in cases 1 and 3, the initial uncertainty in the $x_1(\cdot) - x_2(\cdot)$ direction being null (and the associated equation deterministic), the solution of the DRE along the same direction will be null. Correspondingly, the solution of the DRE written for the new state variables will have a null element on the $(2, 2)$ position.

In case 2, the solution of the DRE will tend to zero in the direction $x_1(\cdot) - x_2(\cdot)$. In case 4, the equation governing $x_1(\cdot) - x_2(\cdot)$ is unstable so that its solution will have a diverging variance. As for the solution of the DRE, it will diverge or converge to a finite value depending on the output transformation. To be precise, if the output transformation is such that the state is observable, then the solution of the DRE will converge.

In order to describe the action of "corruption" of the state due to noise $v_1(\cdot)$, we write the state equation in the form

$$x(t+1) = Fx(t) + G_v\xi(t).$$

Suppose that the initial state be known for example

$$x(1) = 0.$$

Then

$$x(2) = G_v\xi(1).$$

Denoting with $\xi_1(t), \xi_2(t), \ldots, \xi_m(t)$ the entries of $\xi(t)$, and with $g^1, g^2, \ldots, g^m$ the columns of G_v, we have

$$x(2) = \xi_1(1)g^1 + \xi_2(1)g^2 + \cdots + \xi_m(1)g^m.$$

State $x(2)$ is therefore a linear combination of the columns of matrix G_v. The coefficients of such combination are the entries of $\xi(t)$, which are random variables uncorrelated to each other. Hence, $x(2)$ can describe all possible linear combinations of the G_v columns.

Passing to time $t = 3$, we have

$$x(3) = FG_v\xi(1) + G_v\xi(2).$$

$\xi(\cdot)$ being a white noise, $\xi(1)$ and $\xi(2)$ are uncorrelated, so that the noise samples $\xi(1)$ and $\xi(2)$ have influence on all and only those states that can be obtained as combinations of the columns of G_v and FG_v.

By iterating, we come to the conclusion that the noise samples at instants $1, 2, \ldots, j$ have influence on the linear combinations of the columns of G_v, FG_v, F^2G_v, ..., up to $F^{j-1}G_v$. Stated another way, the state variables corrupted by noise are those generated as linear combinations of the columns of matrix

$$\mathscr{R}^j = [G \vdots FG \vdots F^2G \vdots \cdots \vdots F^{j-1}G].$$

Such set is also called *reachability subspace*. This is an hyperplane in the state space with dimension determined by the linearly independent columns of the matrix.

Obviously, such subspace is larger and larger as j increases as more and more columns appear in $\mathscr{R}^j$. However, as shown in Appendix A, its dimension cannot increase for $j \geq n$. This leads to the following definition.

Definition 9.6 (Reachability matrix) The matrix

$$\mathscr{R}^n = [G \vdots FG \vdots F^2G \vdots \cdots \vdots F^{n-1}G]$$

is named reachability matrix of the system (F, G).

The linear combinations of the columns of $\mathscr{R}^n$ is the subspace of the state space where noise carries out its influence.

The system is said to reachable (or the pair (F, G_v) is reachable) when such subspace extends to the whole state space. From the above discussion, we can conclude the following.

Proposition 9.3 (Reachability test) (F, G_v) *is reachable if and only if the reachability matrix* $\mathscr{R}^n$ *has rank n.*

We are now in a position to state the following.

Proposition 9.4 (Second convergence theorem for the DRE) *Suppose that the data generation mechanism is such that* (F, G_v) *is reachable (where* G_v *is any matrix such that* $G_vG_v' = V_1$*) and* (F, H) *is observable. Then,*

1) *For any (positive semi-definite) initial condition, the solution of the DRE asymptotically converges to*

$$\lim_{t \to \infty} P(t) = \bar{P}.$$

$\bar{P}$ *is the same matrix whatever the initial condition* $P(1) = P_1$ *(provided that* $P_1 \geq 0$*).*

2) $\bar{P}$ is the only positive semi-definite solution of the ARE (9.60).
3) The steady-state predictor is stable (all eigenvalues of matrix $F - F\bar{P}H'(H\bar{P}H' + V_2)^{-1}H$, see Eq. (9.61), are inside the unit disk).
4) $\bar{P}$ is positive definite.

Remark 9.13 Matrix V_1 may have various "square roots," for instance there may exists two different matrices G_v and $\tilde{G}_v$ such that $G_v G'_v = \tilde{G}_v \tilde{G}'_v = V_1$. It can be shown that, if (F, G_v) is reachable, then $(F, \tilde{G}_v)$ is also reachable. So, the statement above holds whichever "square root" of V_1 is considered.

Remark 9.14 If the system is subject to an exogenous variable $u(t)$, i.e. $x(t + 1) = Fx(t) + Gu(t) + G_v\xi(t)$, the reachability of (F, G) has no relevance for the convergence of the solution of the Riccati equation. Indeed, matrix G does not even enter the Riccati equation. What matters is the reachability from the noise, namely the reachability of the pair (F, G_v).

9.7.2.5 General Convergence Result
The two propositions above lead to the same conclusions under different assumptions. This is not in contrast with the nature of such results, since they are just sufficient conditions. As a matter of fact, the two propositions are corollaries of a general result, presented in this paragraph.

This study requires the introduction of two new properties, strictly related to the so-called *canonical decomposition* of a dynamical system. Such a decomposition, discussed in Appendix A, can be concisely outlined as follows.

Starting from the usual system

$$x(t + 1) = Fx(t) + v_1(t)$$
$$y(t) = Hx(t) + v_2(t), \qquad (9.69)$$

to study the effect of $v_1(t)$, we introduce the output noise free system as

$$x(t + 1) = Fx(t) + G_v\xi(t)$$
$$y(t) = Hx(t), \qquad (9.70)$$

where G_v is such that $G_v G'_v = V_1$ and $\xi(t) \sim WN(0, I)$.

With a suitable change of basis, this system (as any linear system), can be decomposed into four parts:

r. o.	reachable and observable part
r. no.	reachable and unobservable part
nr. o.	unreachable and observable part
nr. no.	unreachable and unobservable part

as depicted in Figure 9.11. The input signal is effective on the reachable part (composed by two subparts, the reachable and observable one and the

Figure 9.11 Canonical decomposition of system (9.70).

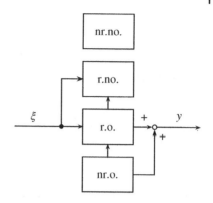

reachable and unobservable one), whereas it does not have influence on the unreachable part. The output signal is informative on the observable part (composed of two subparts, the reachable and observable one and the unreachable and observable one), whereas it does not bring information on the unobservable part.

Accordingly with the canonical decomposition of the system, we can also decompose the Riccati equation into subequations. In view of *second convergence theorem for the DRE*, we expect that the solution of the Riccati subequation associated with the reachable and observable part is asymptotically convergent. This implies that an odd phenomena such as the divergence of the solution or its possible non-uniqueness must originate in the remaining subequations. To overcome such phenomena, the unreachable and unobservable parts have to be stable.

These intuitive considerations can be given a mathematical form by introducing the following definitions.

Definition 9.7 (Detectability) A system is said to be detectable when its unobservable part is stable.

Definition 9.8 (Stabilizability) A system is said to be stabilizable when its unreachable part is stable.

Obviously,

a) if a system is stable, then both its unobservable part and unreachable part are stable, and therefore, the system is also detectable and stabilizable.
b) if a system is reachable and observable, then there is no unreachable part and no unobservable part; hence, it is detectable and stabilizable.

We are now in a position to state the general convergence theorem.

Proposition 9.5 (General convergence theorem for the DRE) *Consider system (9.69) and suppose that the pair (F, H) is detectable and the pair (F, G_v) is stabilizable (G_v is any matrix such that $G_v G_v' = V_1$). Then*

a) *For any positive semi-definite initialization, the solution of the DRE is asymptotically convergent to the limit matrix $\bar{P}$ (always the same whatever the positive semi-definite initial condition be).*

b) *The steady-state Kalman predictor (associated with the limit matrix $\bar{P}$) is stable.*

Proof:
The proof consists of the following points:

1) The detectability assumption implies that, for any positive semi-definite initial condition, $P(1) \geq 0$, the solution of the DRE is bounded. Note that this statement about the non-divergence of $P(t)$ does not guarantee its convergence. As a matter of fact, there are detectable systems with a solution $P(t)$, which is oscillating.

2) Given two initial conditions $\tilde{P}(1) \geq 0$ and $\hat{P}(1) \geq 0$, if $\tilde{P}(1) \geq \hat{P}(1)$, then $\tilde{P}(t) \geq \hat{P}(t), \forall t$, where $\tilde{P}(t)$ and $\hat{P}(t)$ are the solutions of the DRE with initial conditions $\tilde{P}(1)$ and $\hat{P}(1)$, respectively.

3) The solution of the DRE associated with the null initial condition $P(1) = 0$, is monotonically increasing in t.

4) A corollary of points (1) and (3) is that the solution of the DRE with null initial condition, $P(1) = 0$, is asymptotically convergent. The limit matrix is denoted as $\bar{P}$.

5) Thanks to the stabilizability assumption, we prove that the steady-state predictor associated with $\bar{P}$ is stable.

6) When the initial condition $P(1)$ is a generic positive semi-definite matrix, the difference between the solution of the DRE and the solution associated with initial condition $P(1) = 0$ tends to zero.

Proof of point 1:
The detectability assumption means that the unobservable part is stable. We let our readers prove that, under that assumption, there exists a stabilizing gain K such that $F - KH$ is stable. Now, the predictor associated to such gain

$$x(t + 1|t) = Fx(t|t - 1) + K(y(t) - Hx(t|t - 1))$$

has as prediction error $v(t) = x(t) - \hat{x}(t|t - 1)$ governed by the equation

$$v(t + 1) = (F - KH)v(t) + v_1(t) - Kv_2(t).$$

This is a discrete time system with $F - KH$ as dynamical matrix, fed by inputs $v_1(t)$ and $v_2(t)$, two stationary processes. Being a stable system with stationary inputs, the output $v(t)$ has bounded variance.

The Kalman predictor is an optimal predictor. Therefore, the associated variance of the prediction error (namely matrix $P(t)$, solution of the Riccati equation) must be lower than the variance associated to a generic gain. This leads to the conclusion that the solution of the DRE must be bounded.

Remark 9.15 (Oscillating solution of the DRE) The above conclusion does not allow to claim that the solution of the DRE must converge under the detectability assumption. Indeed, consider the second-order system:

$$x(t+1) = Fx(t)$$
$$y(t) = Hx(t) + v_2, \quad v_2 \sim \text{WN}(0, 1),$$

with

$$F = \begin{bmatrix} 0 & f_1 \\ f_2 & 0 \end{bmatrix}, \quad f_1 \neq 0; \quad f_2 \neq 0,$$

$$H = \begin{bmatrix} 1 & 0 \end{bmatrix}.$$

By constructing the 2 × 2 observability matrix, it is easy to see that the system is observable and therefore detectable. We will now show that the solution of the corresponding DRE may be oscillating for a suitable choice of the system parameters. For, assume that, at a certain time instant,

$$P(t) = \begin{bmatrix} p & 0 \\ 0 & 0 \end{bmatrix}, \quad p > 0.$$

A simple computation leads to

$$P(t+1) = \begin{bmatrix} 0 & 0 \\ 0 & f_2 p/(1+p) \end{bmatrix}.$$

If instead

$$P(t) = \begin{bmatrix} 0 & 0 \\ 0 & p \end{bmatrix}, \quad p > 0,$$

then

$$P(t+1) = \begin{bmatrix} f_1^2 p & 0 \\ 0 & 0 \end{bmatrix}.$$

Hence, by posing

$$f_1 = 1$$

and by choosing f_2 such that

$$f_2 p/(1+p) = p$$

the solution associated to the initial condition

$$P(1) = \begin{bmatrix} p & 0 \\ 0 & 0 \end{bmatrix}, \quad p > 0$$

oscillates between

$$\begin{bmatrix} 0 & 0 \\ 0 & p \end{bmatrix}$$

and

$$\begin{bmatrix} p & 0 \\ 0 & 0 \end{bmatrix}.$$

Intuitively, this oscillation can be explained by noticing that the state of the system is governed by equations

$$x_1(t+1) = f_1 x_2(t)$$
$$x_2(t+1) = f_2 x_1(t).$$

This is a deterministic system, so that, if, at a time point t, state $x_1(t)$ is known without uncertainty (namely element $(1,1)$ of matrix P is null), then, being $x_2(t+1) = f_2 x_1(t)$, at the subsequent time point $t+1$, it is state $x_2(t+1)$ to become known without uncertainty (namely the element $(2,2)$ of matrix P is null).

Proof of point 2:
Recall expression (9.40) of the DRE:

$$P(t+1) = (F - K(t)H)P(t)(F - K(t)H)' + V_1 + K(t)V_2 K(t)', \qquad (9.71)$$

where

$$K(t) = FP(t)H'[HP(t)H' + V_2]^{-1}, \qquad (9.72)$$

see Eq. (9.38). With reference to the second member of the DRE equation above, we define

$$\Delta(P, K) = (F - KH)P(F - KH)' + V_1 + KV_2 K'. \qquad (9.73)$$

Given an $n \times n$ matrix P and an $n \times p$ matrix K, $\Delta(\cdot, \cdot)$ supplies an $n \times n$ matrix. If, in place of K, we pose

$$K = K(P) = FPH'(HPH' + V_2)^{-1},$$

we obtain a transformation $\Pi(P)$ defined as

$$\Pi(P) = \Delta(P, K(P))$$

mapping an $n \times n$ matrix into another $n \times n$ matrix. Note that, if $P \geq 0$, then $\Pi(P) \geq 0$ as well.

Thanks to the defined transformation $\Pi(P)$, the DRE and the ARE can be written as

$$\text{DRE}: \quad P(t+1) = \Pi(P(t)),$$

$$\text{ARE}: \quad P = \Pi(P).$$

We now show that the map $\Pi(\cdot)$ preserves ordering, namely, given two $n \times n$ positive semi-definite matrices A and B, if $A \geq B$, then $\Pi(A) \geq \Pi(B)$. Indeed,

$$\Pi(A) = \Delta(A, K(A)).$$

On the other hand, from expression (9.73), it is obvious that, if K is kept constant, Δ is monotonic in P, so that

$$A \geq B \Rightarrow \Delta(A, K(A)) \geq \Delta(B, K(A)).$$

Now $\Delta(B, K(B))$ is the second member of the ARE $B = \Pi(B)$. Therefore, it is the variance of the prediction error of an optimal predictor. As such, it must be lower than the one achievable with any other gain. Hence,

$$\Delta(B, K(A)) \geq \Delta(B, K(B)).$$

From the last two formulas, we have

$$A \geq B \Rightarrow \Delta(A, K(A)) \geq \Delta(B, K(B)),$$

namely $\Pi(A) \geq \Pi(B)$.

Proof of point 3:
Consider now the evolution of the solution of the DRE with null initial condition, $P(1) = 0$. We will denote such solution as $P^{(0)}(\cdot)$. In particular, focus on the value taken by such solution at time 2, namely $P^{(0)}(2)$. As for any solution of the Riccati equation, this matrix must be positive semi-definite, $P^{(0)}(2) \geq 0$. Consequently, $P^{(0)}(2) \geq P^{(0)}(1)$. By posing $A = P^{(0)}(2)$ and $B = P^{(0)}(1)$, we can apply the property seen at the previous point to conclude that $\Pi(P^{(0)}(2)) \geq \Pi(P^{(0)}(1))$, i.e. $P^{(0)}(3) \geq P^{(0)}(2)$. Proceeding in this way, we conclude that the solution of the DRE for null initial condition is monotonically increasing: $P^{(0)}(t + 1) \geq P^{(0)}(t), \forall t$.

Proof of point 4:
It is well known that, when a sequence of real numbers is monotonically increasing and bounded, the sequence is convergent. The same result holds true in the matrix case. Hence, from points 1 and 3, we can say that $\lim_{t \to \infty} P^{(0)}(t)$ is convergent, and we shall denote by $\bar{P}$ the limit

$$\lim_{t \to \infty} P^{(0)}(t) = \bar{P}.$$

Remark 9.16 (Mid proof comment) So far we proved that

a) If we solve the DRE with null initial condition, the solution is convergent to a limit matrix $\bar{P}$.
b) Each solution of the DRE (associated to any feasible initial condition) is bounded.

The only assumption used up to now is detectability.

Proof of point 5:
$\bar{P}$ must satisfy the ARE

$$\bar{P} = (F - \bar{K}H)\bar{P}(F - \bar{K}H)' + \bar{K}V_2\bar{K}' + V_1,$$

where $\bar{K}$ is the optimal gain associated with $\bar{P}$. By contradiction suppose that the optimal predictor is not stable. This means that there exists an x, with $x \neq 0$, such that, for a λ such that $|\lambda| \geq 1$,

$$(F - \bar{K}H)'x = \lambda x. \tag{9.74}$$

Now pre-multiply the two members of the ARE by the conjugate transpose x^* of x and post-multiply the two members by x. We have

$$x^*\bar{P}x = |\lambda|^2 x^*\bar{P}x + x^*\bar{K}V_2\bar{K}'x + x^*V_1x,$$

namely

$$(1 - |\lambda^2|)x^*\bar{P}x = x^*\bar{K}V_2\bar{K}'x + x^*V_1x.$$

The right-hand side of such equality is non-negative since both $\bar{K}V_2\bar{K}$ and V_1 are positive semi-definite matrices. On the opposite, the left-hand side is non-positive since $x^*\bar{P}x \geq 0$ and $(1 - |\lambda^2|) \leq 0$. Consequently, the two members can coincide only if they are null. In turn, being $x^*\bar{K}V_2\bar{K}'x + x^*V_1x$ the sum of two non-negative elements, this entails that both $x^*\bar{K}V_2\bar{K}'x$ and x^*V_1x must be null. Hence, it must be

$$V_1x = 0, \tag{9.75}$$

and, assuming V_2 non-singular,

$$\bar{K}'x = 0.$$

From this last equation and Eq. (9.74), we can say that

$$F'x = \lambda x. \tag{9.76}$$

In view of the PBH test (see Appendix A), the two conditions (9.75) and (9.76) violate the stabilizability assumption.

Proof of point 6:
Among the many ways in which the DRE can be written, we will now refer to expression (9.42), here rewritten:

$$P(t+1) = FP(t)F' - K(t)HP(t)F' + V_1. \tag{9.77}$$

The second member must return a symmetric matrix, so that $K(t)HP(t)F'$ must coincide with its transpose, $(K(t)HP(t)F' = FP(t)H'K(t)')'$. Hence, Eq. (9.77) can also be written in the form

$$P(t+1) = FP(t)(F - K(t)H)' + V_1 \tag{9.78}$$

or in the form

$$P(t + 1) = (F - K(t)H)P(t)F' + V_1.$$

Of course, the same equation is valid for the solution $\bar{P}$ of the ARE:

$$\bar{P} = (F - \bar{K}H)\bar{P}F' + V_1. \tag{9.79}$$

By subtracting Eq. (9.79) from Eq. (9.78), we have

$$P(t + 1) - \bar{P} = FP(t)(F - K(t)H)' - (F - \bar{K}H)\bar{P}F'$$
$$= (F - \bar{K}H)(P(t) - \bar{P})(F - K(t)H)'$$
$$+ \bar{K}HP(t)(F - K(t)H)' - (F - \bar{K}H)\bar{P}H'K(t)'.$$

The last two terms at the right-hand side in this expression cancel each other. Therefore, the time evolution of $\Delta P(t) = P(t) - \bar{P}$ is governed by the equation

$$\Delta P(t + 1) = (F - \bar{K}H)\Delta P(t)(F - K(t)H)'$$

the solution of which is

$$\Delta P(t) = (F - \bar{K}H)^{t-1}\Delta P(1)\psi(t)',$$

where

$$\psi(t) = (F - K(t - 1)H)(F - K(t - 2)H)\cdots(F - K(1)H).$$

We invite our readers to show that $\psi(t)$ is bounded. Then, since $(F - \bar{K}H)^{t-1}$ tends to zero with t, we come to the conclusion that $\Delta P(t)$ must tend to zero.

Example 9.7   Consider the system

$$x(t + 1) = \frac{\sqrt{2}}{2}x(t) + v_1(t)$$
$$y(t) = 2x(t) + v_2(t),$$

where $v_1(t) \sim WN(0, \frac{3}{4})$ and $v_2(t) \sim WN(0, 4)$ are white noises uncorrelated to each other and uncorrelated with the system initial condition.
The DRE is

$$P(t + 1) = \frac{1}{2}P(t) + \frac{3}{4} - \frac{2P(t)^2}{4P(t) + 4},$$

namely

$$P(t \mid 1) = \frac{5P(t) + 3}{4P(t) + 4}.$$

The system is stable; therefore, the solution of the DRE (with positive semi-definite initial condition) will converge to a (positive semi-definite) limit value $\bar{P}$. The limit value must satisfy the ARE:

$$P = \frac{5P + 3}{4P + 4}.$$

This algebraic equation has two solutions, 1 and −0.75, the second of which is unfeasible since it is negative. Therefore, starting from any positive semi-definite initial condition $P(1)$ of the DRE, $P(t)$ will tend to $\bar{P} = 1$.

The Kalman gain, given by

$$K(t) = \frac{\sqrt{2}P(t)}{4P(t) + 4},$$

will therefore tend to

$$\bar{K} = \frac{\sqrt{2\bar{P}}}{4\bar{P} + 4} = \frac{\sqrt{2}}{8}.$$

Hence, the steady-state one-step-ahead Kalman predictor is

$$\hat{x}(t+1|t) = \frac{\sqrt{2}}{2}\hat{x}(t|t-1) + \frac{\sqrt{2}}{8}[y(t) - 2\hat{x}(t|t-1)] . \tag{9.80}$$

Note that this predictor, which can also be written as

$$\hat{x}(t+1|t) = \frac{\sqrt{2}}{4}\hat{x}(t|t-1) + \frac{\sqrt{2}}{8}y(t) ,$$

is stable.

In conclusion, the one-step-ahead steady-state predictor is stable and the variance of the error $v(t) = x(t) - \hat{x}(t|t-1)$ is $\bar{P} = 1$.

We now pass to the two-steps-ahead predictor. As we know from Section 9.5,

$$\hat{x}(t+2|t) = F\hat{x}(t+1|t) .$$

Correspondingly, the prediction error variance is

$$\bar{P}^{(2)} = F\bar{P}F' + V_1 = \frac{5}{4}.$$

The fact that this variance is greater than $\bar{P}$ is expected since the prediction is more uncertain when the prediction horizon increases.

Finally, we determine the optimal steady-state filter $\hat{x}(t|t)$ and the variance $\bar{P}^{(0)}$ of the associated steady-state filtering error $x(t) - \hat{x}(t|t)$. To this purpose, remind formula (9.56) relating the optimal predictor to the optimal filter, i.e.

$$\hat{x}(t+1|t) = F\hat{x}(t|t) .$$

In our case, F, given by $F = \frac{\sqrt{2}}{2}$, is invertible. Therefore,

$$\hat{x}(t|t) = F^{-1}\hat{x}(t+1|t) .$$

Therefore, in view of (9.80),

$$\hat{x}(t|t) = \frac{2}{\sqrt{2}}\left[\frac{\sqrt{2}}{2}\hat{x}(t|t-1)\right] + \frac{2}{\sqrt{2}}\frac{\sqrt{2}}{8}[y(t) - 2\hat{x}(t|t-1)]$$

$$= \hat{x}(t|t-1) + \frac{1}{4}[y(t) - 2\hat{x}(t|t-1)]$$

$$= \frac{\sqrt{2}}{2}\hat{x}(t-1|t-1) + \frac{1}{4}[y(t) - \sqrt{2}\hat{x}(t|t)].$$

As we see, the gain of the optimal filter $\bar{K}^{(0)}$ appearing here is

$$\bar{K}^{(0)} = \frac{1}{4}$$

and this is in agreement with the general expression (9.55) as our readers can easily check.

According to the formula, the variance of the filtering error, $P^{(0)}(N)$, can be computed as

$$\bar{P}^{(0)} = \bar{P} - \bar{P}H'(H\bar{P}H' + V_2)^{-1}H\bar{P} = \frac{1}{2}.$$

In conclusion, we have, as expected, $\bar{P}^{(0)} < \bar{P} < \bar{P}^{(2)}$.

9.8 Innovation Representation

In this section, we introduce the so-called *innovation representation* of a stationary process. Such representation supplies a conceptual bridge between the Kolmogorov–Wiener theory seen at the beginning of the book and the Kalman theory we are now studying. In this section, as well as in the next two, Sections 9.9 and 9.10, we focus on systems with a scalar output $y(t)$.

Consider again the system

$$x(t+1) = Fx(t) + v_1(t) \tag{9.81a}$$

$$y(t) = Hx(t) + v_2(t), \tag{9.81b}$$

with uncorrelated disturbances. Under the stability assumption, output $y(\cdot)$ is a stationary process. Thanks to the first convergence theorem, we also know that the predictor tends to the steady-state one:

$$\hat{x}(t+1|t) = F\hat{x}(t|t-1) + \bar{K}e(t),$$

$$\hat{y}(t+1|t) = H\hat{x}(t+1|t),$$

$$e(t) = y(t) - \hat{y}(t|t-1). \tag{9.82}$$

By posing

$$s(t) = \hat{x}(t|t-1),$$

we can rewrite the previous equations as follows:

$$s(t+1) = Fs(t) + \bar{K}e(t) \tag{9.83a}$$

$$y(t) = Hs(t) + e(t). \tag{9.83b}$$

Note that, thanks to the assumption that $y(t)$ is scalar, here the innovation is a scalar signal. Thus, (9.83) is a single-input single-output dynamical system with $e(t)$ as input, $y(t)$ as output, and $s(t) = \hat{x}(t|t-1)$ as state.

If compared with the original equations (9.81), system (9.83) offers an alternative description of signal $y(t)$, the so-called *innovation representation*.

9.9 Innovation Representation Versus Canonical Representation

In this book, we have seen two main representations of a stationary process, the *canonical representation* of Section 1.16 and the *innovation representation* of the previous section. A natural question is how we can compare these two descriptions.

We start by observing that, while (9.81) supplies $y(\cdot)$ as the output of a system fed by $v_1(\cdot)$ and $v_2(\cdot)$, for a total of $n + 1$ input signals, Eq. (9.83) gives the process as the output of a system with a unique input signal, namely $e(\cdot)$.

The transfer function of the SISO system (9.83) is

$$\hat{K}(z) = H(zI - F)^{-1}\bar{K} + 1. \tag{9.84}$$

$\hat{K}(z)$ has peculiar properties. To explain them, let's denote by $N(z)$ and $D(z)$ its numerator and its denominator, $\hat{K}(z) = N(z)/D(z)$.

- Assuming that there is no pole-zero simplification, $N(z)$ and $D(z)$ are both monic polynomials with degree n, the dimension of the state space. Indeed, considering how a determinant is computed (see Appendix B), it is obvious that $D(z) = \det(zI - F)$ is a monic polynomial of degree n. At the numerator, there is $N(z) = H(zI - F)^{-1}\bar{K} + 1$. If one reflects on how the inverse of matrix is computed (see again Appendix B), we conclude that the numerator of $H(zI - F)^{-1}\bar{K}$ must have a degree lower than or equal to $n - 1$. Therefore, the sum $H(zI - F)^{-1}\bar{K} + 1$ results in a monic polynomial of degree n.
- The singularities of both $N(z)$ and $D(z)$ are all inside the unit disk. Indeed, since we assumed that system (9.81) is stable, matrix F is stable; therefore, the denominator $D(z) = \det(zI - F)$ of $\hat{K}(z)$ must have its singularities inside the unit disk. As for the zeros, we observe that, for the first convergence theorem, predictor (9.82) is stable. Now, the input of such predictor is $y(t)$. As output we can choose any signal; in particular, we can take $e(t)$ as output. In light of the first convergence theorem, we can claim that such a system (from $y(t)$ to $e(t)$) must be stable. On the other hand, this system has a transfer function, which is the inverse of that from $e(t)$ to $y(t)$, i.e. $(\hat{K}(z))^{-1}$. In other words, it is given by $D(z)/N(z)$. Hence, in light of the first convergence theorem, we can say that $N(z)$ must have its singularities inside the unit disk. Stated another way, the zeros of $\hat{K}(z)$ must lie inside the unit disk too.

The properties of $\hat{K}(z)$ coincide with those of the canonical spectral factor $\hat{W}(z)$ of Kolmogorov-Wiener theory. So our conclusion is as follows:

The innovation representation in Kalman theory is the state space version of the canonical representation of Kolmogorov–Wiener theory:

$$\hat{W}(z) = \hat{K}(z).$$

9.10   K-Theory Versus K–W Theory

We now compare Kalman theory and Kolmogorov–Wiener theory. No doubt Kalman theory is much more general than Kolmogorov-Wiener one. The first and most important advantage of Kalman approach is that one can estimate a variable different from the measured one, whereas in Kolmogorov–Wiener theory, at least in the form we presented it in this book, one can only estimate the future evolution of the same variable that was measured. Second, Kolmogorov–Wiener theory makes reference to stationary processes; in a state space framework, this means dealing with time-invariant stable systems. On the contrary, Kalman theory is also applicable to unstable and possibly time-varying systems.

The natural question is then, when both theories are applicable, how can we compare their performances? We anticipate the main result:

When both theories are applicable, they lead to coincident predictors.

We verify this statement by comparing the r step ahead Kolmogorov–Wiener predictor (from the remote white noise) with the r-steps ahead Kalman predictor (from innovation); to be precise, we will show that such predictors have the same transfer function.

We start by observing that, from (9.82), the transfer matrix from $e(t)$ to state $s(t) = \hat{x}(t|t-1)$ is $[zI - F]^{-1}\bar{K}$. Therefore, the transfer matrix from $e(t)$ to $\hat{x}(t+1|t)$ is

$$\hat{K}(z) = z[zI - F]^{-1}\bar{K} = [I - Fz^{-1}]^{-1}\bar{K}.$$

Recalling that the optimal r-steps ahead Kalman predictor is given by $\hat{x}(t+r|t) = F^{r-1}\hat{x}(t+1|t)$, we have that the transfer matrix from $e(t)$ to $\hat{x}(t+r|t)$ is $F^{r-1}[I - Fz^{-1}]^{-1}\bar{K}$. Therefore, the transfer function from $e(t)$ to $\hat{y}(t+r|t)$ is

$$\hat{K}_r(z) = HF^{r-1}[I - Fz^{-1}]^{-1}\bar{K}. \tag{9.85}$$

We now pass to the Kolmogorov–Wiener approach. As previously seen, the canonical spectral factor is given by $\hat{W}(z) = \hat{K}(z)$. To find the optimal predictor in the K–W framework, we apply the long division technique. Recalling the power series expansion

$$[zI - F]^{-1} = z^{-1}[I - Fz^{-1}]^{-1} = z^{-1}[I + Fz^{-1} + F^2z^{-2} + \cdots],$$

we have

$$\hat{W}(z) = \hat{K}(z) = 1 + H\bar{K}z^{-1} + HF\bar{K}z^{-2} + \cdots + HF^{r-2}\bar{K}z^{-r+1}$$
$$+ HF^{r-1}\bar{K}z^{-r} + HF^r\bar{K}z^{-r-1} + \cdots$$
$$= 1 + H\bar{K}z^{-1} + HF\bar{K}z^{-2} + \cdots + HF^{r-2}\bar{K}z^{-r+1}$$
$$+ z^{-r}\{HF^{r-1}[I + Fz^{-1} + F^2z^{-2} + \cdots]\bar{K}\}.$$

Therefore, according to Kolmogorov–Wiener theory, the optimal predictor has the following transfer function:

$$\hat{W}_r(z) = HF^{r-1}[I + Fz^{-1} + F^2 z^{-2} + \cdots]\bar{K}$$
$$= HF^{r-1}[I + Fz^{-1}]^{-1}\bar{K}.$$

This expression coincides with that of $\hat{K}_r(z)$ given by Eq. (9.85).

Remark 9.17 (K vs K–W: Numerical Bottlenecks) For the moment, the comparison between the two approaches to prediction has been performed at a conceptual level. There are however important diversities at the computational level. Indeed, in Kalman theory, the numerical bottleneck is the Riccati equation, an equation where the unknown is a square matrix of dimension $n \times n$, where n is the number of state variables. On the contrary, in Kolmogorov–Wiener theory, even for complex systems, the predictor can be found by means of easy polynomial divisions. Therefore, when both theories are applicable, the K–W approach is preferable for its computational simplicity.

Example 9.8 Consider the autoregressive model of order 1:

$$y(t) = ay(t-1) + \eta(t), \quad \eta(t) \sim (0, \lambda^2) \quad |a| < 1.$$

By applying Kolmogorov–Wiener theory, we know that the optimal predictor is

$$\hat{y}(t+1|t) = ay(t), \tag{9.86}$$

with error variance

$$\text{Var}[v(t+1|t) - v(t)] = \lambda^2.$$

We tackle now the prediction problem with Kalman theory. A state space realization of the AR difference equation is

$$x(t+1) = Fx(t) + G\eta(t)$$
$$y(t) = Hx(t) + \eta(t),$$

with

$$F = a, \quad G = a, \quad H = 1.$$

Indeed, the transfer function of such system is

$$W(z) = H(zI - F)^{-1}G + 1 = \frac{a}{z-a} + 1 = \frac{z}{z-a},$$

which coincides with the transfer function of the given AR(1) model.
The disturbances here acting are

$$v_1(t) = a\eta(t), \quad v_2(t) = \eta(t)$$

with variances

$$V_1 = \mathrm{Var}[v_1(t)] = a^2\lambda^2, \quad V_2 = \mathrm{Var}[v_2(t)] = \lambda^2.$$

Note that $v_1(t)$ and $v_2(t)$ are correlated with cross-variance

$$V_{12} = E[v_1(t)v_2(t)] = a\lambda^2.$$

According to the general expression of the Kalman gain, see Eq. (9.46), we have

$$K(t) = [F(t)P(t)H(t)' + V_{12}(t)][H(t)P(t)H(t)' + V_2(t)]^{-1}$$

$$= \frac{aP(t) + a\lambda^2}{P(t) + \lambda^2} = a. \tag{9.87}$$

Hence, the Kalman predictor is

$$\hat{x}(t+1|t) = F\hat{x}(t|t-1) + K(t)(y(t) - \hat{x}(t|t-1))$$
$$= a\hat{x}(t|t-1) + a(y(t) - \hat{x}(t|t-1)) = ay(t),$$
$$\hat{y}(t+1|t) = \hat{x}(t+1|t) = ay(t).$$

As can be seen, this prediction law coincides with the Kolmogorov–Wiener predictor (9.86). We observe that, in view of (9.47), the DRE is

$$P(t+1) = F(t)P(t)F(t)' + V_1(t) - K(t)\{H(t)P(t)H(t)' + V_2(t)\}K(t)'$$
$$= a^2 P(t) + a^2\lambda^2 - a^2 P(t) - a^2\lambda^2 = 0.$$

This means that the state can be predicted without error. It is easy to understand the reason of such conclusion. Indeed, consider again the state space equations

$$x(t+1) = ax(t) + a\eta(t),$$
$$y(t) = x(t) + \eta(t).$$

From the output equation, we see that $\eta(t) = y(t) - x(t)$. Substituting in the state equation, we obtain

$$x(t+1) = ax(t) + a(y(t) - x(t)) = y(t).$$

This implies that, knowing the output y at time t, we can perfectly guess the value of the state x at time $t+1$.

As for the output, the prediction is not error free since $y(t+1) = x(t+1) + \eta(t)$. While $x(t+1)$ is predictable without any error, $y(t)$ is not.

Example 9.9 Consider the system with two state variables $x_1(t)$ and $x_2(t)$ governed by the equations

$$x_1(t+1) = \alpha x_1(t) + v_{11}(t)$$
$$x_2(t+1) = \beta x_2(t) + v_{12}(t)$$
$$y(t) = x_2(t) + v_2(t), \tag{9.88}$$

where white processes $v_{11} \sim WN(0, 1)$, $v_{12} \sim WN(0, 1)$, and $v_2 \sim WN(0, 1)$ are uncorrelated with each other. The state prediction error covariance $P(t)$ will be a 2×2 matrix:

$$\begin{bmatrix} p_{11}(t) & p_{12}(t) \\ p_{21}(t) & p_{22}(t) \end{bmatrix}.$$

Here, $p_{11}(t)$ is the variance of the prediction error of the first state variable, while $p_{22}(t)$ is the variance of the prediction error of the second state variable. The off-diagonal terms $p_{12}(t) = p_{21}(t)$ are the cross-variances.

In order to discuss how the estimation accuracy depends upon the values of parameters α and β, a possibility is to find the solution of the associated Riccati equation. However, we can say something without any computation.

Indeed, a careful analysis of the given system shows that it is composed of two subsystems, one governed by state $x_1(t)$ and the other by state $x_2(t)$; they operate separately since $x_2(t)$ does not appear in the $x_1(t)$ equation and $x_1(t)$ does not appear in the $x_2(t)$ equation; moreover, the disturbances $v_{11}(t)$ and $v_{12}(t)$ are uncorrelated to each other. As for the output equation, we see that only $x_2(t)$ affects $y(t)$.

Thus, we can define the subsystem associated to state variable $x_2(t)$ as

$$x_2(t + 1) = \beta x_2(t) + v_{12}(t)$$
$$y(t) = x_2(t) + v_2(t), \tag{9.89}$$

and the subsystem associated with state variable $x_1(t)$ as

$$x_1(t + 1) = \alpha x_1(t) + v_{11}(t). \tag{9.90}$$

System (9.89) is reachable from the noise $v_{12}(t)$ and observable. Hence, the corresponding state uncertainty, $p_{22}(t)$, is convergent:

$$p_{22}(t) \longrightarrow \bar{p}_{22}.$$

$\bar{p}_{22}$ can be found by solving the ARE associated with (9.89).

As for state variable $x_1(t)$, it has no influence on the output, so that the data are of no use to estimate it and the only reasonable estimate is that of the mean value $\hat{x}(t + 1|t) = 0$. Therefore, the variance of its prediction error will coincide with the variance of the state, i.e. $p_{11}(t) = \text{Var}[x_1(t)]$. Obviously, such variance will diverge if $|\alpha| > 1$. If instead $|\alpha| < 1$, system (9.90) is stable and the variance of its solution will converge. To be precise, by noting that (9.90) is an AR(1),

$$p_{11}(t) \longrightarrow \frac{1}{1 - \alpha^2}, \quad |\alpha| < 1.$$

Our readers are invited to verify these results by solving the 2×2 Riccati equation associated with the original system (9.88).

9.11   Extended Kalman Filter – EKF

The *extended Kalman filter* (EKF) is a technique aiming at the estimation of the state of nonlinear system via Kalman filtering ideas. It is obtained via a heuristic procedure as follows.

Considering a system of the type

$$x(t + 1) = f(x(t)) + v_1(t) \tag{9.91a}$$
$$y(t) = h(x(t)) + v_2(t), \tag{9.91b}$$

let's define as *nominal state* the solution of the non-noisy equations

$$\bar{x}(1) = E[x(1)], \tag{9.92a}$$
$$\bar{x}(t + 1) = f(\bar{x}(t)). \tag{9.92b}$$

Correspondingly, the *nominal output* is

$$\bar{y}(t) = h(\bar{x}(t)).$$

Note that $\bar{x}(\cdot)$ and $\bar{y}(\cdot)$ can be computed *a priori* by solving a deterministic difference equation, in absence of any data.

By introducing the state and output variations

$$\Delta x(t) = x(t) - \bar{x}(t),$$
$$\Delta y(t) = y(t) - \bar{y}(t),$$

the linearization of Eqs. (9.91a) and (9.91b) leads to

$$\Delta x(t + 1) = \bar{F}(t)\Delta x(t) + v_1(t) \tag{9.93a}$$
$$\Delta y(t) = \bar{H}(t)\Delta x(t) + v_2(t), \tag{9.93b}$$

where

$$\bar{F}(t) = \left. \frac{\partial f(x)}{\partial x} \right|_{x=\bar{x}(t)},$$

$$\bar{H}(t) = \left. \frac{\partial h(x)}{\partial x} \right|_{x=\bar{x}(t)}.$$

Kalman theory can be applied to the linear system so obtained, obtaining the predicted estimate for the variations

$$\widehat{\Delta x}(t + 1|t) = \bar{F}(t)\widehat{\Delta x}(t|t - 1) + \bar{K}(t)e(t) \tag{9.94a}$$

$$e(t) = \Delta y(t) - \widehat{\Delta y}(t|t - 1) = \Delta y(t) - \bar{H}(t)\widehat{\Delta x}(t|t - 1). \tag{9.94b}$$

Here, $\bar{K}(t)$ is computed with the usual expression of the Kalman gain by means of the solution of the Riccati equation computed with matrices $\bar{F}(t)$ and $\bar{H}(t)$.

On the other hand, since

$$\widehat{\Delta x}(t+1|t) = \hat{x}(t+1|t) - \bar{x}(t+1),$$

we also have

$$\hat{x}(t+1|t) = \bar{x}(t+1) + \widehat{\Delta x}(t+1|t)$$
$$= f(\bar{x}(t)) + \bar{F}(t)\widehat{\Delta x}(t|t-1) + \bar{K}(t)e(t)$$
$$= f(\bar{x}(t)) + \bar{F}(t)\Delta x(t)$$
$$+ \bar{K}(t)[y(t) - \bar{y}(t) - \bar{H}(t)\widehat{\Delta x}(t|t-1)]. \tag{9.95}$$

Now $f(\bar{x}(t)) + \bar{F}(t)\Delta x(t)$ can be approximated as

$$f(\hat{x}(t|t-1)).$$

Analogously, $\bar{y}(t) - \bar{H}(t)\widehat{\Delta x}(t|t-1)]$ can be approximated as

$$h(\bar{x}(t|t-1)).$$

In conclusion,

$$\hat{x}(t+1|t) = f(\hat{x}(t|t-1)) + \bar{K}(t)[y(t) - h(\hat{x}(t|t-1))] \tag{9.96}$$

can be seen as the approximated predicted estimate. Analogously, the approximated filtered estimate is given by

$$\hat{x}(t|t) = f(\hat{x}(t|t)) + \bar{K}^{(0)}(t)y[(t) - h(\hat{x}(t-1|t-1))]. \tag{9.97}$$

These formulas are the nonlinear fashion replica of the Kalman filter equations. They could be postulated from the very beginning, without any computation, as an extension of linear Kalman theory to the nonlinear case.

Predictor (9.96) and filter (9.97) are known as *linearized Kalman predictor* and *linearized Kalman filter*. Gains $K(t)$ and $K^{(0)}(t)$ are given by the expressions worked out in the previous sections.

Such estimators have been found by approximating nonlinear functions $f(x)$ and $h(x)$ around the nominal state. However, such state trajectory is computed without taking into consideration any data. Therefore, the nominal state may progressively diverge from the true state; correspondingly, the performance of the predictor may deteriorate in the long run.

Better results are obtained by linearizing $f(x)$ and $h(x)$ around the last estimate of the state, i.e.

$$\hat{F}(t|t-1) = \frac{\partial f(x,t)}{\partial x}\bigg|_{x=\hat{x}(t|t-1)},$$

$$\hat{H}(t|t-1) = \frac{\partial h(x,t)}{\partial x}\bigg|_{x=\hat{x}(t|t-1)}.$$

Of course, this entails that the computation of the solution of the Riccati equation must be performed *in real time*, after $\hat{x}(t|t-1)$ becomes available.

The corresponding estimators are named *extended Kalman predictor* and EKF. Note that, while in the standard linear Kalman problem, the Riccati equation can be solved off-line once forever, in the nonlinear framework, matrices $\hat{F}(t|t-1)$ and $\hat{H}(t|t-1)$ can be computed only after the last state estimate is formed. This entails that the Riccati equation must be solved in real time.

In Chapter 11, we will discuss the tuning of the EKF for a specific problem, that of estimating the frequency of a sinusoid.

9.12 The Robust Approach to Filtering

The filtering problems can be posed in a deterministic setting. This is achieved with a paradigm shift in the very position of the problem, as we will now see.

Consider a data generation system of the type

$$x(t+1) = Fx(t) + Gw(t) \tag{9.98a}$$
$$y(t) = Hx(t) + Ew(t). \tag{9.98b}$$

Here, signal $w(t)$ is a generic disturbance, not necessarily a stochastic process. We deal with the usual problem of estimating the state from the observation of the output. Precisely, we focus on the filtering problem, where to estimate $x(t)$ one can rely on the data sequence, $y(t), y(t-1), y(t-2), \dots$ Such filter is denoted with the same symbol used in the Kalman approach, $\hat{x}(t|t)$.

As estimator, we assume for the very beginning the following structure:

$$\tilde{x}(t+1) = A\tilde{x}(t) + By(t) \tag{9.99a}$$
$$\hat{x}(t|t) = C\tilde{x}(t) + Dy(t), \tag{9.99b}$$

where the state $\tilde{x}(\cdot)$ has the same dimension as the unknown state $x(t)$.

If matrices A, B, C, and D are given, then one can derive the filtering error as

$$v_F(t) = x(t) - \hat{x}(t|t). \tag{9.100}$$

The overall system generating $v_F(t)$ is constituted by three subsystems:

- The data generation mechanism (9.98), fed by $w(t)$ and generating $y(t)$.
- Filter (9.99), elaborating signal $y(t)$ to produce the estimate $\hat{x}(t|t)$.
- Error equation (9.100), which makes use of both (9.98) and (9.99).

This composite system is linear, with $w(t)$ as input and $v_F(t)$ as output. We shall denote it by S and its transfer matrix by $N(z)$.

As usual, the objective is to minimize the estimation error. In the previous sections of this chapter, we have posed the problem in a Bayesian frame and developed the Kalman filter. However, the basic requirement of reducing as much as possible the effect of disturbance $w(t)$ on error $v_F(t)$ can also be tackled as an optimization problem aiming at choosing matrices A, B, C and D

so as to minimize – so to say – the "size" of the overall composite system S. This can be precisely stated as the problem of *minimizing the norm of S*. In this way, one can disregard the nature of disturbance $w(t)$, be it stochastic or deterministic.

Remark 9.18 By comparing the model used in Kalman theory, namely (9.1a) and (9.1b), with the model (9.98a) and (9.98b) now adopted, we see that one can reduce one model to the other one by posing

$$v_1(t) = Gw(t), \quad v_2(t) = Ew(t).$$

Hence, if $w(t)$ is assumed to be a white noise with unitary variance, then the variance and cross-variance matrices of noises $v_1(t)$ and $v_2(t)$ are

$$V_1 = GG', \quad V_2 = EE', \quad V_{12} = GE'. \tag{9.101}$$

Note that the case of uncorrelated state and output disturbances in Kalman's approach corresponds now to condition $GE' = 0$. Moreover, the condition of positive definiteness of the matrix variance of the output disturbance corresponds to imposing $EE' > 0$.

This circle of ideas leads us to open a parenthesis to define the norm of a signal and the norm of a system.

9.12.1 Norm of a Dynamic System

Given a discrete time real function $f(t)$, its norm, or more precisely its two-norm, is defined as

$$\|f(t)\|_2 = \sqrt{\sum_t^{+\infty} f(t)^2},$$

if $f(t)$ is a scalar or as

$$\|f(t)\|_2 = \sqrt{\sum_t^{+\infty} f(t)'f(t)},$$

if $f(t)$ is a vector. In any case, the norm is a non-negative real number.

The set of functions $f(t)$ for which this quantity is finite is named l_2 *space*.

In the scalar case, if $f(t) = \text{imp}(t)$, then $\|f(t)\|_2 = 1$. If instead $f(t)$ is the step function (namely the function that is null, $\forall t < 0$ and equal to 1, $\forall t \geq 0$), then $\|f(t)\|_2 = \infty$. Hence, the impulse belongs to l_2, while the step does not. Incidentally, the example of the step function shows that there exist bounded signals with unbounded l_2 norm.

We now introduce the two-norm and the ∞-norm of a stable system with input $u(t)$ and output $y(t)$ by focusing on the single-input single-output case. In other words, both $u(t)$ and $y(t)$ are scalar signals. Note that, thanks to stability, if $u(t) \in l_2$, then $y(t) \in l_2$ as well. In particular, $y(t) \in l_2$ when $u(t) = imp(t)$.

Definition 9.9 (Two-norm of a system) The two-norm, $\|S\|_2$, of a generic system S is given by

$$\|S\|_2 = \|h(t)\|_2,$$

where $h(t)$ is the impulse response.

Roughly speaking, the squared two-norm is the "energy" of its impulse response.

Passing now to ∞-norm, we start by noting that, as mentioned above, a stable system subject to an input $u(t) \in l_2$ generates an output that is also a signal in l_2. The quotient between the norm of the output and that of the input can be seen as an amplification coefficient, characterizing the entity of the effect of the input on the output. When the input is the impulse, this coefficient is the two-norm $\|S\|_2$. However, when the input is another signal, the amplification coefficient is different. To describe the effect of a *set of input signals*, rather than that of a unique special signal (the impulse), we introduce the following notion.

Definition 9.10 (∞-Norm of a system) The quantity

$$\|S\|_\infty = \sup_{u(\cdot)\in l_2,\, u(\cdot)\neq 0} \frac{\|y(t)\|_2}{\|u(t)\|_2}$$

is named ∞-norm of system S.

We can interpret the ∞-norm as the largest input-output amplification coefficient of the system for a set of input signals constituted by all functions $u(\cdot)$ with limited two-norm.

For a linear system with transfer matrix $N(z)$, these norms are also denoted with the symbols

$$\|N(z)\|_2, \quad \|N(z)\|_\infty,$$

respectively.

Remark 9.19 The above definitions can be generalized to the multi-input multi-output case. For example, if there are p outputs and m inputs, we can consider an experiment with input $u(t)$ having as entries $u_i(t) = imp(t)$ and $u_j(t) = 0$, $\forall j \neq i$, where $i = 1, 2, \ldots, m$. The corresponding output (a vector with p entries) is denoted as $h^{(i)}(t)$. The two-norm is then defined as

$$\|S\|_2 = \sqrt{\sum_{i}^{p} \|h^{(i)}(t)\|_2{}^2}.$$

The two-norm and the ∞-norm admit a frequency-domain interpretation, thanks to a celebrated result known as *Plancherel theorem* (or sometimes *Parseval* theorem).

To state it, we begin by recalling the notion of *frequency response*, see Section A.5 of Appendix A. Given a stable discrete-time scalar system with transfer function $N(z)$ subject to a sinusoidal input $u(t) = A \sin(\bar{\omega}t + \alpha)$, the output tends to a sinusoid with the same frequency of the input: $y(t) = B \sin(\bar{\omega}t + \beta)$. The amplitude B is given by

$$B = A|N(e^{j\bar{\omega}})|.$$

$|N(e^{j\omega})|$, seen as a function of angular frequency ω, is called the *frequency response*. It supplies the input-output amplification for sinusoidal signals at various frequencies.

The *Plancherel theorem* provides an expression of the two-norm and the ∞-norm in terms of frequency response. To be precise

$$\|N(z)\|_2 = \left[\frac{1}{2\pi} \int_{-\pi}^{+\pi} |N(e^{j\omega})|^2 d\omega \right]^{\frac{1}{2}},$$

$$\|N(z)\|_\infty = \left[\sup_{-\pi \leq \omega \leq +\pi} |N(e^{j\omega})|^2 \right]^{\frac{1}{2}}.$$

Thus, the ∞-norm is the ordinate of the peak of the frequency response, whereas the two-norm depends upon the area beneath the entire frequency response.

For multi-input multi-output systems,

$$\|N(z)\|_2 = \left[\frac{1}{2\pi} \int_{-\pi}^{+\pi} \operatorname{tr}(N(e^{j\omega})'N(e^{-j\omega})) d\omega \right]^{\frac{1}{2}},$$

$$\|N(z)\|_\infty = \left[\sup_{-\pi < \omega < +\pi} \lambda_{\text{MAX}}(N(e^{j\omega})'N(e^{-j\omega})) \right]^{\frac{1}{2}},$$

where $\lambda_{\text{MAX}}(\cdot)$ denotes maximum eigenvalue.

9.12.2 Robust Filtering

We are now in a position to deal with the filtering problem as the problem of minimizing the norm of system S, relating disturbance $w(t)$ to the estimation error $v_F(t)$.

If we adopt the two-norm, then it can be shown that the optimal filter coincides with that we have seen in the Bayesian approach, provided that as variance and cross-variance of the state and output disturbances v_1 and v_2 we adopt the values given in formulas (9.101).

If instead we adopt the ∞-norm, recalling Definition 9.10, the problem is to *minimize* the *maximum effect* of disturbance $w(\cdot)$ on error $v_F(\cdot)$ *for all* possible signals $w(\cdot) \in l_2$. The worst $w(\cdot)$ is that producing the maximum norm of $v_F(\cdot)$. This is a *min-max* (or *minimax*) problem.

As a matter of fact, a wise way of posing such problem is to find a filter (if any) such that the ∞-norm is "sufficiently small," lower than a given threshold, say γ:

$$\|S\|_\infty < \gamma. \tag{9.102}$$

The value of γ is a fundamental choice in this framework. Indeed, condition (9.102) imposes a limit to the impact of all possible disturbance signals (in l_2) on the estimation error. Obviously, the smaller the value of γ, the more demanding the problem becomes. Indeed, choosing a smaller γ means imposing a higher degree of precision in the estimation.

A basic result can be stated as follows. Assuming $GE' = 0$ and $EE' > 0$, suppose that there exists a positive definite solution of the equation

$$P = F \left[I + P \left(H'(EE')^{-1}H - \frac{1}{\gamma^2}I \right) \right]^{-1} PF' + GG' \tag{9.103}$$

such that inequality

$$P^{-1} + H'(EE')^{-1}H - \frac{1}{\gamma^2}I > 0 \tag{9.104}$$

holds and matrix

$$F \left[P^{-1} + H'(EE')^{-1}H - \frac{1}{\gamma^2}I \right]^{-1} P^{-1} \tag{9.105}$$

is stable. This last condition is the stability condition of the filtering error system associated with the worst-case disturbance, while (9.104) is named feasibility condition.

Then, it can be shown that a robust filter satisfying inequality (9.102) exists and is given by

$$\tilde{x}(t+1) = F\tilde{x}(t) + K_1(y(t) - H\tilde{x}(t)), \tag{9.106a}$$
$$\hat{x}(t|t) = \tilde{x}(t) + K_2(y(t) - H\tilde{x}(t)), \tag{9.106b}$$

with gains K_1 and K_2 given by

$$K_2 = PH'(HPH' + EE')^{-1}, \quad K_1 = FK_2. \tag{9.107}$$

The theory of optimal filtering in the H_∞ context is named *robust filtering*; correspondingly, Eq. (9.103) is referred to as *robust Riccati equation*.

The analogy between Eqs. (9.106) of the robust filter and Eqs. (9.54) for the optimal filter is striking.

Remark 9.20 (Robust filtering for $\gamma \to \infty$)   When $\gamma \to \infty$, algebraic equation (9.103) becomes

$$P = F[I + PH'(EE')^{-1}H]^{-1}PF' + GG'. \tag{9.108}$$

On the other hand, thanks to the *matrix inversion lemma* (see Appendix B),

$$(I + PH'(EE')^{-1}H)^{-1} = I - PH'(EE' + HPH')^{-1}H,$$

so that expression (9.108) can be equivalently written as

$$P = FPF' - FPH'(HPH' + EE')^{-1}HPF' + GG'.$$

This is precisely the ARE of the Kalman approach (provided we set $V_1 = GG'$ and $V_2 = EE'$ as in (9.101)).

Remark 9.21 (On the feasibility condition)   The left-hand side of the feasibility condition (9.104) is the sum of a positive definite matrix, $P^{-1} + H'(EE')^{-1}H$, and a negative definite matrix, $-\frac{1}{\gamma^2}I$. For high values of γ, we expect that the feasibility and stability conditions are satisfied; the minimum value of γ is identified as the largest γ for which one of the two conditions of feasibility and stability begins to be violated.

Remark 9.22 (Partial state estimation)   In our formulation, the state estimation was posed as the problem of determining the state $x(t)$ of a system from the observation of the output. More in general, one can consider the problem of determining some entries of the state vector or a combination of them. A possibility is then to define as unknown

$$x_r(t) = Lx(t),$$

where L is a given matrix. Obviously, when $L = I$, the full state estimation problem is recovered.

In the framework of classical Kalman filter theory, the optimal filter of $x_r(t) = Lx(t)$ is simply given by $\hat{x}_r(t|t) = L\hat{x}(t|t)$, where $\hat{x}(t|t)$ is the full state estimate. However, if one approaches the problem in the ∞-norm framework, then also the Riccati equation has to be modified. To be precise, in place of (9.103) one has to consider

$$P = F\left[I + P\left(H'(EE')^{-1}H - \frac{1}{\gamma^2}L'L\right)\right]^{-1}PF' + GG'$$

as robust Riccati equation, while the feasibility condition becomes

$$P^{-1} + H'(EE')^{-1}H - \frac{1}{\gamma^2}L'L > 0.$$

Finally, one has to impose the stability of

$$F\left[P^{-1} + H'(EE')^{-1}H - \frac{1}{\gamma^2}L'L\right]^{-1}P^{-1}. \tag{9.109}$$

The circle of ideas on robust filtering, concisely presented in this section, can be widely extended to include also modeling uncertainties. For example, one can build a theory for systems with a matrix F whose entries are uncertain (with some norm constraints). For such extensions, the readers are referred to the literature, in particular to the books Basar and Bernhard, (1991), and Lewis et al. (2008).

10

Parameter Identification in a Given Model

10.1 Introduction

In this chapter, we deal with the problem of estimating an unknown parameter in a given dynamical model describing a real system. Such a problem is very frequently encountered in a huge variety of applications. Here, we investigate how to resort to the observation of the measurable external variables affecting the systems to find the value for the missing parameter.

In analogy with previous chapters, the unknown parameter vector is denoted by θ and its dimension by q. The observations of the input and output signals, $u(t)$ and $y(t)$, over the interval $[1, 2, \ldots, N]$ form the set of data: $D^N = \{y(1), u(1), y(2), u(2), \ldots, y(N), u(N)\}$.

This situation is summarized in Figure 10.1, where $P(\theta)$ is the system affected by the unknown parameter θ and $e(t)$ denotes the set of non-measurable exogenous signals influencing the system.

We assume that an exact mathematical model (and a corresponding simulator) for $P(\theta)$ is available; only parameter θ is unknown and its value must be retrieved from data. This framework is known as *white box identification*.

Conceptually, the estimation problem is solved by introducing a suitable function $\widehat{f} : \mathbb{R}^{2N} \to \mathbb{R}^q$ mapping the observations D^N into an estimate $\widehat{\theta} := \widehat{f}(y(1), u(1), \ldots, y(N), u(N))$, for the true parameter θ.

Finding such function explicitly, however, is an extremely hard problem. In the sequel, we will see some methods and assess their validity via a simulation example.

10.2 Kalman Filter-Based Approaches

To estimate the unknown parameter, a possibility is to resort to Kalman filtering: The basic idea is to deal with parameter θ as a new state variable. This can be achieved by introducing an additional state equation of the type:

Model Identification and Data Analysis, First Edition. Sergio Bittanti.
© 2019 John Wiley & Sons, Inc. Published 2019 by John Wiley & Sons, Inc.

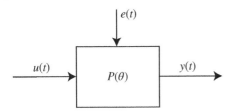

Figure 10.1 The parameter estimation problem.

$\theta(k+1) = \theta(k) + w(k)$, where $w(\cdot)$ is white noise with suitable variance. Such fake noise is introduced to let the estimate $\hat{\theta}$ of the parameter more responsive to observed data and improve convergence.

When the parameter estimation problem is reformulated as a state prediction problem, function $\hat{f}$ mapping the data into the estimate is *implicitly* defined by the Kalman filter equations.

This procedure is illustrated in the following example.

Example 10.1 (Parameter estimation as a state estimation problem) Consider the following data-generation mechanism:

$$x_1(t+1) = \theta \cdot x_1(t) + v_{11}(t) \tag{10.1a}$$

$$x_2(t+1) = x_1(t) + \theta^2 \cdot x_2(t) + v_{12}(t) \tag{10.1b}$$

$$y(t) = \theta \cdot x_1(t) + x_2(t) + v_2(t), \tag{10.1c}$$

where v_{11}, v_{12}, and v_2 are uncorrelated white noises with zero mean value.
θ is an unknown real parameter belonging to the range $[-0.9, 0.9]$.

Note that Eqs. (10.1a) and (10.1b) constitute a linear system with eigenvalues θ and θ^2. Hence, for each $\theta \in [-0.9, 0.9]$, such system is stable, and signals $x_1(\cdot)$, $x_2(\cdot)$, and $y(\cdot)$ are stationary processes (with zero mean value).

We can pose the problem of estimating θ in a state space framework by rewriting the original system (10.1) as

$$x_1(t+1) = x_3(t) \cdot x_1(t) + v_{11}(t) \tag{10.2a}$$

$$x_2(t+1) = x_1(t) + x_3(t)^2 \cdot x_2(t) + v_{12}(t) \tag{10.2b}$$

$$x_3(t+1) = x_3(t) + w(t) \tag{10.2c}$$

$$y(t) = x_3(t) \cdot x_1(t) + x_2(t) + v_2(t), \tag{10.2d}$$

where $x_3(t)$ is an additional state variable representing parameter θ and the fake noise $w(t)$ has characteristics to be chosen by the designer.

This system has an extended state vector:

$$x_E(t) = [x_1(t) \, x_2(t) \, x_3(t)]'.$$

By means of the available N data, one can find the optimal Kalman predictor $\hat{x}_E(N|N-1)$ or the optimal Kalman filter $\hat{x}_E(N|N)$; the third entry of such vector will supply the parameter estimate $\hat{x}_3$.

Observe that, even if the original system (10.1) was linear, the new system (10.2) contains the product of state variables and therefore is *nonlinear*. The state estimation problem calls therefore for a nonlinear estimation method such as the extended Kalman filter (EKF) introduced in Section 9.11.

The study of the performance of the EKF as parameter estimator is discussed in various papers. In particular, Reif and Unbehauen (1999), it is shown that the estimation error remains bounded assuming that the system satisfies a nonlinear observability condition and the initial estimation error as well as the disturbing noise terms are small enough. In other words, only local convergence holds. In general, the obtained estimate may be far from the correct value as shown in the following example.

Example 10.2 (Example 10.1 continued)   Consider again the data-generation mechanism of the previous example and let $v_{11} \sim \mathrm{WN}(0,1)$, $v_{12} \sim \mathrm{WN}(0,1)$, $v_2 \sim \mathrm{WN}(0,0.01)$ be mutually uncorrelated.

For the estimation, the EKF technique was adopted. To assess the performance of this approach, 800 values for the parameter θ were extracted, with a uniform distribution in the interval $[-0.9, 0.9]$. For each extracted value, $N = 1000$ observations of the output variable $y(\cdot)$ were generated by simulation. Each time, the $N = 1000$ observations were made available to the estimation algorithms that returned an estimate of the corresponding θ. Thus, 800 estimates $\hat{\theta}$ were obtained and compared with the corresponding 800 true values of θ.

To set up the EKF technique, main issues are (i) tuning the variance of the fake noise and (ii) choosing the initialization of the Riccati equation.

In the performed simulations, many possibilities have been considered. Herein we focus on two cases. In both of them, $w(t) \sim \mathrm{WN}(0, 10^{-6})$ was taken as fake noise, while the initialization of the difference Riccati equation (DRE) was

$$P(1) = \begin{bmatrix} 10 & 0 & 0 \\ 0 & 10 & 0 \\ 0 & 0 & 2 \end{bmatrix} \tag{10.3}$$

in the first case and

$$P(1) = \begin{bmatrix} 1 & 0 & 0 \\ 0 & 1 & 0 \\ 0 & 0 & 10^{-2} \end{bmatrix} \tag{10.4}$$

in the second one.

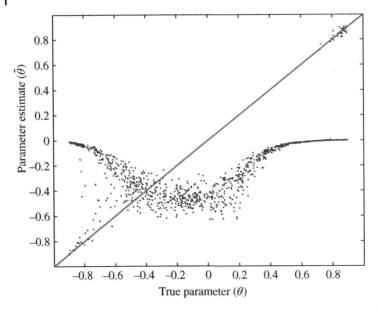

Figure 10.2 Estimates of the unknown parameter with the EKF method. Example 10.2 with initialization of the Riccati equation given by Eq. (10.3)

As parameter estimate $\hat{\theta} = \hat{x}_3(1001|1000)$ was adopted.

In Figures 10.2 and 10.3, the estimates obtained this way are depicted. As it is apparent, the estimation error $\hat{\theta} - \theta$ may be large.

A number of variants on the EKF technique have been studied to ensure better convergence properties. One of them is the so-called *particle filter* (PF) method, where unknown θ is seen as a random vector described with its probability distribution. According to such distribution, a cloud of possible parameter values are chosen. By letting the system equation evolve with such parameter values, the *a posteriori* probability distribution of θ can be constructed. The procedure is repeated with the new distribution, and then iterated up to convergence. In the article by Crisan and Doucet (2002), it is shown that, under suitable assumptions, the PF estimate converges indeed to the true parameter, and this is one reason for the increasing popularity of the PF filter. On the other hand, this estimation algorithm requires an intensive simulation effort, and it is, therefore, computationally demanding.

10.3 Two-Stage Method

The *two-stage* method, first proposed in the paper (Garatti and Bittanti 2008), operates by constructing the function $\hat{f}$, which maps input/output data

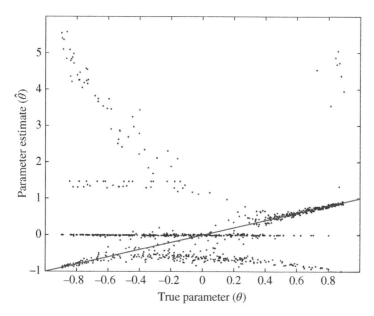

Figure 10.3 Estimates of the unknown parameter with the EKF method. Example 10.2 with initialization of the Riccati equation given by Eq. (10.4).

sequences D^N into the corresponding parameter θ affecting the model, namely $\widehat{f} : D^N \rightarrow \theta$. Then, the value of the unknown parameter corresponding to a certain set $\bar{D}^N$ of observations can be immediately computed as $\widehat{f}(\bar{D}^N)$.

To determine $\widehat{f}$, the map $\widehat{g}$ associating to each value of the parameter a sequence of data is preliminarily determined by intensive simulation trials (using the plant model): $\widehat{g} : \theta \rightarrow D^N$.

Function $\widehat{f}$ is the inverse of $\widehat{g}$. The *two-stage* method enables to determine $\widehat{f}$ from $\widehat{g}$ with an intelligent inversion procedure articulated in two phases as described in the sequel.

10.3.1 First Stage – Data Generation and Compression

The simulator is extensively used to generate a large number of input/output data each of which is associated to a certain value of the unknown parameter θ. For each simulation, the corresponding input–output data sequences are recorded and set into a table called *simulated data chart*. To be precise, we collect N measurements

$$D_1^N = \{y^1(1), u^1(1), \dots, y^1(N), u^1(N)\}$$

for $\theta = \theta_1$; N measurements

$$D_2^N = \{y^2(1), u^2(1), \dots, y^2(N), u^2(N)\}$$

Table 10.1 The simulated data chart as the starting point of the two-stage method.

θ_1	$D_1^N = \{y^1(1), u^1(1), \dots, y^1(N), u^1(N)\}$
θ_2	$D_2^N = \{y^2(1), u^2(1), \dots, y^2(N), u^2(N)\}$
$\vdots$	$\vdots$
θ_m	$D_m^N = \{y^m(1), u^m(1), \dots, y^m(N), u^m(N)\}$

for $\theta = \theta_2$; and so on. By repeated simulation experiments, one can work out a set of, say m, pairs $\{\theta_i, D_i^N\}$ as summarized in Table 10.1, the *simulated data chart*.

This chart provides m samples of map $\hat{g}$, namely $\{\hat{g}(\theta_1), \hat{g}(\theta_2), \dots, \hat{g}(\theta_m)\}$.

In principle, from the simulated data chart, the map associating a parameter to a set of data, $\hat{f} : \mathbb{R}^{2N} \to \mathbb{R}^q$, can be reconstructed by solving the optimization problem:

$$\hat{f} \leftarrow \min_f \frac{1}{m} \sum_{i=1}^{m} \left\| \theta_i - f(y^i(1), u^i(1), \dots, y^i(N), u^i(N)) \right\|^2. \tag{10.5}$$

However, solving (10.5) is a critical optimization problem when the entries of function $f(\cdot)$, namely the number of data, is very large, as often happens.

To pass to an affordable problem, the information conveyed by sequences D_i^N is compressed into sequences $\tilde{D}_i^n$ of reduced dimensionality. This can be achieved by fitting a simple model to each sequence D_i^N and then taking the parameters of this model, say $\alpha_1^i, \alpha_2^i, \dots, \alpha_n^i$, as compressed sequences, i.e. $\tilde{D}_i^n = \{\alpha_1^i, \dots, \alpha_n^i\}$. In this way, a new data chart can be worked out, constituted by the pairs $\{\theta_i, \tilde{D}_i^n\}$, $i = 1, \dots, m$, as in Table 10.2. This is named *compressed artificial data chart*.

While in D_i^N the information on the unknown parameter θ_i is scattered in a long sequence of N samples, in the new compressed data $\tilde{D}_i^n$, such information is condensed in a short sequence of n samples ($n \ll N$).

Table 10.2 The compressed artificial data chart.

θ_1	$\tilde{D}_1^n = \{\alpha_1^1, \dots, \alpha_n^1\}$
θ_2	$\tilde{D}_2^n = \{\alpha_1^2, \dots, \alpha_n^2\}$
$\vdots$	$\vdots$
θ_m	$\tilde{D}_m^n = \{\alpha_1^m, \dots, \alpha_n^m\}$

Remark 10.1 (Turing and the Enigma machine) The information compression described above is reminiscent of the celebrated problem of deciphering the *Enigma* code. The Enigma was a type of encryption machine used by the German armed forces to transmit secret messages during World War II. An operator typed in a message, then scrambled it by means of some notched wheels, which automatically displayed different letters of the alphabet. By knowing the exact setting of these wheels, the receiver was able to reconstruct the original message. The UK intelligence assigned Alan Turing (1912–1954) the crucial mission of cracking the code. Turing set up a special computer (called *Bombe*) for that purpose. However, deciphering seemed an impossible task, since the number of characters that made up each message allowed for an astronomical number of combinations and thus a nearly limitless number of possible phrases. Turing's winning idea was to focus on a portion of the message. As the portion was made up of a limited number of characters, the number of possible combinations of words was enormously lower. Concision of information was the key to the solution.

10.3.2 Second Stage – Compressed Data Fitting

Once the compressed artificial data chart has been worked out, problem (10.5) is reformulated in a new one, that of finding a map $\widehat{h} : \mathbb{R}^n \to \mathbb{R}^q$, which fits the set of m compressed artificial observations to the corresponding parameter vectors, i.e.

$$\widehat{h} \leftarrow \min_h \frac{1}{m} \sum_{i=1}^{m} \left\| \theta_i - h(\alpha_1^i, \dots, \alpha_n^i) \right\|^2. \tag{10.6}$$

n being small, the determination of the optimal function $\widehat{h}$ is no more an issue.

As for the choice of h, one can select a linear function: $h(\alpha_1^i, \dots, \alpha_n^i) = c_1 \alpha_1^i + \cdots + c_n \alpha_n^i$, $c_i \in \mathbb{R}^q$, i.e. each component of h is just a linear combination of the compressed artificial data $\alpha_1^i, \dots, \alpha_n^i$. As in any linear regression, the parameters c_i appearing here can be computed via least squares, at a low computational cost. Of course such a way of parameterizing h is computationally cheap but possibly loose. Better results are obtained by choosing a class of complex (nonlinear) functions.

The splitting of the map determination into two phases is the key to transform a numerically intractable problem into an affordable one.

Remark 10.2 (Physical interpretation of the artificial data) While $P(\theta)$ is a mathematical description of a real plant, the simple model class selected to produce the compressed artificial data does not have physical meaning; this class plays a purely instrumental and intermediary role in the process of bringing into light the hidden relationship between the unknown parameter and the collected data. In this connection, we observe that the choice of the ARX model order is not a critical issue. Anyhow, one can resort to appropriate complexity selection methods such as the FPE, AIC, or MDL criteria introduced in Chapter 6.

Remark 10.3 (Nonlinearity in estimation)   Suppose that both in the first stage and in the second one, a linear parametrization is used. In other words: in the first stage, the simple class of models is the ARX one, and in the second stage, a linear regression of the compressed artificial data sequences is used. Even in such a case, the final estimation rule is nonlinear. Indeed, the generation of the compressed artificial data in the first stage requires the use of the LS algorithm applied to the simulated data sequences $D^N i$, and this is by itself a nonlinear manipulation of data.

As a matter of fact, in some cases, such nonlinearity limited to the first stage of elaboration suffices for capturing the relationship between the unknown θ and the data $y(1), u(1), \dots, y(N), u(N)$. In other cases, instead, introducing also a nonlinearity in the second stage (namely, taking h as a nonlinearly parameterized function of the compressed artificial data) is advisable to achieve better global results.

Remark 10.4 (Multivalue estimation)   The two-stage approach relies on intensive simulations of the plant model and this fact can be computationally demanding. Yet, differently from other approaches, all these simulations have to performed once for all, off line. The result then is the relation $\widehat{f}$, which can be easily applied over and over, for estimating all possible values of θ without any supervision from a human operator. This is quite different from the estimation process to be faced when using the EKF method for parameter estimation. Indeed, the EKF approach requires the construction of a special state equation to transform the unknown parameter into a state of an extended dynamical system. This special equation is affected by a special noise (fake noise), the variance of which is normally unknown. This leads to the problem of tuning the fake noise variance, which is a main problem in the EKF framework. Even more so, such variance has to be suitably re-tuned whenever the model is called to operate with a different parametrization θ. Thus, the two-stage approach is better suited for multivalue estimation.

Example 10.3 (Example 10.1 further continued)   The two-stage method has been applied to system (10.1) by considering $m = 500$ *new* values of θ uniformly extracted from the interval $[-0.9, 0.9]$; correspondingly, 500 sequences of 1000 output values were simulated so as to construct the *simulated data chart*.

For each sequence $y^i(1), \dots, y^i(1000)$, $i = 1, \dots, 500$, the corresponding compressed artificial data set was obtained by identifying through the least

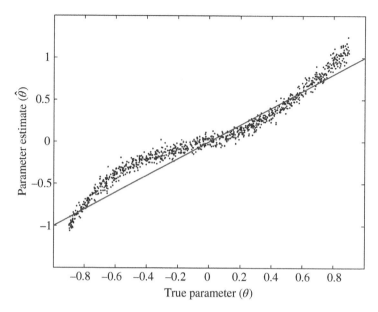

Figure 10.4 Estimates of the unknown parameter of system (10.1) with the two-stage method.

squares algorithm the coefficients $\alpha_1^i, \dots, \alpha_5^i$ of an autoregressive model of order 5 ($y^i(t) = \alpha_1^i y(t-1) + \cdots + \alpha_5^i y(t-5)$).

Estimator $\widehat{h}(\alpha_1^i, \dots, \alpha_5^i)$ was computed by resorting to a linear parametrization ($h = c_1 \alpha_1^i + \cdots + c_5 \alpha_5^i$), with coefficients $c_1, \dots, c_5$ estimated again by the least squares algorithm.

The so-achieved results are depicted in Figure 10.4, clearly showing the improvement with respect to the Kalman filtering method used in the previous example. Even better results are achievable by resorting to a more complex (nonlinear, neural networks-based) function $h(\cdot)$, as discussed in Garatti and Bittanti (2012).

11

Case Studies

11.1 Introduction

The methods of prediction and identification explained in this book have a huge variety of applications.

Here, we focus on two case studies. The first one is the analysis of the data of an earthquake that took place in Kobe (Japan) in 1995. The data cover various phases and can be distinguished in three segments; the first one corresponds to a normal seismic activity, while the third one is the earthquake phase. In between there is a transition phase on which our attention focuses in order to find some feature hidden in data to detect the occurrence of the earthquake. In this case study, input–output models will be used, with focus on parameter estimation and model complexity selection.

The second case study concerns the estimation of the frequency of a sinusoid from noisy measurements. The problem will be tackled both with a prediction error identification method based on input–output models and by extended Kalman filtering state space techniques.

11.2 Kobe Earthquake Data Analysis

Kobe earthquake occurred on 16 January 1995, at 20:46:49 (UTC) and measured 6.8 on the moment magnitude scale. Our analysis is based on the data collected by a seismograph located at the University of Tasmania, Hobart, Australia. Measurements began at 20:56:51 (UTC), with a sampling interval $T = 1$s, and lasted for 3000 seconds (about 51 minutes). The corresponding time series is depicted in Figure 11.1, consisting of the measurements of the unbiased vertical acceleration in nm s^{-2} of the earth ground.

As it appears from the data, seismic waves showed up 1700 seconds after the beginning of data recording, that is, at time 21:25:10 (UTC). The 40-minute delay with respect to the earthquake occurrence at 20:46:49 (UTC) due to the

Model Identification and Data Analysis, First Edition. Sergio Bittanti.
© 2019 John Wiley & Sons, Inc. Published 2019 by John Wiley & Sons, Inc.

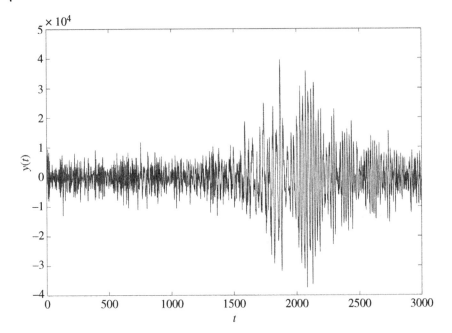

Figure 11.1 The time series $y(t)$ of Kobe earthquake.

8600 km distance between Kobe and Hobart. It corresponds to a propagation speed of about 3580 m s^{-1}, a typical velocity of seismic waves.

In the following, the time series will be denoted by $y(t)$, $t = 1, 2, \ldots, 3000$. See Figure 11.1 for a plot of $y(t)$.

Not surprisingly, due to the earthquake effect, the time series cannot be thought of as a realization of a stationary stochastic process (variance at least is time varying). For this reason, the time series has been partitioned into three segments as shown in Figure 11.2. The first segment for $t = 1, 2, \ldots, 1200$ is labeled as *normal seismic activity*, since seismic waves have not arrived yet, while the third segment for $t = 1601, 1602, \ldots, 3000$ is labeled as *earthquake* for the very opposite reason. The second segment for $t = 1201, 1202, \ldots, 1600$, instead, represents a *transition phase* between the normal seismic activity and the earthquake phase.

As said at the beginning, the measured vertical acceleration y is unbiased (zero mean); hence, the empirical variances in the three segments can be estimated as

- first segment (*normal seismic activity*)

$$\frac{1}{1200} \sum_{t=1}^{1200} y(t)^2 = 1.15 \times 10^7 (\text{nm s}^{-2})^2,$$

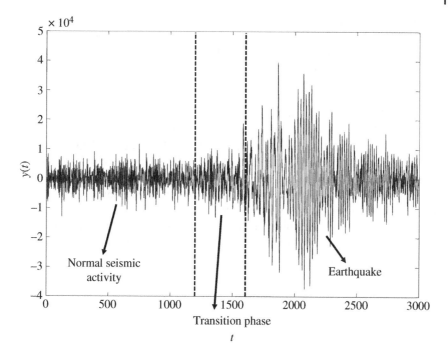

Figure 11.2 Partition of the time series $y(t)$ into three segments.

- second segment (*transition phase*)

$$\frac{1}{400} \sum_{t=1201}^{1600} y(t)^2 = 2.5 \times 10^7 (\text{nm s}^{-2})^2,$$

- third segment (*earthquake*)

$$\frac{1}{1400} \sum_{t=1601}^{3000} y(t)^2 = 1.12 \times 10^8 (\text{nm s}^{-2})^2.$$

Thus, we see that, during the transition phase, the magnitude of the oscillations is not very different from that relative to the normal seismic activity phase. During the earthquake segment, instead, the variance is 1 order of magnitude bigger.

The main question we want to address is whether it is possible to discover some special feature during the transition phase, which can be taken as a warning sign of the earthquake occurrence. As a matter of fact, we will show that the transition phase and the normal seismic activity period are radically different in terms of the underlying generation mechanism. In fact, after performing model identification for the normal activity, we will see that the identified model is subject to a degradation of its prediction capabilities during the transition phase. This fact clearly denotes a deviation of the transition snapshots from normal seismic oscillations. Such feature is a much more clear indicator than the modest diversity in the intensity of the oscillations during the two segments.

The study is in three steps. First, by prediction error techniques, we identify a model for the normal seismic activity segment. Then, we validate such model by checking its performance over the earthquake segment. Finally, we apply the model to the transition data and we will see a clear degradation of the model performance during such phase. This permits one to discern the transition phase from the normal seismic activity 200–300 seconds prior to the beginning of the earthquake. Hence, this degradation could be seen as a warning sign (albeit with very little notice).

This case study on the Kobe earthquake was reported in the paper (Garatti and Bittanti, 2014). Some figures are published here upon permission of the conference Chairman Prof. Ignacio Rojas.

11.2.1 Modeling the Normal Seismic Activity Data

For the first segment of data, Figure 11.3 displays the periodogram

$$\hat{\Gamma}_y(\omega) = \frac{1}{1200} \left| \sum_{t=1}^{1200} y(t) e^{-j\omega t} \right|^2,$$

the empirical covariance function

$$\hat{\gamma}_y(\tau) = \frac{1}{1200 - \tau} \sum_{t=1}^{1200-\tau} y(t) y(t+\tau),$$

and the partial covariance function denoted as PARCOV(k) (computed with the Durbin Levinson algorithm).

As can be seen, both the empirical covariance function and the partial covariance function are not null after a finite number of time lags, not even approximately. This suggests that autoregressive or moving average models are not appropriate. We then consider the time series as the realization of an ARMA(n_a, n_c) stochastic linear model:

$$y(t) = \frac{C(z)}{A(z)} e(t) = \frac{1 + c_1 z^{-1} + \cdots + c_{n_c} z^{-n_c}}{1 + a_1 z^{-1} + \cdots + a_{n_a} z^{-n_a}} e(t), \quad e(t) \sim \text{WN}(0, \lambda^2).$$

$$(11.1)$$

For fixed model orders n_a, n_c, the model parameters are estimated via the prediction error minimization rationale. That is, letting

$$\theta = [a_1 \ \cdots \ a_{n_a} \ c_1 \cdots \ c_{n_c}]',$$

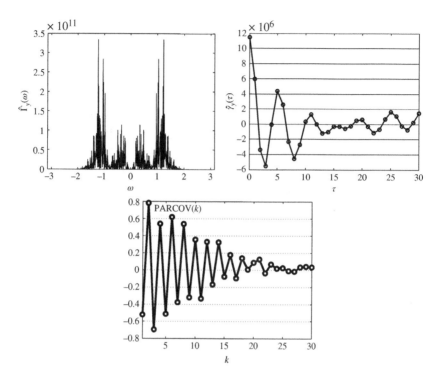

Figure 11.3 Nonparametric properties of $y(t)$ over the normal seismic activity segment.

the optimal parameter vector is obtained as the minimizer of

$$\hat{\theta}_{1200}^{n_a,n_c} := \arg\min_{\theta} \frac{1}{1200} \sum_{t=1}^{1200} (y(t) - \hat{y}(t|t-1,\theta))^2, \tag{11.2}$$

where

$$\hat{y}(t|t-1,\theta) = \frac{C(z) - A(z)}{C(z)} y(t)$$

is the one-step optimal predictor for model (11.1). The value $\hat{\theta}_{1200}^{n_a,n_c}$ can be actually computed based on the Newton-like minimization technique explained in Section 5.4.

As for the optimal model orders, we let n_a and n_c range from 1 up to 12 and compute $\hat{\theta}_{1200}^{n_a,n_c}$ for the various n_a and n_c according to (11.2). For each identified model, the minimum description length (MDL) indicator is computed as

$$\text{MDL}(n_a, n_c) = \ln(1200)\frac{n_a + n_c}{1200} + \ln\left(\frac{1}{1200} \sum_{t=1}^{1200} \left(y(t) - \hat{y}\left(t|t-1, \hat{\theta}_{1200}^{n_a,n_c}\right)\right)^2\right).$$

The optimal n_a, n_c are those returning the lowest value for MDL.

$n_a\backslash n_c$	1	2	3	4	5	6	7	8	9	10	11	12
1	14 967	14 153	13 635	13 538	13 252	13 365	13 547	13 860	13 631	14 110	14 357	14 018
2	14 010	13 317	13 087	13 262	13 306	13 462	13 640	13 840	13 929	14 150	14 357	14 541
3	13 563	13 162	13 082	13 050	13 308	13 378	13 658	13 759	13 987	14 204	14 383	14 458
4	13 378	13 145	13 105	13 100	13 133	13 405	13 538	13 791	13 982	14 212	14 389	14 241
5	13 214	13 193	13 236	13 249	13 192	13 202	13 568	13 841	14 016	14 226	14 413	14 432
6	13 104	13 394	13 210	13 359	13 315	13 475	13 682	13 847	14 041	14 222	14 079	14 153
7	13 212	13 576	13 417	13 516	13 481	13 712	13 576	13 807	12 124	14 166	12 330	14 070
8	13 370	13 695	13 671	13 821	13 775	13 735	13 781	12 130	12 137	12 141	14 086	12 160
9	12 301	13 640	12 132	12 136	12 141	12 129	12 120	12 139	12 142	12 152	12 144	12 160
10	12 253	12 245	12 144	12 148	12 152	12 118	12 143	12 137	12 134	12 136	12 156	12 160
11	12 236	12 198	12 151	12 139	12 150	12 137	12 142	12 144	12 136	12 149	12 154	12 162
12	12 232	12 180	12 159	12 159	12 155	12 160	12 165	12 169	12 174	12 147	12 154	12 160

Figure 11.4 MDL(n_a, n_c) for $n_a = 1, \ldots, 12$ and $n_c = 1, \ldots, 12$.

From the computed values of MDL reported in Figure 11.4, one can conclude that the minimum is achieved for $n_a = 7$ and $n_c = 9$. The identified model, corresponding to $\widehat{\theta}^{7,9}_{1200}$, is given by

$$
\begin{aligned}
A(z) &= 1 - 2.5z^{-1} + 2.8z^{-2} - 1.6z^{-3} + 0.7z^{-5} - 0.6z^{-6} + 0.2z^{-7}, \\
C(z) &= 1 + 1.1z^{-1} - 1.6z^{-2} - 1.7z^{-3} + 1.6z^{-4} + 1.2z^{-5} - 1.5z^{-6} \\
&\quad - 1.0z^{-7} + 0.6z^{-8} + 0.4z^{-9}, \\
\lambda^2 &= 1.59 \times 10^5 (\text{nm s}^{-2})^2.
\end{aligned}
\tag{11.3}
$$

Interestingly enough, the spectrum

$$
\Gamma_y(\omega) = \frac{|C(e^{-j\omega}, \widehat{\theta}^{7,9}_{1200})|^2}{|A(e^{-j\omega}, \widehat{\theta}^{7,9}_{1200})|^2} \lambda^2
$$

of the ARMA stochastic process generated by (11.3), depicted in Figure 11.5, is in full agreement with the periodogram of Figure 11.3.

The singularities of (11.3) are reported in Figure 11.5 (with poles as crosses and zeros as circles).

11.2.2 Model Validation

With the identified model (11.3), the predictor $\widehat{y}(t|t - 1, \widehat{\theta}^{7,9}_{1200})$ can be computed. Figure 11.6a depicts $y(t)$ versus $\widehat{y}(t|t - 1, \widehat{\theta}^{7,9}_{1200})$ over the normal seismic activity segment, i.e. for $t = 1, \ldots, 1200$. The corresponding prediction error $\epsilon(t, \widehat{\theta}^{7,9}_{1200}) = y(t) - \widehat{y}(t|t - 1, \widehat{\theta}^{7,9}_{1200})$ has empirical variance $\frac{1}{1200} \sum_{t=1}^{1200} \epsilon(t, \widehat{\theta}^{7,9}_{1200})^2 = \lambda^2 = 1.59 \times 10^5 (\text{nm s}^{-2})^2$, which is 2 order of magnitude smaller than the empirical variance of $y(t)$. Figure 11.6b depicts the

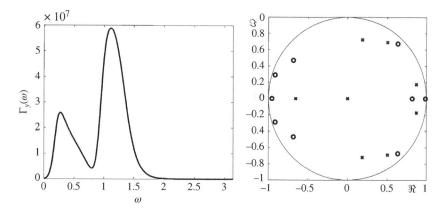

Figure 11.5 Spectrum and poles and zeros of the identified stochastic model.

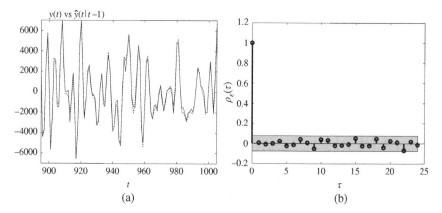

Figure 11.6 Prediction error and Anderson's whiteness test.

correlation coefficient $\hat{\rho}_\epsilon(\tau) = \frac{\hat{\gamma}_\epsilon(\tau)}{\hat{\gamma}_\epsilon(0)}$, along with the 95% confidence interval for the Anderson's whiteness test, introduced in Section 2.7.

Since all the displayed $\hat{\rho}_\epsilon(\tau)$, $\tau = 1, \ldots, 24$ are within the confidence interval, the whiteness test is passed and $\epsilon(t, \hat{\theta}_{1200}^{n_a, n_c})$ can be presumed to be a white noise.

Thus, altogether, the conclusion is drawn that the identified model is an appropriate descriptor of the data during the normal seismic activity.

We pass then to check whether the obtained model is also valid for the earthquake segment (ranging from $t = 1601$ to $t = 3000$). It is straightforward to verify that the answer is negative. Indeed, both the plot of $y(t)$, $t = 1601, \ldots, 3000$ (Figure 11.7a) and that of the corresponding periodogram (Figure 11.7b) are remarkably different from the analogous plots relative to the normal seismic

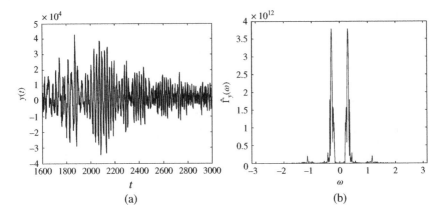

Figure 11.7 The earthquake phase segment (a) along with the corresponding periodogram (b).

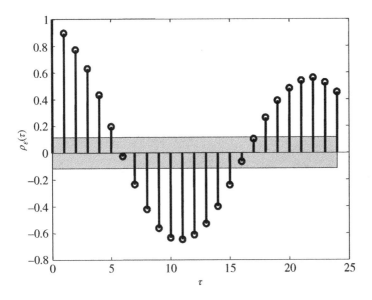

Figure 11.8 Whiteness test for the earthquake phase.

activity. This is confirmed by the whiteness test applied to the prediction error $\epsilon(t, \hat{\theta}_{1200}^{n_a, n_c})$ for $t = 1601, \ldots, 3000$, the result of which is shown in Figure 11.8 as obtained.

In conclusion, model (11.3) is not suitable for the earthquake phase. In other words, the earthquake phase differs from the normal seismic one not only because of the different amplitude of the oscillations measured by the seismograph but also because of a drastic change in dynamics of the underlying data generation mechanism.

This conclusion is negative and positive at the same time. It is negative because it states that model (11.3) cannot be used in the earthquake phase. But it is also positive in the sense that, with model (11.3), one can likely discern between the normal seismic activity and the earthquake activity. We will better elaborate this aspect in the next point.

11.2.3 Analysis of the Transition Phase via Detection Techniques

We now address the following main question: Is it possible to build some indicator to envisage the earthquake insurgence already in the transition phase?

We have seen that the amplitude of oscillations during the transition phase is not very different from that of the normal seismic activity. So, a premonition indicator based on the oscillation intensity is not so reliable.

We probe then an alternative procedure, borrowed from system detection technique ideas. Let's apply the normal activity model to the transition phase, by partitioning such phase into the four windows each of which is 100 seconds long (see Figure 11.9). Performing the whiteness test to each window, we see that the test is passed just in the first time window. Instead, the whiteness test is not passed in the other time windows, and this becomes clearer and clearer as windows get closer to the earthquake phase. The results are shown in Figure 11.10.

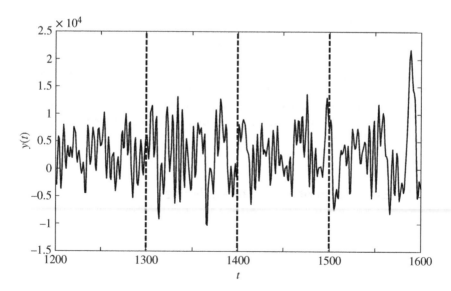

Figure 11.9 The transition phase segment and its further partition.

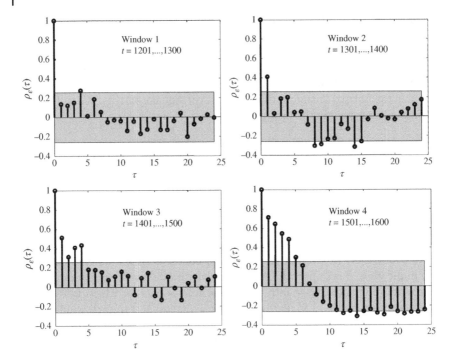

Figure 11.10 Whiteness test for each time window in the transition segment.

11.2.4 Conclusions

In the time series of data concerning the Kobe earthquake of 1995, it is possible to discern the transition phase from the normal seismic activity phase about 200–300 seconds before the beginning of the earthquake phase. This is achieved by monitoring the prediction capabilities of the model identified during the normal seismic activity when applied to the transition phase segment. In particular, the whiteness test for the prediction error exhibits a clear degradation of performance.

11.3 Estimation of a Sinusoid in Noise

Suppose to observe

$$y(t) = s(t) + n(t), \tag{11.4}$$

where

$$s(t) = A \cos(\Omega^0 t + \phi) \tag{11.5}$$

is a sinusoidal signal and $n(t)$ is a noise. How can we find Ω^o given a set $y(1), y(2), \ldots, y(N)$ of snapshots ? Moreover, if Ω^o is time varying, can we track it?

This question can be posed as follows: design a dynamical system (filter) providing at its output an estimate $\hat{\omega}$ of ω^o when fed by sequence $y(\cdot)$. We will tackle it by means of two possible rationales:

i) Choose the structure of the filter from the very beginning, a structure apt to detect a frequency peak in data. A typical choice is the so-called *notch filter*. Then, by resorting to prediction error identification methods, tune the parameters of the filter for the best identification of the angular frequency.
ii) Describe a sinusoidal signal with a state space model. The unknown angular frequency becomes a parameter in such a model. Then resort to the extended Kalman filter method to estimate the state.

11.3.1 Frequency Estimation by Notch Filter Design

Consider the transfer function

$$G(z, \Omega, \rho) = \frac{1 - 2\rho \cos(\Omega)z^{-1} + \rho^2 z^{-2}}{1 - 2\cos(\Omega)z^{-1} + z^{-2}}, \quad 0 < \rho < 1. \tag{11.6}$$

The two poles are on the unit circle at points $e^{\pm j\Omega}$. The zeros, at $\rho e^{\pm j\Omega}$, are radially aligned with the poles but lie inside the unit circle. Such peculiar location of the singularities produces a frequency response exhibiting a peak around Ω. The width of the peak is determined by the value of ρ: The frequency response becomes more and more peaked in the extent to which ρ is closer and closer to 1.

On the contrary, the inverse transfer function

$$F(z, \Omega, \rho) = G(z, \Omega, \rho)^{-1} = \frac{1 - 2\cos(\Omega)z^{-1} + z^{-2}}{1 - 2\rho\cos(\Omega)z^{-1} + \rho^2 z^{-2}} \tag{11.7}$$

has two zeros on the unit circle (see Figure 11.11), so that the corresponding frequency response is null at Ω, creating a sort of notch, as shown in Figure 11.12. We will call (11.6) the *notch model* and (11.7) the *notch filter*.

To estimate the frequency of the sinusoid embedded in noise, the basic idea is to see the signal as generated by system (11.6) when fed by white noise. The estimation is then posed as the problem of properly tuning the model parameters Ω and ρ.

This idea can be put into practice in a prediction error minimization framework as follows. Let

$$\hat{y}(t, \Omega, \rho)$$

be the one-step-ahead predictor of the output of model (11.6) with a white noise input. Correspondingly,

$$\epsilon(t, \Omega, \rho) = y(t) - \hat{y}(t, \Omega, \rho)$$

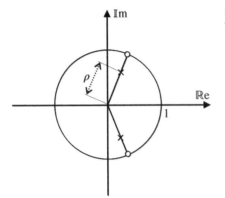

Figure 11.11 Poles and zeros of a notch filter.

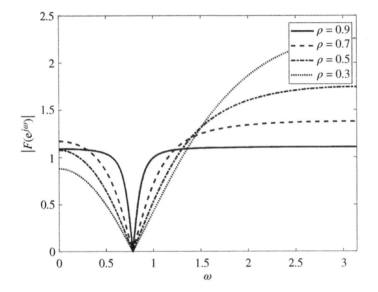

Figure 11.12 Frequency response of notch filter.

is the prediction error. The loss function

$$J(\Omega) = \lim_{N \to \infty} \frac{1}{N} \sum_{t}^{N} E[\epsilon(t, \Omega, \rho)^2]$$

is taken as an indicator of the resemblance of $\hat{y}$ to y: the lower the $J(\Omega)$, the better the adherence of the model to data. For a given ρ, the frequency estimate $\hat{\Omega}$ is obtained by minimizing $J(\Omega)$.

This rationale can be implemented by defining a loss function $J(\Omega)$ as a sample variance of the error. Such variance can be derived from the spectrum $\Gamma_{\epsilon\epsilon}(\omega)$

of $\epsilon(t, \Omega, \rho)$ as

$$J(\Omega) = \frac{1}{2\pi} \int_{-\pi}^{+\pi} \Gamma_{\epsilon\epsilon}(\omega) d\omega$$

(see Remark 5.3). In light of expression (11.4), $y(t)$ is constituted by a sinusoidal signal and a stationary stochastic process. Hence, its spectrum is given by the sum of two spectra:

$$\Gamma_{yy}(\omega) = \Gamma_{yy,s}(\omega) + \Gamma_{yy,n}(\omega).$$

The sinusoid (11.5) can be described in spectral terms by an impulsive function (delta function) located at $\pm\Omega^o$: $\frac{A}{2}\delta(\omega + \Omega^o) + \frac{A}{2}\delta(\omega - \Omega^o)$; the noise contribution is determined by its spectrum, a constant in case of a white noise.

Turn now to $\epsilon(t, \Omega, \rho)$. The error can be obtained by filtering the data through the notch filter (11.7):

$$\epsilon(t, \Omega, \rho) = F(z, \Omega, \rho)y(t).$$

Therefore,

$$J(\Omega) = J_s(\Omega) + J_n(\Omega), \tag{11.8}$$

with

$$J_s(\Omega) = \frac{A^2}{2\pi}|F(e^{j\Omega_o}, \Omega, \rho)|^2, \tag{11.9}$$

$$J_n(\Omega) = \frac{1}{2\pi} \int_{-\pi}^{+\pi} |F(e^{j\omega}, \Omega, \rho)|^2 \Gamma_{yy,n}(\omega) d\omega. \tag{11.10}$$

Now, from (11.7) and (11.9), an easy computation shows that

$$J_s(\Omega) = \alpha(\Omega, \rho)(\cos(\Omega^o) - \cos(\Omega))^2, \tag{11.11}$$

where the non-negative coefficient $\alpha(\Omega, \rho)$ does not depend on Ω_o. Hence, we conclude that the minimum of $J_s(\Omega)$ is in Ω^o.

Turning to $J_n(\Omega)$, from (11.10), we see that it does not depend upon Ω_o. Therefore, should it be constant (with respect to Ω), the minimum of the global loss function $J(\Omega) = J_s(\Omega) + J_n(\Omega)$ would coincide with the minimum of $J_s(\Omega)$, namely Ω_o. If instead $J_n(\Omega)$ is not a constant, the minimum of $J(\Omega)$ is pushed away from that of $J_s(\Omega)$.

If the noise $n(t)$ is white,

$$n(t) \sim WN(0, \lambda^2)$$

then, by resorting to the Rugizka algorithm introduced in paragraph (1.12), one can compute (11.10) to obtain

$$J_n(\Omega) = \frac{\lambda^2}{2\pi} \frac{2(1 + 2\cos(\Omega)^2 + 2\rho^2 \cos(\Omega)^2 - \rho^4 - 8\rho\cos(\Omega)^2 + 4\rho^2 \cos(\Omega)^2)}{1 + \rho^2 - \rho^4 - \rho^6 - 4\rho^2 \cos(\Omega)^2 + 4\rho^4 \cos(\Omega)^2}. \tag{11.12}$$

This is not a constant with respect to Ω, so that the point of minimum of the overall loss function (11.8) does not coincide with the true frequency Ω^o: The estimator is biased.

By studying this function, it can be seen that $J_n(\Omega)$ is more and more flat in the extent to which parameter ρ is closer and closer to 1. In the limit, for $\rho = 1$, it is indeed a constant. So, one could consider this as a remedy against bias. However, when ρ is close to 1, finding the minimum of the loss function from a finite number of snapshots by minimizing

$$J(\Omega) = \frac{1}{N} \sum_{1}^{N} \epsilon(t, \Omega, \rho)^2$$

becomes more impervious due to the presence of false minima. Such issues are highlighted in the simulation reported in Figure 11.13, where the frequency of the unknown signal $s(t)$ is subject to a step perturbation at $t = 2000$:

$$s(t) = \begin{cases} \sin((\pi/2)t), & t \leq 2000, \\ \sin((\pi/3)t), & t > 2000. \end{cases}$$

In order to obtain an asymptotically unbiased estimator, a possibility is to remove the condition that the zeros of the notch model (11.6) be radially aligned

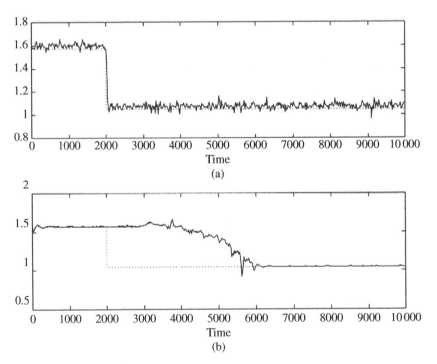

Figure 11.13 Tracking performance of the notch filter: (a) $\rho = 0.70$ and (b) $\rho = 0.99$.

with the poles, and choose for them an optimal positioning aiming at a flat $J_n(\Omega)$. To this purpose, expression (11.7) can be replaced with

$$F(z, \Omega, \rho) = G(z, \Omega, \rho)^{-1} = \frac{1 - 2\cos(\Omega)z^{-1} + z^{-2}}{1 - 2\rho f(\Omega)z^{-1} + \rho^2 z^{-2}}, \tag{11.13}$$

where $f(\Omega)$ is a design function enabling the optimization of the positioning of the zeros.

To see how such optimization can be achieved, consider again the case when the noise is white, $n(t) \sim WN(0, \sigma^2)$. Then, expression (11.9) is still valid, while the recomputation of (11.10) leads to

$$J_n(\Omega, f(\Omega)) = \frac{\lambda^2}{2\pi}$$
$$\times \frac{2(1 + 2\cos(\Omega)^2 + 2\rho^2\cos(\Omega)^2 - \rho^4 - 8\rho f(\Omega)\cos(\Omega) + 4\rho^2 f(\Omega)^2)}{1 + \rho^2 - \rho^4 - \rho^6 - 4\rho^2 f(\Omega)^2 + 4\rho^4 f(\Omega)^2}.$$
$$\tag{11.14}$$

Now, function $J_n(\Omega, b)$ can be minimized with respect to b so obtaining as optimal choice

$$f_m(\Omega) = \frac{(1 + \rho^2)\cos(\Omega)}{2\rho}.$$

Notch filters have been studied in a large number of contributions. The results presented here are based on the articles Bittanti etal., (1997), and Savaresi etal., (2003). In this last paper, the case when the noise is colored is also investigated in detail.

11.3.2 Frequency Estimation with EKF

We now pose the problem of frequency estimation in a state space framework, and then address it with EKF. We start by writing the equations of the discrete time oscillator by introducing the second-order system:

$$x_1(t + 1) = \cos(\Omega)x_1(t) - \sin(\Omega)x_2(t),$$
$$x_2(t + 1) = \sin(\Omega)x_1(t) + \cos(\Omega)x_2(t). \tag{11.15}$$

A solution of such system is

$$x_1(t + 1) = \cos(\Omega t),$$
$$x_2(t + 1) = \sin(\Omega t).$$

Indeed, by substituting these expressions for $x_1(t)$ and $x_2(t)$ into the right-hand sides of (11.15), we have

$$\cos(\Omega)\cos(\Omega t) - \sin(\Omega)\sin(\Omega t),$$
$$\sin(\Omega)\cos(\Omega t) + \cos(\Omega)\sin(\Omega t).$$

By means of the Werner formulas

$$\cos(\alpha)\cos(\beta) = \frac{1}{2}[\sin(\alpha + \beta) + \sin(\alpha - \beta)],$$

$$\sin(\alpha)sin(\beta) = \frac{1}{2}[\cos(\alpha - \beta) - \cos(\alpha + \beta)],$$

it can be seen that

$$\cos(\Omega)\cos(\Omega t) - \sin(\Omega)\sin(\Omega t) = \cos(\Omega(t + 1)),$$

$$\sin(\Omega)\cos(\Omega t) + \cos(\Omega)\sin(\Omega t) = \sin(\Omega(t + 1)),$$

so verifying that (11.15) is the description of a harmonic signal of angular frequency Ω.

As for measurements, we assume that the observed signal is

$$y(t) = x_1(t) + v(t), \quad v(t) \sim \text{WN}(0, \lambda^2).$$

To tackle the frequency estimation problem, we introduce the new state variable

$$x_3(t) = \Omega,$$

so obtaining the overall system:

$$x_1(t + 1) = \cos(x_3)x_1(t) - \sin(x_3)x_2(t)$$
$$x_2(t + 1) = \sin(x_3)x_1(t) + \cos(x_3)x_2(t)$$
$$x_3(t + 1) = x_3(t) + w(t)$$
$$y(t) = x_1(t) + v_2(t). \tag{11.16}$$

The disturbance terms are

a) the noise affecting the measurement of state variable $x_1(t)$. Denoted as $v_2(t)$, it is assumed to be white

$$v_2(t) \sim \text{WN}(0, \lambda^2)$$

b) the noise corrupting the angular frequency in the $x_3(t)$ equation

$$w(t) \sim \text{WN}(0, \sigma^2).$$

We also assume that $w(t)$ and $v_2(t)$ are uncorrelated.

In system (11.17), the state is

$$x_E(t) = \begin{bmatrix} x_1(t) \\ x_2(t) \\ x_3(t) \end{bmatrix}.$$

As for the noise vector

$$v_1(t) = \begin{bmatrix} 0 \\ 0 \\ w(t) \end{bmatrix},$$

the covariance matrix is given by

$$V_1 = \lambda^2 \tilde{I},$$

where

$$\tilde{I} = \begin{bmatrix} 0 & 0 & 0 \\ 0 & 0 & 0 \\ 0 & 0 & 1 \end{bmatrix}.$$

If we let

$$f(x_E) = \begin{bmatrix} \cos(x_3)x_1 - \sin(x_3)x_2 \\ \sin(x_3)x_1 + \cos(x_3)x_2 \\ x_3 \end{bmatrix}$$

and

$$H = \begin{bmatrix} 1 & 0 & 0 \end{bmatrix},$$

then the Kalman filter recursion is

$$\hat{x}_E(t+1|t+1) = f(\hat{x}_E(t|t)) + K(t)(y(t) - Hf(\hat{x}_E(t|t))), \tag{11.17a}$$

$$K(t) = P(t)H'(HP(t)H' + \sigma^2)^{-1}, \tag{11.17b}$$

$$P(t+1) = F(t)[P(t) - P(t)H'(HP(t)H' + \sigma^2)HP(t)]F(t)' + \lambda^2 \tilde{I}. \tag{11.17c}$$

Here, $y(t) - Hf(\hat{x}_E(t|t)) = y(t) - \hat{x}_1(t|t)$ is the innovation, the difference between the measured signal and its estimate provided by the filter. The gain $K(t)$ is a vector with three components determining the influence of the innovation on each element of the extended state. The gain depends upon the 3×3 symmetric matrix $P(t)$ solving the Riccati equation (11.17c). In the Riccati equation, the 3×3 matrix

$$F(t) = \left. \frac{\partial f(x)}{\partial x} \right|_{x = \hat{x}(t-1|t-1)}$$

can be computed by performing the derivatives of the 3×1 vector $f(x, t)$ with respect to the 3×1 vector $x_E(t)$, so obtaining

$$F = \begin{bmatrix} \cos(x_3) & -\sin(x_3) & -\sin(x_3)x_1 - \cos(x_3)x_2 \\ \sin(x_3) & \cos(x_3) & \cos(x_3)x_1 - \sin(x_3)x_2 \\ 0 & 0 & 1 \end{bmatrix}.$$

To complete the EKF equation, we add the angular frequency estimation equation, simply given by

$$\hat{\Omega}(t) = \hat{x}_3(t|t).$$

At this point, the EKF algorithm is fully specified and provides a real-time estimate of the unknown angular frequency.

Going through the algorithm equations, we see however two parameters to be still specified, variance λ^2 and variance σ^2. To decide the value of λ^2, one can make reference to the intensity of the disturbance affecting the measurements, while, for the selection of σ^2, one can use some rule of thumb based, e.g. on the expected amplitude of the signal to be estimated. Actually, both λ^2 and σ^2 are often seen as tuning knobs to achieve the best performance in estimation. It is therefore important to investigate their influence on the algorithm performance.

We claim that the performance of the EKF algorithm above is determined by the *quotient* between the two variances.

To prove such claim, we investigate the effect of a rescaling of the two variances, by replacing them by $\tilde{\lambda}^2 = \alpha\lambda^2$ and $\tilde{\sigma}^2 = \alpha\sigma^2$. We first consider the Riccati equation, where both variances appear. It is easy to see that, if no other change is made in the terms of (11.17c), the new solution $\tilde{P}(t)$ is simply given by $\tilde{P}(t) = \alpha P(t)$.

Turn now to the gain $K(t)$. If in (11.17b) we replace $P(t)$ with $\tilde{P}(t) = \alpha P(t)$ and λ^2 with $\tilde{\lambda}^2 = \alpha\lambda^2$, the gain does not change: $\tilde{K}(t) = K(t)$.

Then, Eq. (11.17a) tells us that $\tilde{\hat{x}}_E(t+1|t+1) = \hat{x}_E(t+1|t+1)$, provided that we impose the same initial condition in the two situations (before and after rescaling).

Finally, form the expression of F, we conclude that $\tilde{F} = F$. The invariance of this matrix validates the statement that $\tilde{P}(t) = \alpha P(t)$ since this relation was derived by assuming that no other change was made in the terms of (11.17c), in particular that matrix F was not modified.

This proves our initial claim. In practice, we can set one of the two parameters to 1, for example $\lambda^2 = 1$, and act on the other parameter for the tuning. Expression (11.17b) shows that on decreasing σ^2, the gain $K(t)$ increases, so that the estimate is more sensitive to the innovation (see Eq. (11.17a)). Hence, a lower value of σ^2 is recommended to track angular frequencies that are fast varying. On the other hand, if the frequency is slowly changing, in the limit it is a constant, then a small σ^2 will result in a larger variance in the estimated signal.

This approach to the estimation of the frequency of a sinusoid in noise was proposed in the paper (Bittanti and Savaresi, 2000).

Appendix A

Linear Dynamical Systems

In this appendix, we study linear discrete-time and time-invariant dynamical systems, their basic models, and their fundamental properties.

A.1 State Space and Input–Output Models

The first description we refer to is the *state space* representation:

$$x(t + 1) = Fx(t) + Gu(t) \tag{A.1a}$$

$$y(t) = Hx(t) + Ku(t), \tag{A.1b}$$

where t is discrete time and

- input u is a vector with m components,
- output y has p components,
- state x has n components (n is the *system order*),
- F, G, H, and K are real matrices of suitable dimensions (compatible with the dimensions of vectors u, y, and x).

Equation (A.1a) is the *state space equation*, while Eq. (A.1b) is the *output equation*.

From expression (A.1b), we see that the output is related to the state and to the input via an *algebraic* relation. The time evolution of the system is determined by Eq. (A.1a), supplying the state at time $t + 1$, given the state and the input at time t. Therefore, (A.1a) is a *dynamic* equation, whereas (A.1b) is a *static* (or *algebraic*) link.

We say that the system is *single input single output* (or simply *SISO*) when $m = 1$ and $p = 1$.

A.1.1 Characteristic Polynomial and Eigenvalues

The symbol $\det(\cdot)$ denotes the determinant of a matrix (more on this notion can be found in Section B.3 of Appendix B). The polynomial $\det(zI - F)$ is

Model Identification and Data Analysis, First Edition. Sergio Bittanti.
© 2019 John Wiley & Sons, Inc. Published 2019 by John Wiley & Sons, Inc.

named *characteristic polynomial* of matrix F; correspondingly, the equation $\det(zI - F) = 0$ is named *characteristic equation* of F. The solutions of such equation are the so-called *eigenvalues* of matrix F, also named eigenvalues of system (A.1). F being an $n \times n$ matrix, $\det(zI - F)$ is a polynomial in z of degree n. Therefore, a system with n state variables has n eigenvalues. They can be real or complex; in case an eigenvalue is not real, then among the eigenvalues is its complex conjugate.

A.1.2 Operator Representation

An alternative to the state space description is the *operator representation* of the system. We use the symbol z to denote the *one-step-ahead operator*, i.e. $zx(t)$ is a symbol denoting $x(t + 1)$. Thus, we can rewrite the state equation as $zx(t) = Fx(t) + Gu(t)$, so that $x(t) = (zI - F)^{-1}Gu(t)$ and $y(t) = H(zI - F)^{-1}Gu(t)$; therefore, the input–output operator description is

$$W(z) = H[zI - F]^{-1}G + K. \tag{A.2}$$

Obviously, $W(z)$ is a $p \times m$ matrix, whose entries are polynomial functions of operator z.

A.1.3 Transfer Function

A third possibility is to consider the *transfer function* representation. To be precise, we define the *z-transform* of the input signal $u(t)$ and the output signal $y(t)$ as the complex functions $U(z)$ and $Y(z)$ of the complex variable z defined by

$$U(z) = u(0) + u(1)z^{-1} + u(2)z^{-2} + \cdots ,$$

$$Y(z) = y(0) + y(1)z^{-1} + y(2)z^{-2} + \cdots .$$

It is easy to see that, if $x(0) = 0$, then the relationship between $Y(z)$ and $U(z)$ is

$$Y(z) = W(z)U(z),$$

where $W(z)$ is again given by expression (A.2) introduced above, provided that z is now seen as the complex variable.

A.1.4 Zeros, Poles, and Eigenvalues

If we focus on SISO systems, then $W(z)$ is the quotient between two polynomials in z with real coefficients. The roots of the numerator are called *zeros* and the roots of the denominator are called *poles* of the system.

Recalling the definition of the inverse of a matrix, from expression (A.2), we conclude that the denominator is just the determinant of matrix $zI - F$, a polynomial with real coefficients of degree n (recall the n is the number of state variables so that F is $n \times n$). Therefore, the poles coincide with the system eigenvalues.

This statement comes with a warning: It is possible that the numerator and the denominator of the transfer function have a common root. In that case, after simplification, the denominator degree decreases and there is an eigenvalue that disappears in the simplified transfer function representation. We speak then of *hidden eigenvalue*. Then, the set of poles is a subset of the set of eigenvalues.

A.1.5 Relative Degree

Again with reference to SISO systems, we make another interesting observation. Considering the procedure to compute the inverse of $(zI - F)$, the degree of the numerator of $H(zI - F)^{-1}G$ is strictly lower than the degree of its denominator. Hence, if $K = 0$, the number of zeros is strictly lower than the number of poles, whereas if $K \neq 0$, then the zeros and poles are in equal number. The difference between the degree of the denominator and the degree of the numerator is called *relative degree*. Interestingly enough, the relative degree, say k, is the delay with which the input acts on the output. For instance, suppose that $W(z) = 1/z^2$, so that the relative degree is 2, then $y(t) = z^{-2}u(t) = z^{-1}(z^{-1}u(t))$. z^{-1} being the unit delay operator, $z^{-1}u(t) = u(t - 1)$, so that $y(t) = z^{-1}u(t - 1) = u(t - 2)$, namely the output is a two-step delayed version of the input.

A.1.6 Equilibrium Point and System Gain

Consider system (A.1) subject to a constant input $u(t) = \bar{u}$. Correspondingly, any state $\bar{x}$ such that $x(t + 1) = x(t) = \bar{x}$ is said to be an equilibrium point or equilibrium state (associated with that input). If in (A.1) we set $x(t + 1) = x(t) = \bar{x}$, we have

$$\bar{x} = (I - F)^{-1}G\bar{u}.$$

The output $\bar{y}$ is then

$$\bar{y} = (H(I - F)^{-1}G + K)\bar{u}.$$

Thus, the relation between the input and the output at *steady state*, namely at equilibrium, is determined by quantity $H(I - F)^{-1}G + K$, which is named *system gain*.

From expression (A.2), we see that the gain is given by the transfer function evaluated at $z = 1$:

$$\text{gain} = W(z)|_{z=1}.$$

A.2 Lagrange Formula

If in expression (A.1) we set $u(t) = 0$, then the system becomes

$$x(t + 1) = Fx(t)$$
$$y(t) = Hx(t),$$

the solution of which for an initial state $x(t_0)$ at time t_0 is the so-called *free motion*, denoted as $x_L(\cdot)$ for the state and as $y_L(\cdot)$ for the output. It is easy to see that

$$x_L(t) = F^{t-t_0}x(t_0),$$
$$y_L(t) = HF^{t-t_0}x(t_0).$$

Passing to the case when the system is subject to any input function and the initial state at time t_0 is null, $x(t_0) = 0$, we have

$$x(t_0 + 1) = Gu(t_0),$$
$$x(t_0 + 2) = FGu(t_0) + Gu(t_0 + 1),$$
$$x(t_0 + 3) = F^2Gu(t_0) + FGu(t_0 + 1) + Gu(t_0 + 2),$$
$$\vdots$$

Hence, the so-called *forced motion* is

$$x_F(t) = Gu(t - 1) + FGu(t - 2) + \cdots + F^{t-t_0-1}Gu(t_0).$$

In general, due to the linearity of the system,

$$x(t) = x_L(t) + x_F(t)$$
$$= F^{t-t_0}x(t_0) + \sum_{k=0}^{t-t_0-1} F^kGu(t - k - 1). \tag{A.3}$$

Correspondingly,

$$y(t) = HF^{t-t_0}x(t_0) + \sum_{k=0}^{t-t_0-1} HF^kGu(t - k) + Ku(t). \tag{A.4}$$

Expressions (A.3) and (A.4) are named *Lagrange formulas* in discrete time. Coefficients HF^kG are called *Markov parameters* of the system.

A.3 Stability

The state evolution of a dynamical system associated with a given initial state $x(t_0)$ and a given input signal $u(\cdot)$ is provided by (A.3).

Consider now another initial state given by the previous one subject to a perturbation $\delta x(t_0)$: $x_p(t_0) = \bar{x}(t_0) + \delta x(t_0)$. Assuming that the input sequence is unchanged, the state evolution becomes

$$x_p(t) = F^{t-t_0}(\bar{x}(t_0) + \delta x(t_0)) + \sum_{k=0}^{t-t_0-1} F^k Gu(t - k - 1). \tag{A.5}$$

We call $x_p(t)$ *perturbed state*.

A main question is whether $x_p(t)$ will diverge from $x(t)$ or not.

The state $\bar{x}(t_0)$ is said to be asymptotically stable, or for simplicity, stable, if, for any perturbation $\delta x(t_0)$, $x_p(t) - x(t)$ tends to 0 as $t \to \infty$. A system is said to be asymptotically stable (or, for simplicity, stable) if any state $\bar{x}(t_0)$ is asymptotically stable.

By comparing (A.5) with (A.3), we see that

$$x_p(t) - x(t) = F^{t-t_0} \delta x(t_0).$$

Therefore, this system is stable if and only if

$$\lim_{t \to \infty} F^{t-t_0} = 0,$$

or, equivalently, if and only if

$$\lim_{t \to \infty} F^t = 0. \tag{A.6}$$

If F were a scalar, say $F = f$, then this condition simply means that the discrete time exponential of f tends to 0: $f^t \to 0$. Obviously, this happens if and only if $|f| < 1$. One can extend this result to the matrix case and prove that (A.6) is satisfied if and only if all eigenvalues of matrix F are inside the unit disk in the complex plane, namely

$$|\lambda_i[F]| < 1, \quad \forall i.$$

A.4 Impulse Response

In this section, as well as in the subsequent Section A.5, we make reference to SISO dynamical systems, i.e. systems with ($m = 1, p = 1$). We define as impulse response the output obtained when the initial state (taken, for example at $t_0 = 0$) is null, $x(0) = 0$, and the input is given by the *impulse signal* denoted by imp(t), and defined as follows:

$$u(t) = \text{imp}(t) = \begin{cases} 1, & t = 0, \\ 0, & t \neq 0. \end{cases} \tag{A.7}$$

We will compute the response starting from both a state space representation and an input–output representation.

A.4.1 Impulse Response from a State Space Model

From (A.4), we have

$$x(1) = Gu(0) = G,$$

$$x(2) = Fx(1) + Gu(1) = FG,$$

$$x(3) = Fx(2) + Gu(2) = F^2 G,$$

$$\vdots$$

$$y(0) = Ku(0) = K,$$

$$y(1) = HG,$$

$$y(2) = HFG$$

$$y(3) = HF^2 G,$$

$$\vdots$$

Thus, the impulse response is the sequence of Markov parameters:

$$y(t) = \begin{cases} K, & t = 0, \\ HF^{t-1}G, & t > 0. \end{cases} \tag{A.8}$$

Note that, if the system is asymptotically stable, then $F^t \to 0$. Consequently, the impulse response of a stable system is vanishing on the long run:

$$\lim_{t \to \infty} y(t) = 0. \tag{A.9}$$

A.4.2 Impulse Response from an Input–Output Model

The impulse response can be determined from the input–output representation as well,

$$y(t) = W(z)u(t).$$

Indeed, by considering the expansion of $W(z)$ in negative powers of z:

$$W(z) = w_0 + w_1 z^{-1} + w_2 z^{-2} + \cdots , \tag{A.10}$$

and assuming that the input is the impulse signal (A.7), an easy computation shows that

$$y(0) = w_0,$$

$$y(1) = w_1,$$

$$y(2) = w_2,$$

$$\vdots$$

In other words, the coefficients $\{w_0, w_1, w_2, \dots\}$ of expansion (A.10) constitute the impulse response of the system with transfer function $W(z)$.

A.4.3 Quadratic Summability of the Impulse Response

We have seen above that the impulse response of a stable system vanishes in the long run, see (A.9). We now show that it vanishes sufficiently fast to guarantee the quadratic summability of the coefficients, namely

$$\sum_{i=0}^{+\infty} y(i)^2 \text{ exists finite,}$$

a statement that can be written as

$$\sum_{i=0}^{+\infty} y(i)^2 < \infty.$$

We can prove this by referring to expression (A.11) of the impulse response. In the scalar case, i.e. when matrix F is scalar, $F = f$, we have

$$y(t) = \begin{cases} K, & t = 0, \\ Hf^{t-1}G, & t > 0, \end{cases} \tag{A.11}$$

so that

$$y(0)^2 + y(1)^2 + y(2)^2 + \cdots = K^2 + (HG)^2(f^2 + f^4 + \cdots). \tag{A.12}$$

The series $\sum_{i=1}^{+\infty} y(i)^2$ is the famous geometric series, which is convergent for $|f| < 1$. The same conclusion holds in the matrix case.

As a special important case, we can say that the process generated by a stable ARMA model is equivalent to a well-defined MA(∞) process and therefore is stationary.

A.5 Frequency Response

Focus now on the case when the input is sinusoidal,

$$u(t) = A \, \sin(\bar{\omega}t + \alpha).$$

If the system is stable, for any initial state the output tends to a signal that is sinusoidal as well, with the *same* angular frequency:

$$y(t) = B \, \sin(\bar{\omega}t + \beta).$$

Amplitude B and phase β can be easily computed from the transfer function $W(z)$ as follows:

$$B = A|W(e^{j\bar{\omega}})|,$$

$$\beta = \alpha + \phi,$$

where ϕ is the phase of complex number $W(e^{j\bar{\omega}})$.

Thus, if

$$|W(e^{j\bar{\omega}})| > 1,$$

the input amplitude is amplified, while if

$$|W(e^{j\bar{\omega}})| < 1,$$

the input amplitude is attenuated.

A.6 Multiplicity of State Space Models

In the state space, there are many quadruples (F, G, H, K) that can represent the same system. As we will see, there are two basic reasons for that: (i) change of basis and (ii) redundancy in the system order.

A.6.1 Change of Basis

As an introductory example, consider a system with two state variables x_1 and x_2. By exchanging the two variables, we can define the new basis $\tilde{x}_1 = x_2$ and $\tilde{x}_2 = x_1$. The original and the new basis are related to each other as

$$\begin{bmatrix} \tilde{x}_1 \\ \tilde{x}_2 \end{bmatrix} = \begin{bmatrix} 0 & 1 \\ 1 & 0 \end{bmatrix} \begin{bmatrix} x_1 \\ x_2 \end{bmatrix},$$

namely

$$\tilde{x} = Tx, \quad T = \begin{bmatrix} 0 & 1 \\ 1 & 0 \end{bmatrix}.$$

In general, given a system of any order, the relationship

$$\tilde{x}(t) = Tx(t), \quad \det T \neq 0$$

describes a *change of basis*.

If we refer to (A.1), it is easy to see that

$$\tilde{x}(t+1) = \tilde{F}\tilde{x}(t) + \tilde{G}u(t)$$
$$y(t) = \tilde{H}\tilde{x}(t) + \tilde{K}u(t),$$

with

$$\tilde{F} = TFT^{-1}, \quad \tilde{G} = TG, \quad \tilde{H} = HT^{-1}, \quad \tilde{K} = K.$$

Thus, for any choice of matrix T, one obtains a different quadruple F, G, H, K.

Presumably, such a formal variation in the choice of the basis does not have any consequence on the system intrinsic characteristics. And indeed, there are a number of invariant properties, as we will now see.

First, we analyze the characteristic polynomial of $\tilde{F}$:

$$\det (zI - \tilde{F}) = \det (zI - TFT^{-1}) = \det (zTT^{-1} - TFT^{-1})$$
$$= \det (T(zI - F)T^{-1}).$$

On the other hand, given two square matrices A and B, $\det(AB) = (\det A)(\det B)$, so that

$$\det T(z - F)T^{-1} = \det T \; \det(zI - F) \; \det T^{-1}.$$

Since $\det T \; \det T^{-1} = 1$, the two characteristic polynomials coincide. Second, as an obvious corollary of the above conclusion, we can state that the eigenvalues of F and $\tilde{F}$ are the same.

We now prove that the transfer function does not change:

$$\tilde{G}(z) = \tilde{H}[zI - \tilde{F}]^{-1}\tilde{G} + \tilde{K}$$
$$= HT^{-1}[zTT^{-1} - TFT^{-1}]^{-1}TG + K$$
$$= HT^{-1}T[zI - F]^{-1}T^{-1}TG + K$$
$$= H[zI - F]^{-1}G + K$$
$$= G(z) \tag{A.13}$$

(note that $(AB)^{-1} = B^{-1}A^{-1}$).

This is somewhat expected since $\tilde{x}(t) = Tx(t)$ is a change of basis in the state space. Such transformation is purely formal and cannot modify the input–output relationship.

Summing up, there are infinitely many representations of the same order of a dynamical system, each of which is associated with a different choice of the adopted basis in the state space. All such representations share the same transfer function.

A.6.2   Redundancy in the System Order

There is a second reason of multiplicity in the state space representation: the system order. For instance, the system of order 2

$$x_1(t + 1) = x_1(t) - 3x_2(t) + u(t) \tag{A.14a}$$

$$x_2(t + 1) = -2x_2(t) + u(t) \tag{A.14b}$$

$$y(t) = x_1(t) + x_2(t), \tag{A.14c}$$

has two real eigenvalues, one in $+1$ and one in -2. The associated transfer function is

$$G(z) = 2\frac{z-1}{(z-1)(z+2)} = 2\frac{1}{(z+2)}.$$

Hence, $+1$ is a *hidden eigenvalue* as, after simplification, it does not appear among the poles. Not surprisingly, the system

$$x_1(t+1) = -2x_1(t) + 2u(t) \tag{A.15a}$$

$$y(t) = x_1(t), \tag{A.15b}$$

whose transfer function is

$$G(z) = 2\frac{1}{(z+2)},$$

has identical external behavior of the previous one, even if its order is 1 only. This last reason of multiplicity can be removed by introducing the notion of *realization* and *minimal realization*: given a transfer function $G(z)$, a system described by the quadruple (F, G, H, K) having $G(z)$ as transfer function is said to be a *realization* of $G(z)$; a realization is said to be a *minimal realization* if there does not exist any other realization with lower order.

The number of state variables of a minimal realization is named *Smith–McMillan degree* of the system.

For instance, (A.14) and (A.15) are realizations of the same transfer function; however, only (A.15) is minimal. Correspondingly, the Smith–McMillan degree is 1 and representation (A.14) is redundant.

A.7 Reachability and Observability

We now introduce the concepts of reachability and observability, thanks to which many aspects of the relationship between internal (input–state–output) and external (input–output) representations can be clarified.

A.7.1 Reachability

A state $\bar{x}$ is said to be reachable in the interval $[1, t]$ if there exists an input function $\bar{u}(\cdot)$ such that the system state, assumed to be in the origin at time 1, can be transferred to $\bar{x}$ at time t. The set of reachable states in the interval $[1, t]$ is denoted by X_t.

It is easy to see that, if $\bar{x}_1$ and $\bar{x}_2$ are both reachable in the interval $[1, t]$ (by means of input functions $u_1(\cdot)$ and $u_2(\cdot)$, respectively), any linear combination of them $\bar{x} = \alpha_1 \bar{x}_1 + \alpha_2 \bar{x}_2$, where α_1 and α_2 are any pair of real numbers, is reachable in the interval $[1, t]$ (by input $u(\cdot) = \alpha_1 u_1(\cdot) + \alpha_2 u_2(\cdot)$). Therefore, X_t is a subspace.

Remark A.1 (Subspace) By *subspace* we mean a set of states such that, if x_a and x_b belong to the set, then also any linear combination of them belongs to the set too. Intuitively, a subspace is a plane–or an hyperplane–passing through the origin of the whole space. The *dimension* of a subspace is the number of vectors in the set, which are linearly independent. For instance, a straight line through the origin in a 3D space is a subspace of dimension 1, while a plane passing through the origin is a subspace of dimension 2.

Obviously, $X_{t+1} \supseteq X_t$ since, prolonging the interval from $[1, t]$ to $[1, t + 1]$, there is a further degree of freedom in the choice of the input signal.

Remark A.2 (On the definition of reachability) The definition of reachability can be given a more general form by considering a generic initial point t_1. This would mean defining reachability over the interval $[t_1, t_2]$ in place of $[1, t]$. On the other hand, for time-invariant systems, what matters is only the duration of the time interval. In other words, reachability over $[t_1, t_2]$ is the same as reachability over $[1, t]$ with $t = t_2 - t_1 + 1$.

A simple test of reachability can be obtained as follows. Starting from the state $x(1) = 0$, we have $x(2) = Gu(1)$. Focus on the set of such $x(2)$ achievable by manipulating at will vector $u(1)$. By denoting with $u(1)_1, u(1)_2, \ldots, u(1)_m$ the entries of $u(1)$, it is straightforward to see that $x(2)$ is a linear combination (LC) of the columns of matrix G with coefficients of the combinations given by $u(1)_1, u(1)_2, \ldots, u(1)_m$, respectively. The values $u(1)_1, u(1)_2, \ldots, u(1)_m$ being free of constraints, the set of reachable states in one step, X_1, is the subspace generated by the columns of G:

$$X_1 = \text{LC columns } G.$$

The dimension of such subspace is given by the number of columns of G, which are linearly independent.

Passing to the next time point, we have $x(3) = Fx(2) + Gu(2) = FGu(1) + Gu(2)$. Hence, the set of states $x(3)$, which can be reached in two steps are all and only the LC of the columns of matrices G and FG. In other words, the reachable subspace in a two-step evolution of the system is

$$X_2 = \text{LC columns } G, \text{ and columns } FG.$$

In general, since

$$x(t) = F^{t-1}Gu(1) + F^{t-2}Gu(2) + \cdots + Gu(t - 1),$$

the subspace of reachable states in t steps is

$$X_t = \text{LC columns } G, \text{ columns } FG, \ldots, \text{ columns } F^{t-1}G.$$

This suggests to introduce matrix

$$\mathscr{R}^t = [G \,\vdots\, FG \,\vdots\, F^2G \,\vdots\, \cdots \,\vdots\, F^{t-1}G],$$

known as the *reachability matrix* over the interval from 1 to t. Note that

$$\dim X_t = \text{rank } \mathscr{R}^t,$$

where dim denotes the dimension of the subspace. Obviously, $X_{t+1} \supseteq X_t$ since, prolonging the interval from $[1, t]$ to $[1, t + 1]$, there is a further degree of freedom in the choice of the input signal. Therefore, dim $X_{t+1} \geq$ dim X_t.e. On the other hand, such dimension cannot increase without limits, as it is bounded by the dimension of the state space. A celebrated theorem of matrix theory, proved in Section B.5 of Appendix B, is of help to clarify this issue.

Proposition A.1 (Cayley–Hamilton theorem) *Consider any square matrix A of dimension $n \times n$, and denote by $z^n + \alpha_1 z^{n-1} + \cdots + \alpha_{n-1}z + \alpha_n$ its characteristic polynomial. Then*

$$A^n + \alpha_1 A^{n-1} + \cdots + \alpha_{n-1}A + \alpha_n I = 0.$$

This result implies that the columns of the nth power of any $n \times n$ matrix A are linear combinations of the columns of $A^{n-1}, A^{n-2}, \ldots$. Therefore, the rank of matrix $\mathscr{R}^t$ may increase passing from $t = 1$ to $t = 2$, from $t = 2$ to $t = 3$, and so on but cannot increase for $t \geq n$. Then we come to the following definition.

Definition A.1 (Reachability matrix) The matrix

$$\mathscr{R}^n = [G \,\vdots\, FG \,\vdots\, F^2G \,\vdots\, \cdots \,\vdots\, F^{n-1}G]$$

is named reachability matrix of the system (F, G).

If the rank of such matrix coincides with the dimension n of the state space, then all states are reachable. In such a case, we say that *the system is reachable* (or that the system is completely reachable).

A.7.2 Observability

As seen above, reachability has to do with the effect of the input on the system state. We now pass to study the effect of the state on the output. To make things as simple as possible, we study such effect in absence of an input signal, by considering the system

$$x(t + 1) = Fx(t)$$
$$y(t) = Hx(t). \tag{A.16}$$

Starting from a state $x(1)$, we have

$$y(1) = Fx(1),$$
$$y(2) = HFx(1),$$
$$\vdots$$
$$y(t) = HF^{t-1}x(1). \tag{A.17}$$

Suppose now we want to identify state $x(1)$ from observations $y(1), y(2), \ldots, y(t)$ of the output. This amounts to solving the above system of linear equations with respect to unknown $x(1)$. Depending on the particular pair (F, H), there might be various initial states giving rise to the same sequence $y(1), y(2), \ldots, y(t)$, or a unique $x(1)$. In this latter case, one could determine the state without ambiguity by observing the output signal.

A state $\bar{x}$ is said to be observable over the interval $[1, t]$ if no other state $\tilde{x}, \tilde{x} \neq \bar{x}$, exists such that functions $y(\cdot)$ and $\tilde{y}(\cdot)$ generated by system (A.16) with those states as initial conditions coincide over the interval from 1 to t, $y(1) = \tilde{y}(1), y(2) = \tilde{y}(2), \ldots, y(t) = \tilde{y}(t)$.

A system is said to be observable (or completely observable) if all states are observable.

A simple test of observability is obtained as follows.

Definition A.2 (Observability matrix)   The matrix

$$\mathcal{O}^n = \begin{bmatrix} H \\ \cdots \\ HF \\ \cdots \\ HF^2 \\ \cdots \\ \vdots \\ \cdots \\ HF^{n-1} \end{bmatrix} \tag{A.18}$$

is named observability matrix of the system (F, G).

If the rank of such matrix coincides with the dimension n of the system, then the system is observable.

A.7.3   PBH Test of Reachability and Observability

Another reachability test is the so-called *PBH test*, where the acronym *PBH* comes from the initials of V.M. Popov, V. Belevitch, and M.L.J. Hautus. Such test is based on the so-called *PBH reachability matrix*, defined as

$$\mathcal{R}_{\text{PBH}}(\lambda) = [\lambda I - F \;\vdots\; G],$$

where λ is a complex number.

If λ is not an eigenvalue of matrix F, then $\lambda I - F$ is invertible so that $\lambda I - F$ has rank equal to n. Consequently, in such a situation, matrix $\mathscr{R}_{\text{PBH}}(\lambda)$ has maximum rank: rank $\mathscr{R}_{\text{PBH}}(\lambda) = n$. If instead λ is an eigenvalue of matrix F, then rank$(\lambda I - F) < n$, so that the rank of $\mathscr{R}_{\text{PBH}}(\lambda)$ may be lower than n.

Proposition A.2 (PBH reachability test)   *The pair (A, B) is reachable if and only if rank $\mathscr{R}_{PBH}(\lambda) = n$, $\forall \lambda$.*

Proof: Suppose that $\mathscr{R}_{\text{PBH}}(\lambda) < n$. Then there exists an n-dimensional vector $x, x \neq 0$ such that

$$x' \mathscr{R}_{\text{PBH}}(\lambda) = 0.$$

In view of the definition of $\mathscr{R}_{PBH}(\lambda)$, this is equivalent to

$$\lambda x' = x'F, \tag{A.19a}$$
$$x'G = 0. \tag{A.19b}$$

By multiplying the two sides of Eq. (A.19a) by matrix G on the right, we have

$$\lambda x'G = x'FG. \tag{A.20}$$

On the other hand, from (A.19b), we know that $x'G = 0$; therefore, from (A.20), we conclude that

$$x'FG = 0. \tag{A.21}$$

Multiply now the two members of the first equations in (A.19) by FG, so that

$$\lambda x'FG = x'F^2G. \tag{A.22}$$

From (A.21), we can then say that

$$x'F^2G = 0. \tag{A.23}$$

By iterating, we come to the following conclusion:

$$x'[G \ \vdots \ FG \ \vdots \ F^2G \ \vdots \ \cdots \ \vdots \ F^{n-1}G] = 0,$$

thus violating the reachability test of Section A.7.1.

We are now in a position to introduce the notion of *reachable eigenvalue* and *unreachable eigenvalue*: an eigenvalue λ of F for which rank $\mathscr{R}_{\text{PBH}}(\lambda) = n$ is said to be a *reachable*, whereas an eigenvalue λ for which rank $\mathscr{R}_{\text{PBH}}(\lambda) < n$ is said to be *unreachable*.

We can rephrase Proposition A.2 by saying that a system is reachable if and only if all eigenvalues are reachable.

In an entirely analogous way, we can introduce the *PBH observability matrix* as

$$\mathcal{O}_{PBH}(\lambda)] = \begin{bmatrix} \lambda I - F \\ \cdots \\ H \end{bmatrix},$$

where λ is a complex number. The eigenvalues of F for which rank $\mathcal{O}_{PBH}(\lambda) = n$ are said to be observable, the remaining ones unobservable.

It can be proved that a system is observable if and only if all the eigenvalues are observable in the sense above specified. Equivalently, a system is observable if and only if rank $\mathcal{O}_{PBH}(\lambda) = n$, $\quad \forall \lambda$.

A.8 System Decomposition

A.8.1 Reachability and Observability Decompositions

When a system is not reachable, there are some state variables, or linear combinations of state variables, that are not influenced by the input signal. More precisely, with a suitable change of basis, the original system (A.1), with state x, can be transformed into a system with state

$$\tilde{x} = \begin{bmatrix} \tilde{x}_1 \\ \tilde{x}_2 \end{bmatrix}$$

having the following structure

$$\dot{\tilde{x}}_1 = \tilde{F}_{11}\tilde{x}_1 + \tilde{F}_{12}\tilde{x}_2 + \tilde{G}_1 u \tag{A.24a}$$

$$\dot{\tilde{x}}_2 = \tilde{F}_{22}\tilde{x}_2. \tag{A.24b}$$

These expressions clearly point out that the input u acts on $\tilde{x}_1$, whereas it does not have any effect on $\tilde{x}_2$, as schematically depicted in Figure A.1.

Correspondingly, we speak of reachable and unreachable parts. The reachable part is described by the equation

$$\dot{\tilde{x}}_1 = \tilde{F}_{11}\tilde{x}_1 + \tilde{G}_1 u,$$

where the pair $(\tilde{F}_{11}, \tilde{G}_1)$ is reachable. As for the remaining part with state $\tilde{x}_2$, obviously it is not reachable.

Figure A.1 Decomposition of a system into its reachable and unreachable parts.

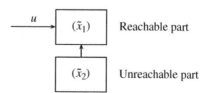

We also observe that the dynamical matrix of system (A.24) is given by

$$\tilde{F} = \begin{bmatrix} \tilde{F}_{11} & \tilde{F}_{12} \\ 0 & \tilde{F}_{22} \end{bmatrix}.$$

This is a block triangular matrix, so that its eigenvalues are those of the diagonal blocks, $\tilde{F}_{11}$ and $\tilde{F}_{22}$:

$$\{\lambda_i(\tilde{F})\} = \{\lambda_i(\tilde{F}_{11})\} \cup \{\lambda_i(\tilde{F}_{22})\}.$$

On the other hand, we know that a change of basis does not modify the eigenvalues:

$$\{\lambda_i(F)\} = \{\lambda_i(\tilde{F}_{11})\} \cup \{\lambda_i(\tilde{F}_{22})\}.$$

In conclusion, the eigenvalues of the original system can be partitioned into *reachable eigenvalues* (those of $\tilde{F}_{11}$) and *unreachable eigenvalues* (those of $\tilde{F}_{22}$).

Turning to observability, we can perform a decomposition into observable and unobservable parts. Indeed, with a suitable change of coordinates $\tilde{x} = Tx$, the original system can be rewritten as

$$\begin{cases} \dot{\tilde{x}}_1 = \tilde{F}_{11}\tilde{x}_1 + \tilde{F}_{12}\tilde{x}_2 \\ \dot{\tilde{x}}_2 = \tilde{F}_{22}\tilde{x}_2 \\ y = \tilde{H}_2\tilde{x}_2 \end{cases}$$

graphically represented in Figure A.2.

As it is clear, observing output y does not bring any information on $\tilde{x}_1$, which is therefore unobservable. On the opposite, the pair $(\tilde{F}_{22}, \tilde{H}_2)$ is observable and $\tilde{x}_2$ is the observable part of the state. The eigenvalues of $(\tilde{F}_{22})$ are the observable eigenvalues.

A.8.2 Canonical Decomposition

Reachability captures the effect of the input on the state of a system, while observability describes the influence of the state on the output. With such concepts, we can clarify the relationship between the state space representation

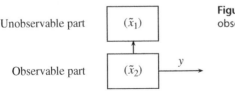

Unobservable part $(\tilde{x}_1)$

Observable part $(\tilde{x}_2)$ y

Figure A.2 Decomposition of a system into observable and unobservable parts.

Figure A.3 Canonical decomposition.

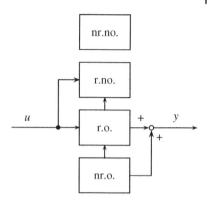

and the input–output description. Indeed, with a change of basis, any system can be decomposed into four parts:

r. o.	reachable and observable part,
r. no.	reachable and unobservable part,
nr. o.	unreachable and observable part,
nr. no.	unreachable and unobservable part.

Correspondingly, one obtains the so-called *canonical decomposition*, schematically represented in Figure A.3.

Such drawing clearly points out that the input–output behavior depends upon the reachable and observable part only.

The following symbols are used to denote the dimensions of the various parts of the decomposition.

$n_{r.o.}$	dimension of the reachable and observable part,
$n_{r.no.}$	dimension of the reachable and unobservable part,
$n_{nr.o.}$	dimension of unreachable and observable part,
$n_{nr.no.}$	dimension of the unreachable and unobservable part.

Obviously, the sum $n_{r.o.} + n_{r.no.} + n_{nr.o.} + n_{nr.no.}$ coincides with the total number, n, of state variables.

The n eigenvalues of the system spread out on the various parts in a number equal to the dimension of each part. In particular, $nr.o.$ eigenvalues are reachable and observable.

In the computation of the transfer function from u to y of a SISO system, only these last eigenvalues survive, the remaining ones either do not appear (those of the unreachable and unobservable part) or cancel out (those of the unreachable or unobservable parts). Therefore, after zero-pole cancellations, the remaining poles are those of the reachable and observable part.

Example A.1 (Decomposition of a second-order system) Consider the system

$$\begin{cases} x_1(t+1) = x_1(t) - 1.5x_2(t) + u & \text{(A.25a)} \\ x_2(t+1) = -0.5x_2(t) + u & \text{(A.25b)} \\ y(t) = x_1(t) + x_2(t), & \text{(A.25c)} \end{cases}$$

whose matrices are

$$F = \begin{bmatrix} 1 & -1.5 \\ 0 & -0.5 \end{bmatrix}, \quad G = \begin{bmatrix} 1 \\ 1 \end{bmatrix}, \quad H = \begin{bmatrix} 1 & 1 \end{bmatrix}, \quad K = [0].$$

The system eigenvalues are $\lambda_1 = 1$ and $\lambda_2 = -0.5$. As for the reachability and observability matrices, they are given by

$$\mathcal{R}^2 = \begin{bmatrix} G & FG \end{bmatrix} = \begin{bmatrix} 1 & -0.5 \\ 1 & -0.5 \end{bmatrix},$$

$$\mathcal{O}^2 = \begin{bmatrix} H \\ HF \end{bmatrix} = \begin{bmatrix} 1 & 1 \\ 1 & -2 \end{bmatrix}$$

The rank of $\mathcal{R}^2$ is 1, whereas the rank of $\mathcal{O}^2$ is 2. Hence, the system is observable but not reachable.

Notice in passing that, being the system SISO, these matrices are square, so that the rank test for reachability or observability can also be performed as a test of non-singularity for the corresponding matrices. In this connection, we see that $\det(\mathcal{O}^2) \neq 0$, whereas $\det(\mathcal{R}^2) = 0$, so that we come again to the conclusions above.

Compute now the transfer function; it is easy to see that

$$G(z) = 2\frac{z-1}{(z-1)(z+0.5)} = \frac{2}{z+0.5}.$$

Thus, eigenvalue $+1$ cancels out and only eigenvalue -0.5 is a pole. Note that the former is associated with the (unreachable, observable) part while the latter comes from the (reachable, observable) part.

Example A.2 (Decomposition of a third-order system) The system

$$\begin{cases} x_1(t+1) = -a_1x_1(t) & +u(t) \\ x_2(t+1) = & -a_2x_2(t) & +u(t) \\ x_3(t+1) = & -a_3x_3(t) & +u(t) \\ y(t) = x_1(t) + x_2(t) + \beta x_3(t) \end{cases} \quad \text{(A.26)}$$

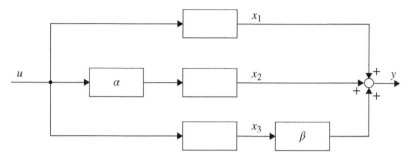

Figure A.4 Block representation of the system in Example A.2.

has an output that is the sum of state variables x_1, x_2, and x_3. Each of them is generated by a scalar system with input u. In other words, system A.26 is just the parallel of three first-order systems, as illustrated in Figure A.4.

Correspondingly, the transfer function is the sum of the transfer functions of the individual first-order systems, i.e.

$$G(z) = \frac{1}{z + a_1} + \frac{\alpha}{z + a_2} + \frac{\beta}{z + a_3}$$
$$= \frac{(z + a_2)(z + a_3) + \alpha(z + a_1)(z + a_3) + \beta(z + a_1)(z + a_2)}{(z + a_1)(z + a_2)(z + a_3)}.$$

We make the assumption that $a_1 \neq a_2 \neq a_3$, and analyze the following possibilities:

1) $\alpha \neq 0; \beta \neq 0$. There is no zero-pole simplification.
2) $\alpha = 0; \beta \neq 0$. At the numerator, the term $(z + a_2)$ is a common factor. Therefore, a simplification occurs and $G(z)$ has two poles only.
3) $\alpha \neq 0; \beta = 0$. At the numerator, the term $(z + a_3)$ is common factor. Again, after simplification, there are two poles.
4) $\alpha = 0; \beta = 0$. The numerator reduces to $(z + a_2)(z + a_3)$, so that $G(z)$ keeps a single pole.

We invite our readers to verify that

1) The system is reachable and observable.
2) The system is observable but not reachable.
3) The system is reachable but not observable.
4) The system is neither reachable nor observable.

This can be seen by constructing the reachability and observability matrices. Alternatively, one can proceed by inspection: The state equations are indeed decoupled with each other. Hence, if $\alpha = 0$, it is obvious that u acts on x_1 and x_3 but has no effect on x_2. Hence, the system is not reachable; to be precise, variables x_1 and x_3 are reachable, whereas x_2 is not reachable (in other words, a point $\bar{x} = (\bar{x}_1, \bar{x}_2, \bar{x}_3)$ of the state space is reachable if and only if $\bar{x}_2 = 0$). If instead $\beta = 0$, while $\alpha \neq 0$, then x_3 has no influence on y and the system is not observable.

When a zero-pole simplification occurs, the system is not a minimal representation. Indeed, one can construct a realization of order 2 (cases 2 and 3) or a realization of order 1 (case 4).

We can conclude this section with the following main statements.

Proposition A.3 (Minimal representation)   *A realization is minimal if and only if the system is reachable and controllable.*

Proposition A.4 (Smith–McMillan degree)   *The Smith–McMillan degree of a system is the order of its reachable and observable part.*

A.9   Stabilizability and Detectability

Consider the system

$$x(t+1) = Fx(t) + Gu(t) \tag{A.27}$$

subject to the state feedback

$$u(t) = Lx(t). \tag{A.28}$$

The overall system, governed by the equation

$$x(t+1) = (F + GL)x(t) \tag{A.29}$$

is named *closed-loop system*. By contrast, the original system is called *open-loop system*.

In open loop, the system eigenvalues are those of matrix F; in closed loop, they are those of matrix $(F + GL)$.

Two main problems are

i) *Stabilization problem*: Assume the open-loop system to be unstable. With a suitable choice of matrix L, is it possible to obtain a stable closed-loop system?

ii) *Eigenvalue assignment problem*: With a suitable choice of matrix L, is it possible to locate the closed-loop poles in desired positions in the complex plane?

Of course, the stabilization problem is a special case of the eigenvalue assignment problem, since stabilization just requires a suitable positioning of the eigenvalues (to be all moved inside the unit desk of the complex plane).

It can be seen that the eigenvalue assignment problem can be solved if the system is reachable. If the system is not reachable, then only the eigenvalues of the reachable part can be arbitrarily assigned, while the unreachable eigenvalues cannot be modified. This fact is rather intuitive, since the input cannot affect the unreachable part of the system. In conclusion, the stabilization problem can be solved if the system is reachable, or if its unreachable part is stable.

This leads to the following notion:

Stabilizability:

A system is said to be *stabilizable* if it is reachable or if its unreachable part is stable.

Analogously, we can introduce the dual notion of

Detectability:

A system is said to be *detectable* if it is observable or if its unobservable part is stable.

Reachability and observability are named *structural properties*. Stabilizability and detectability are known as *extended structural properties*.

Appendix B

Matrices

B.1 Basics

A set of numbers ordered in a column as

$$x = \begin{bmatrix} a_1 \\ a_2 \\ \vdots \\ a_n \end{bmatrix}$$

is named vector. The entries a_i may be real or complex. n is the dimension of the vector.

The vector with all entries equal to 0 is also denoted by the symbol 0.

Given a scalar α, the product αx is the vector with entries αa_i.

If $x_1, x_2, \dots, x_k$ are vectors of the same dimension, for any set of scalar coefficients $\alpha_1, \alpha_2, \dots, \alpha_k$, the vector $\alpha_1 x_1 + \alpha_2 x_2 + \cdots + \alpha_k x_k$ is a *linear combination* of $x_1, x_2 \cdots x_k$. Vectors $x_1, x_2, \dots, x_k$ are said to be *linearly independent* if $\alpha_1 x_1 + \alpha_2 x_2 + \cdots + \alpha_k x_k = 0$ only if $\alpha_i = 0, \forall i$. Otherwise, they are said to be linearly dependent. For example, in a two-dimensional framework, the column vectors

$$x_1 = \begin{bmatrix} 3 \\ 4 \end{bmatrix}, \quad x_2 = \begin{bmatrix} 2 \\ 1 \end{bmatrix}$$

are linearly independent. Indeed, the system $\alpha_1 x_1 + \alpha_2 x_2 = 0$ has the only solution $\alpha_1 = 0$ and $\alpha_2 = 0$. On the opposite, given

$$x_1 = \begin{bmatrix} 8 \\ 4 \end{bmatrix}, \quad x_2 = \begin{bmatrix} 2 \\ 1 \end{bmatrix}$$

the system $\alpha_1 x_1 + \alpha_2 x_2 = 0$ has the solution $\alpha_2 = -4\alpha_1$. Therefore, they are linearly dependent.

As anticipated in Appendix A, the set of all linear combinations of a given bunch of vectors is named *subspace*, the subspace *generated* by those vectors.

Model Identification and Data Analysis, First Edition. Sergio Bittanti.
© 2019 John Wiley & Sons, Inc. Published 2019 by John Wiley & Sons, Inc.

By *dimension* of a subspace we mean the number of independent vectors generating it. For example, consider the vectors

$$x_1 = \begin{bmatrix} 3 \\ 0 \\ 5 \end{bmatrix}, \quad x_2 = \begin{bmatrix} 2 \\ 1 \\ 4 \end{bmatrix}, \quad x_3 = \begin{bmatrix} 5 \\ 1 \\ 9 \end{bmatrix}.$$

x_1 and x_2 are linearly independent. As for x_3, we note that it is the sum of the previous two, $x_3 = x_1 + x_2$. Therefore, x_1, x_2, and x_3 generate a subspace of dimension 2, i.e.

$$\{\alpha_1 x_1 + \alpha_2 x_2 \; ; \forall \alpha_1, \forall \alpha_2\}.$$

Note that the pairs $\{x_1, x_3\}$ and $\{x_2, x_3\}$ are also linearly independent. Therefore, the same subspace may be generated as

$$\{\alpha_1 x_1 + \alpha_3 x_3 \; ; \forall \alpha_1, \; \forall \alpha_3\}$$

or

$$\{\alpha_2 x_2 + \alpha_3 x_3; \; \forall \alpha_2, \; \forall \alpha_3\}.$$

A matrix A is an array of entries organized in rows and columns as

$$A = \begin{bmatrix} a_{11} & a_{12} & \cdots & a_{1n} \\ a_{21} & a_{22} & \cdots & a_{2n} \\ \vdots & \vdots & \vdots & \vdots \\ a_{m1} & a_{m2} & \cdots & a_{mn} \end{bmatrix},$$

where a_{ij} are real or complex numbers.

We say that the matrix has dimension $(m \times n)$ to point out the number of rows (m) and columns (n).

Obviously, when $n = 1$, we obtain a vector, also called *column vector*. When $m = 1$, the matrix is named *row vector*. If $m = 1$ and $n = 1$, the matrix is a scalar.

If $m = n$, we say that the matrix is square.

The *transpose* A' is obtained by exchanging rows with columns, i.e. the (i, j) entry of A' is entry (j, i) of A. Hence, the dimension of A' is $(n \times m)$.

The *adjoint matrix* is the $(n \times m)$ matrix A^* with entries given by the complex conjugate of the entries of the transpose.

Entries a_{ii} constitute the *principal diagonal* or simply *diagonal* of A.

A square matrix is said to be *symmetric* if $a_{ij} = a_{ji}$, $\forall i, j$.

A matrix is *lower triangular* if $a_{ij} = 0$, $\forall i < j$, *upper triangular* if $a_{ij} = 0$, $\forall i > j$. A matrix that is either lower or upper triangular is said to be *triangular*.

A matrix is *diagonal* when $a_{ij} = 0$, $\forall i \neq j$. Such matrix is denoted by diag $[a_{ii}]$, where the a_{ii} are the elements along the diagonal. A diagonal matrix is also triangular.

The product of a matrix A by a scalar α is the matrix whose entries are given by the entries of A multiplied by α:

$$\alpha A = [\alpha\ a_{ij}].$$

For two matrices of the same dimension, their sum is

$$A + B = [a_{ij} + b_{ij}].$$

Given two matrices A and B with dimensions $(m \times p)$ and $(p \times n)$, the product AB is the $(m \times n)$ matrix whose (i, j) entry is given by

$$\{AB\}_{ij} = \sum_{k}^{p} a_{ik} b_{kj}.$$

This is the celebrated rule of the *rows by columns product*.

By multiplying a square matrix A by itself a number of times, we obtain the power A^n.

If there exists an integer n such that $A^n = 0$, then A is said to be *nilpotent*.

Example B.1 (Nilpotent matrix)   The various powers of the 5×5 matrix

$$N = \begin{bmatrix} 0 & 1 & 0 & 0 & 0 \\ 0 & 0 & 1 & 0 & 0 \\ 0 & 0 & 0 & 1 & 0 \\ 0 & 0 & 0 & 0 & 1 \\ 0 & 0 & 0 & 0 & 0 \end{bmatrix}$$

are given by

$$N^2 = \begin{bmatrix} 0 & 0 & 1 & 0 & 0 \\ 0 & 0 & 0 & 1 & 0 \\ 0 & 0 & 0 & 0 & 1 \\ 0 & 0 & 0 & 0 & 0 \\ 0 & 0 & 0 & 0 & 0 \end{bmatrix}, \quad N^3 = \begin{bmatrix} 0 & 0 & 0 & 1 & 0 \\ 0 & 0 & 0 & 0 & 1 \\ 0 & 0 & 0 & 0 & 0 \\ 0 & 0 & 0 & 0 & 0 \\ 0 & 0 & 0 & 0 & 0 \end{bmatrix},$$

$$N^4 = \begin{bmatrix} 0 & 0 & 0 & 0 & 1 \\ 0 & 0 & 0 & 0 & 0 \\ 0 & 0 & 0 & 0 & 0 \\ 0 & 0 & 0 & 0 & 0 \\ 0 & 0 & 0 & 0 & 0 \end{bmatrix}, \quad N^5 = \begin{bmatrix} 0 & 0 & 0 & 0 & 0 \\ 0 & 0 & 0 & 0 & 0 \\ 0 & 0 & 0 & 0 & 0 \\ 0 & 0 & 0 & 0 & 0 \\ 0 & 0 & 0 & 0 & 0 \end{bmatrix}.$$

Therefore, N is nilpotent.

Remark B.1 (Various types of zeros)   We warn our readers that the symbol 0 used herein may have various meanings. It may be a scalar, a vector, or a matrix. For instance, in the condition of linear independence, $\alpha_1 x_1 + \alpha_2 x_2 + \cdots + \alpha_k x_k = 0$, the 0 is a vector with all entries equal to the scalar 0. On the other hand, the zero appearing in the definition of nilpotent matrix, $A^n = 0$, is a matrix of the same dimension of A with all entries equal to the scalar 0.

The square matrix defined by

$$a_{ij} = \begin{cases} 0, & i \neq j \\ 1, & i = j \end{cases}$$

is named *identity matrix*, usually denoted with the symbol I:

$$I = \begin{bmatrix} 1 & 0 & 0 & \cdots & 0 \\ 0 & 1 & 0 & \cdots & 0 \\ 0 & 0 & 1 & \cdots & 0 \\ \vdots & \vdots & \vdots & \ddots & \vdots \\ 0 & 0 & 0 & \cdots & 1 \end{bmatrix}.$$

It is straightforward to verify that

$$AI = IA = A.$$

A square matrix A is said to be *invertible* when there exists a square matrix B such that

$$AB = I. \tag{B.1}$$

If it exists, B is called *inverse* of A and denoted by A^{-1}. Note that, if B is the inverse of A, A is the inverse of B.

A diagonal matrix $\text{diag}[a_{ii}]$ admits inverse if and only if $a_{ii} \neq 0, \forall i$. In that case, the inverse is $\text{diag}[a_{ii}^{-1}]$.

The sum is commutative, i.e. $A + B = B + A$. The product, in general, is not, i.e. AB may be different than BA.

As for transposition,

$$(A + B)' = A' + B',$$
$$(AB)' = B'A'.$$

The same property holds for the adjoint matrix.

The *trace* of a square matrix is defined as the sum of the entries along the principal diagonal

$$\text{trace}[A] = \sum_i a_{ii}.$$

A remarkable property is that, if both the products AB and BA make sense, then the trace is the same

$$\text{trace}[AB] = \text{trace}[BA], \tag{B.2}$$

even if $AB \neq BA$.

Example B.2 (Trace of identity) Obviously, the trace of the identity matrix is its dimension:

$$\text{trace}[I] = n.$$

Example B.3 (The diad) Given a vector y of dimension $n > 1$,

$$y = \begin{bmatrix} y_1 \\ y_2 \\ \vdots \\ y_n \end{bmatrix},$$

the $n \times n$ symmetric matrix $A = yy'$ is named *diad*.

Note that the first column of yy' is vector y multiplied by the scalar y_1, the second column is vector y multiplied by the scalar y_2, and so on. Hence, all columns linearly depend upon the unique vector y.

Its trace can be computed as

$$\text{trace}[yy'] = \text{trace}[y'y].$$

On the other hand, matrix $y'y$ is the scalar, given by the sum of the squares of its entries. In conclusion,

$$\text{trace}[yy'] = y_1^2 + y_2^2 + \cdots + y_n^2.$$

B.2 Eigenvalues

In Appendix A, we defined an *eigenvalue* of a matrix A as any λ such that $\det(\lambda I - A) = 0$.

The definition we introduce here is: λ is an *eigenvalue* if there exists a vector x whose entries are not all null such that

$$Ax = \lambda x, \quad x \neq 0.$$

The two definitions are obviously equivalent to each other. The last definition enables one to introduce a new concept, that of *eigenvector*: Vector x is named *eigenvector* of A associated to eigenvalue λ.

An $n \times n$ matrix has n eigenvalues, some of which can be coincident.

One can give a simple geometric interpretation of the notion of eigenvalue and eigenvector by considering the relation $y = Ax$. In general, vector y has a direction different from that of vector x. The only exception takes place when x is an eigenvector (associated to a certain eigenvalue λ): in that case, y has the same direction of x. The eigenvalue determines the proportionality coefficient between the two vectors y and x.

It is apparent that, if x is an eigenvector, αx is also an eigenvector for any scalar α; moreover, if $x_1, x_2, \ldots, x_k$ are eigenvectors associated with the same eigenvalue, then any linear combination of them is an eigenvector too. In other words, the set of eigenvectors associated with a given eigenvalue form a subspace.

Example B.4 (Eigenvalues and eigenvectors of a reflection matrix) The eigenvalues and eigenvectors of the 2×2 matrix

$$A = \begin{bmatrix} 0 & 1 \\ 1 & 0 \end{bmatrix}$$

can be determined by solving the system

$$\begin{bmatrix} 0 & 1 \\ 1 & 0 \end{bmatrix} \begin{bmatrix} x_1 \\ x_2 \end{bmatrix} = \lambda \begin{bmatrix} x_1 \\ x_2 \end{bmatrix}.$$

It is easy to see that the eigenvalues are $\lambda_1 = 1$ and $\lambda_2 = -1$. The eigenvectors associated to $\lambda_1 = 1$ are the set of all vectors in the $x_1 - x_2$ plane on the bisector of the first and third quadrants (i.e. all vectors with $x_2 = x_1$). The eigenvectors associated to $\lambda_2 = -1$ are the set of all vectors belonging to the bisector of the second and fourth quadrants (i.e. all vectors with $x_2 = -x_1$).

Note that this matrix, when applied to a vector x, gives rise to $y = Ax$, which is x reflected with respect to the bisector of the first and third quadrants. This is why A is named *reflection matrix*.

Example B.5 (Eigenvalues and eigenvectors of the identity matrix) For the computation of the eigenvalues and eigenvectors of the 2×2 identity matrix

$$I = \begin{bmatrix} 1 & 0 \\ 0 & 1 \end{bmatrix},$$

we have to deal with the system

$$\begin{bmatrix} 1 & 0 \\ 0 & 1 \end{bmatrix} \begin{bmatrix} x_1 \\ x_2 \end{bmatrix} = \lambda \begin{bmatrix} x_1 \\ x_2 \end{bmatrix}.$$

The only λ solving this system is $\lambda = 1$. For this eigenvalue, all vectors x are eigenvectors.

Example B.6 (Eigenvalues and eigenvectors of the diad) If we multiply the diad $A = yy'$ by y, we obtain

$$Ay = [yy'] \, y = y \, [y'y] = y \, [y_1^2 + y_2^2 + \cdots + y_n^2].$$

Therefore, by letting

$$\lambda = y_1^2 + y_2^2 + \cdots + y_n^2,$$

we have

$$Ay = \lambda y.$$

Hence, λ is an eigenvalue of the diad, with y as an associated eigenvector. Our readers are invited to show that the set of all eigenvectors associated to λ is the line $\{\alpha y \mid \forall \alpha\}$.

The eigenvalues of a triangular matrix (and therefore of a diagonal matrix) are the diagonal entries.

B.3   Determinant and Inverse

The *determinant* of a square matrix is defined by an iteration over the dimension of the matrix as follows. For a 2×2 matrix

$$A = \begin{bmatrix} a_{11} & a_{12} \\ a_{21} & a_{22} \end{bmatrix},$$

the determinant is defined as

$$\det[A] = a_{11}a_{22} - a_{12}a_{21}.$$

For a matrix of any dimension $n \times n$, the definition is

$$\det[A] = a_{i1} \ \det[\tilde{A}_{i1}] + a_{i2} \ \det[\tilde{A}_{i2}] + \cdots + a_{in} \ \det[\tilde{A}_{in}],$$

where, for each $j = 1, 2, \ldots, n$,

$$\tilde{A}_{ij} = (-1)^{i+j} A_{ij}. \tag{B.3}$$

A_{ij} being the $(n-1) \times (n-1)$ matrix obtained from A by erasing row i and column j. Note that the result of the computation of the determinant is the same whatever i, $i = 1, 2, \ldots, n$, be chosen.

Example B.7 (Determinant of a 3×3 matrix)   For the 3×3 matrix

$$A = \begin{bmatrix} a_{11} & a_{12} & a_{13} \\ a_{21} & a_{22} & a_{23} \\ a_{31} & a_{32} & a_{33} \end{bmatrix},$$

by taking $i = 1$, the determinant is given by

$$\det[A] = a_{11} \ \det[A_{11}] - a_{12} \ \det[A_{12}] + a_{13} \ \det[A_{13}],$$

with

$$A_{11} = \begin{bmatrix} a_{22} & a_{23} \\ a_{32} & a_{33} \end{bmatrix}, \quad A_{12} = \begin{bmatrix} a_{21} & a_{23} \\ a_{31} & a_{33} \end{bmatrix}, \quad A_{13} = \begin{bmatrix} a_{21} & a_{22} \\ a_{31} & a_{32} \end{bmatrix}.$$

We say that a matrix is *singular* when the determinant is null, *non-singular* or *invertible* in the opposite.

It is easy to see that the determinant of a triangular matrix (and therefore of a diagonal matrix) is given by the product of its diagonal entries. In particular, the identity matrix has unit determinant

$$\det[I] = 1.$$

The determinant of the transpose coincides with the determinant of the original matrix

$$\det[A'] = \det[A]$$

and the determinant of the product is the product of the determinants

$$\det[AB] = \det[A] \, \det[B]. \tag{B.4}$$

From this last property, it follows that if a matrix is singular (null determinant), then its inverse does not exist. Indeed, Eq. (B.1) entails that

$$\det[AB] = 1.$$

Hence,

$$\det[A] \, \det[B] = 1$$

so that, when $\det[A] = 0$, this identity would lead to the absurd conclusion that $0 = 1$. Hence, a matrix is invertible if and only if it is non-singular.

Obviously, assuming that $\det[A] \neq 0$,

$$\det[A^{-1}] = \frac{1}{\det[A]}.$$

The determinant is strictly related to the eigenvalues. Indeed, it can be shown that the product of all eigenvalues is the determinant

$$\det[A] = \Pi_i \lambda_i.$$

This implies that a matrix is singular, $\det[A] = 0$, if and only if A has a null eigenvalue.

Remark B.2 (Geometrical interpretation of the determinant) Consider the vectors

$$x^{(1)} = \begin{bmatrix} x_1^{(1)} \\ x_2^{(1)} \end{bmatrix}, \quad x^{(2)} = \begin{bmatrix} x_1^{(2)} \\ x_2^{(2)} \end{bmatrix}$$

and the matrix

$$X = \begin{bmatrix} x_1^{(1)} & x_1^{(2)} \\ x_2^{(1)} & x_2^{(2)} \end{bmatrix}.$$

Figure B.1 The parallelogram associated to a 2 × 2 matrix.

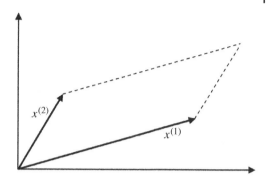

We leave to our readers to prove that the absolute value of the determinant of matrix X is the area of the parallelogram formed by vectors $x^{(1)}$ and $x^{(2)}$ (Figure B.1).

Consider now a 2×2 matrix A. When applied to matrix X, we obtain the matrix

$$Y = AX,$$

whose columns are

$$y^{(1)} = Ax^{(1)}, \quad y^{(2)} = Ax^{(2)}.$$

In the same way as $|\det[X]|$ is the area of the parallelogram defined by columns $x^{(1)}$ and $x^{(2)}$, $|\det[Y]|$ is the area of the parallelogram defined by columns $y^{(1)}$ and $y^{(2)}$.

Relationship (B.4) then leads to the conclusion that $|\det[A]|$ is the amplification/attenuation coefficient linking the area of the parallelogram associated with matrix Y to the area of the parallelogram associated with matrix X.

Note that this geometrical interpretation of the determinant can be extended to 3×3 matrices; the absolute value of the determinant is then the volume of the parallelepiped defined by the three columns of the matrix.

Assuming that $\det[A] \neq 0$, the entries b_{ij} of the inverse B of A can be computed as follows:

$$b_{ij} = \frac{1}{\det[A]} \det[\tilde{A}_{ji}],$$

where $\tilde{A}_{ij}$ was defined in (B.3). Note that, in this formula, b_{ij} and $\tilde{A}_{ji}$ have interchanged indexes.

The inverse of a product of two invertible matrices can be computed as

$$(AB)^{-1} = B^{-1}A^{-1}.$$

A real matrix is said to be *orthogonal* when $A'A = I$, or equivalently $A' = A^{-1}$. For such matrices:

- any pair of rows x and y of A are orthogonal, i.e. $x'y = 0$;
- any pair of columns are orthogonal;
- the product of two orthogonal matrices is an orthogonal matrix;
- the inverse of an orthogonal matrix is orthogonal.

The following result provides a very useful inversion formula, the *matrix inversion lemma*.

Proposition B.1 (Matrix inversion lemma) *Consider four matrices A, B, C, and D of suitable dimensions such that the square matrix $A + BCD$ can be formed according to the matrix multiplication rule. Suppose that A and C are invertible. Then*

$$(A + BCD)^{-1} = A^{-1} - A^{-1}B(C^{-1} + DA^{-1}B)^{-1}DA^{-1}.$$

In particular, when we add a diad yy' to a square matrix A, we have

$$(A + yy')^{-1} = A^{-1} - \frac{1}{1 + y'A^{-1}y}A^{-1}yy'A^{-1}.$$

Proof: Indeed, we can write A^{-1} as

$$A^{-1} = (A + BCD)^{-1}(A + BCD)A^{-1}$$

$$= (A + BCD)^{-1}(I + BCDA^{-1})$$

$$= (A + BCD)^{-1} + (A + BCD)^{-1}BCDA^{-1}.$$

On the other hand, our readers can verify that

$$(A + BCD)^{-1}BC = A^{-1}B(C^{-1} + DA^{-1}B)^{-1}.$$

This formula, together with the previous one, leads to the lemma.

B.4   Rank

For a $m \times n$ matrix A, we define as *column rank* the maximum number of linearly independent columns and as *row rank* the maximum number of linearly independent rows.

Perhaps surprisingly, the column rank and the row rank coincide. This leads to the definition of *rank* of a (non-necessarily square) matrix as the number of linearly independent columns (or equivalently the number of linearly independent rows).

The rows (columns) of the transpose A' of a matrix A are the columns (rows) of A. Thus,

$$\text{rank}[A'] = \text{rank}[A].$$

Obviously, rank$[A] \leq \min(m, n)$. We say that the matrix is *maximum rank* when rank$[A] = \min(m, n)$.

An $n \times n$ matrix is invertible if and only if its rank is maximum, rank$[A] = n$.

Example B.8 The 2×3 matrix

$$A = \begin{bmatrix} 8 & 6 & 4 \\ 4 & 3 & 2 \end{bmatrix}$$

has rank 1 since the first row is the double of the second one.

On the opposite, the 2×3 matrix

$$A = \begin{bmatrix} 8 & 6 & 4 \\ 4 & 2 & 2 \end{bmatrix}$$

has rank 2 since the two rows are linearly independent.

Consider now the product of two matrices, AB where A is $m \times p$ and B is $p \times n$. Its rank satisfies the inequality

$$\text{rank}[AB] \leq \min(\text{rank}[A], \text{rank}[B]). \tag{B.5}$$

Indeed, in view of the definition of matrix product, the rows of AB are linear combinations of the rows of B, so that the number of linearly independent rows of AB cannot be greater than the number of linear independent rows of B. Analogously, the columns of AB are linear combinations of the columns of A so that the number of independent columns cannot be greater than the number of independent columns of A.

Another important inequality concerning the rank of a product is

$$\text{rank}[AB] \geq \text{rank}[A] + \text{rank}[B] - p, \tag{B.6}$$

where p is the number of columns of A (same as the number of rows of B).

From (B.5) and (B.6), the so-called *Sylvester inequality* follows:

$$\text{rank}[A] + \text{rank}[B] - p \leq \text{rank}[AB] \leq \min(\text{rank}[A], \text{rank}[B]).$$

If rank$[A] = \text{rank}[B] = p$, then rank$[AB] = p$.

By applying such result to the product of two invertible matrices of dimensions $n \times n$, we have rank$[AB] = n$. When $B = A$, and rank$[A] = n$, rank$[A^2] = \text{rank}[A] = n$. However, if A is singular, then rank$[A^2]$ may be strictly lower than rank$[A]$. For instance, matrix N^2 in Example B.1 is of rank 3 while rank$[N] = 4$.

When applied to a diad yy', with $y \neq 0$, the Sylvester inequality leads to

$$\text{rank}[yy'] = 1,$$

a conclusion easily achievable by a direct analysis of matrix yy'.

We finally observe that the rank of the sum $yy' + zz'$ of two diads is 1 if z and y are linearly dependent. If instead z and y are linearly independent, it is equal to 2.

B.5 Annihilating Polynomial

Up to now, in this appendix, we have considered matrices as arrays of numbers. We now pass to consider matrices as *arrays of functions*.

Given an $n \times n$ matrix A, we can introduce the $n \times n$ matrix

$$\lambda I - A, \tag{B.7}$$

where I is the $n \times n$ identity matrix. The entries of (B.7) are scalar functions of variable λ. We can compute the determinant of such matrix in the same way we defined the determinant for a matrix of numbers. This leads to

$$\Delta(\lambda) = \det[\lambda I - A].$$

$\Delta(\lambda)$ is a scalar function of λ. To be precise, it is a polynomial of degree n in λ, the *characteristic polynomial* of A.

Correspondingly,

$$\det[\lambda I - A] = 0$$

is called *characteristic equation* of A.

We know that any eigenvalue satisfies the characteristic equation, and vice versa, a solution of the characteristic equation is an eigenvalue. In other words, *the roots of the characteristic equation are all and only the eigenvalues of A.*

The fundamental theorem of algebra states that a polynomial with real coefficients of degree n has precisely n roots, which may be real or complex. If there is a complex root, then its complex conjugate is also a root. Hence, an nxn matrix has n eigenvalues. If there is a complex eigenvalue, then its complex conjugate is also an eigenvalue.

The eigenvalues of a symmetric matrix are all real.

We have previously seen that the product of the eigenvalues is the determinant of the matrix. We now observe that sum of the eigenvalues is its trace:

$$\text{trace}[A] = \sum_i \lambda_i. \tag{B.8}$$

Example B.9 (Characteristic polynomial of the diad) Consider again the diad introduced in Example B.6: $A = yy'$.

If $n = 2$, then $\det[\lambda I - A] = \lambda^2 - (y_1^2 + y_2^2)\lambda$. Hence, the eigenvalues are $\lambda = y_1^2 + y_2^2$ and $\lambda = 0$.

In general, for any $n \geq 2$, $n - 1$ eigenvalues of a diad are null; the only non-null eigenvalue is $\lambda = y_1^2 + y_2^2 + \cdots + y_n^2$. This means that the characteristic polynomial is $\lambda^n - (y_1^2 + y_2^2 + \cdots + y_n^2)\lambda^{n-1}$.

It is easy to verify that $\text{trace}[A] = y_1^2 + y_2^2 + \cdots + y_n^2$ so that it coincides with the (only non-null) eigenvalue of the matrix.

Sometimes, given a polynomial in λ, say $\Theta(\lambda)$, it is useful to replace λ with matrix A. For instance, if $\Theta(\lambda) = \lambda^2 + 3\lambda + 5$, then, replacing λ with A, we obtain matrix $\Delta(A) = A^2 + 3A + 5I$. Note that the known term $5 = 5\lambda^0$ is replaced by $5A^0 = 5I$.

Polynomial $\Theta(\lambda)$ is said to be *annihilating* for matrix A if by replacing λ with matrix A, we obtain the null matrix, $\Theta(A) = 0$.

A special case is that of the characteristic polynomial; thanks to the Cayley–Hamilton theorem stated in Proposition A.1, we can say that, if $\Delta(\lambda)$ is the characteristic polynomial of A, then $\Delta(A) = 0$, namely the characteristic polynomial is annihilating.

Obviously, if $\Theta(\lambda)$ is annihilating, $\alpha\Theta(\lambda)$ is also annihilating $\forall\alpha$. To avoid such type of multiplicity, one often imposes that $\Theta(\lambda)$ is monic (i.e. the coefficient of the leading term is 1).

Given a matrix A, the annihilating monic polynomial of minimal degree is called *minimal polynomial*, denoted as $\nabla(\lambda)$.

Example B.10 (Characteristic polynomial of a diagonal matrix) The eigenvalues of an $n \times n$ diagonal matrix

$$
\begin{bmatrix}
p_1 & 0 & 0 & 0 \\
0 & p_2 & 0 & 0 \\
0 & 0 & \ddots & 0 \\
0 & 0 & 0 & p_n
\end{bmatrix}
$$

are the elements along the diagonal, i.e. $p_1, p_2, \ldots, p_n$. The characteristic polynomial is $\Delta(\lambda) = (\lambda - p_1)(\lambda - p_2)\cdots(\lambda - p_n)$.

In particular, the identity matrix of dimension n has n coincident eigenvalues, all equal to 1, and its characteristic polynomial is $\Delta(\lambda) = (\lambda - 1)^n$.

Example B.11 (Characteristic and minimal polynomials of certain matrices) Consider the following 4×4 matrices

$$
K_0 = \begin{bmatrix}
p & 0 & 0 & 0 \\
0 & p & 0 & 0 \\
0 & 0 & p & 0 \\
0 & 0 & 0 & p
\end{bmatrix}, \quad
K_1 = \begin{bmatrix}
p & 1 & 0 & 0 \\
0 & p & 0 & 0 \\
0 & 0 & p & 0 \\
0 & 0 & 0 & p
\end{bmatrix},
$$

$$
K_2 = \begin{bmatrix}
p & 1 & 0 & 0 \\
0 & p & 1 & 0 \\
0 & 0 & p & 0 \\
0 & 0 & 0 & p
\end{bmatrix}, \quad
K_3 = \begin{bmatrix}
p & 1 & 0 & 0 \\
0 & p & 1 & 0 \\
0 & 0 & p & 1 \\
0 & 0 & 0 & p
\end{bmatrix}.
$$

For all $i = 0, 1, 2, 3$, the matrix $\lambda I - K_i$ is upper triangular, with identical diagonal entries, all given by $\lambda - p$. Since the determinant of a triangular matrix is the product of the diagonal entries, all the above matrices have the same characteristic polynomial, given by

$$\Delta(\lambda) = (\lambda - p)^4.$$

Consider now polynomial $(\lambda - p)$. Correspondingly,

$$K_0 - pI = \begin{bmatrix} 0 & 0 & 0 & 0 \\ 0 & 0 & 0 & 0 \\ 0 & 0 & 0 & 0 \\ 0 & 0 & 0 & 0 \end{bmatrix}.$$

We let our readers verify that the following polynomials are all annihilating:

$$\Theta(\lambda)_a = (\lambda - p)^4,$$

$$\Theta(\lambda)_b = (\lambda - p)^3,$$

$$\Theta(\lambda)_c = (\lambda - p)^2,$$

$$\Theta(\lambda)_d = (\lambda - p)^1.$$

Here, the first polynomial $\Theta(\lambda)_a$ is the characteristic polynomial of K_0, $\Delta(\lambda)$, while the last one, $\Theta(\lambda)_d$, is the minimal polynomial $V(\lambda)$.

Passing to K_1, we have

$$K_1 - pI = \begin{bmatrix} 0 & 1 & 0 & 0 \\ 0 & 0 & 0 & 0 \\ 0 & 0 & 0 & 0 \\ 0 & 0 & 0 & 0 \end{bmatrix}.$$

Hence, $(\lambda - p)$ is not annihilating for K_1. In this case, the minimal polynomial is

$$V(\lambda) = (\lambda - p)^2.$$

Indeed,

$$(K_1 - pI)^2 = \begin{bmatrix} 0 & 1 & 0 & 0 \\ 0 & 0 & 0 & 0 \\ 0 & 0 & 0 & 0 \\ 0 & 0 & 0 & 0 \end{bmatrix}^2 = \begin{bmatrix} 0 & 0 & 0 & 0 \\ 0 & 0 & 0 & 0 \\ 0 & 0 & 0 & 0 \\ 0 & 0 & 0 & 0 \end{bmatrix}.$$

We let our readers compute the minimal polynomial for K_2 and K_3. In conclusion,

$$K_0 \to V(\lambda) = (\lambda - p), \quad K_1 \to V(\lambda) = (\lambda - p)^2,$$

$$K_2 \to V(\lambda) = (\lambda - p)^3, \quad K_3 \to V(\lambda) = (\lambda - p)^4.$$

As illustrated in the previous example, the minimal polynomial may coincide with the characteristic polynomial or may be a factor of it.

B.6 Algebraic and Geometric Multiplicity

Given a square matrix A, an eigenvalue has been defined as any number λ such that $Ax = \lambda x$ for some $x \neq 0$. Vector x is an eigenvector associated with that eigenvalue.

Obviously, if x is an eigenvector, for all scalars α, αx is also an eigenvector. Moreover, if x_1 and x_2 are eigenvectors associated with the same eigenvalue λ, any linear combination of them $\alpha_1 x_1 + \alpha_2 x_2$ is also an eigenvector.

In general, as already pointed out, the set of eigenvectors associated with a given eigenvalue λ is a subspace.

The number of linearly independent eigenvectors associated with an eigenvalue λ (i.e. the dimension of the subspace of eigenvectors associated with that λ) is named *geometric multiplicity* of λ.

We also know that any eigenvalue is a solution of the characteristic equation. The *algebraic multiplicity* is the multiplicity of the eigenvalue as solution of the characteristic equation.

Example B.12 (Example B.11 continued)   Consider again matrices K_i of the previous Example B.11. All matrices have $\Delta(\lambda) = (\lambda - p)^4$ as characteristic polynomial. Hence, eigenvalue p has algebraic multiplicity equal to 4.

Passing to the geometric multiplicity we note that, for K_0, the vectors

$$\begin{bmatrix} 1 \\ 0 \\ 0 \\ 0 \end{bmatrix} \quad \begin{bmatrix} 0 \\ 1 \\ 0 \\ 0 \end{bmatrix} \quad \begin{bmatrix} 0 \\ 0 \\ 1 \\ 0 \end{bmatrix} \quad \begin{bmatrix} 0 \\ 0 \\ 0 \\ 1 \end{bmatrix}$$

are eigenvectors associated to p. Since they are linearly independent, the geometric multiplicity of p is 4. We let our readers show that the geometric multiplicity of K_1, K_2, and K_3 is 3, 2, and 1, respectively.

The geometric multiplicity of an eigenvalue is lower than or equal to its algebraic multiplicity.

B.7 Range and Null Space

A $m \times n$ real matrix A can be seen as a transformation from a space of dimension n to a space of dimension m. Indeed, given a vector x with n entries, the vector $y = Ax$ has m entries. We write

$$A : \mathbb{R}^n \to \mathbb{R}^m.$$

The set of all vectors obtained as $y = Ax$ when x ranges in the whole space $\mathbb{R}^n$ is named *range* of A, denoted as $\mathscr{R}[A]$

$$\mathscr{R}[A] = \{Ax | x \in \mathbb{R}^n\}.$$

By letting x_i be the ith entry of x and $a^{(i)}$ the ith column of A, it is apparent that

$$Ax = \sum_i x_i a^{(i)}.$$

This implies that the *range* of A is the linear combination of the columns of A, namely the subspace in $\mathbb{R}^m$ generated by the columns of A.

The dimension of the range of A is given by

$$\dim \mathscr{R}[A] = \text{rank}[A].$$

The *null space* of A, denoted by $\mathscr{N}[A]$, is defined as the set of vectors x such that $Ax = 0$:

$$\mathscr{N}[A] = \{x | Ax = 0\}.$$

The *null space* is a subspace of $\mathbb{R}^n$. Its dimension is called *nullity*. It turns out that

$$\text{rank}[A] + \text{nullity}[A] = n.$$

Example B.13 The 2×3 matrix

$$A = \begin{bmatrix} 8 & 6 & 4 \\ 4 & 3 & 2 \end{bmatrix}$$

is a transformation from $\mathbb{R}^2$ to $\mathbb{R}^3$. The readers are invited to verify that the *null space* is a plane (2D subspace) in $\mathbb{R}^3$, whereas the range is a line (1D subspace) in $\mathbb{R}^2$. Thus, the nullity is 2 and the rank 1. The sum of nullity and rank is 3, the number of columns of A.

B.8 Quadratic Forms

In this section, we deal with a symmetric real matrix A of dimension $n \times n$. Recall that the eigenvalues of such a matrix are real. Consider now the function

$$f(x) = x'Ax, \tag{B.9}$$

where vector x is n dimensional with entries $x_1, x_2, \ldots, x_n$. Function $f(x)$ is named *quadratic form*.

Example B.14 (Quadratic form of a diagonal matrix) For $n = 2$, if

$$A = \begin{bmatrix} a & 0 \\ 0 & b \end{bmatrix},$$ (B.10)

then

$$f(x) = x'Ax = ax_1^2 + bx_2^2.$$ (B.11)

A is said to be

- *positive semi-definite* (p.s.d.) if $x'Ax \geq 0, \forall x$;
- *positive definite* (p.d.) if it is p.s.d. and $x'Ax = 0$ only for $x = 0$;
- *negative semi-definite* (n.s.d.) if $-A$ is p.s.d.;
- *negative definite* (n.d.) if $-A$ is p.d.

Note that there exist matrices that are *indefinite*, i.e. do not enjoy any of the above properties. For example matrix

$$A = \begin{bmatrix} +1 & 0 \\ 0 & -1 \end{bmatrix}$$

is indefinite.

Often the following symbols are used:

- $A \geq 0$, positive semi-definite matrix,
- $A > 0$, positive definite matrix,
- $A \leq 0$, negative semi-definite matrix,
- $A < 0$, negative definite matrix.

The eigenvalues of a positive semi-definite matrix are non-negative. Indeed, assume that, by contradiction, a positive semi-definite matrix A admits a negative eigenvalue, $\lambda < 0$. By the definition of eigenvalue, there must exist a vector $x \neq 0$ such that $Ax = \lambda x$. Then, by pre-multiplying by x', one obtains $x'Ax = \lambda x'x$. Since $\lambda x'x < 0, \forall x \neq 0$, we come to the absurd conclusion $x'Ax < 0$.

In general, the following properties hold true:

- $A \geq 0$, if and only if all eigenvalues are non-negative,
- $A > 0$, if and only if all eigenvalues are positive,
- $A \leq 0$, if and only if all eigenvalues are non-positive,
- $A < 0$, if and only if all eigenvalues are negative.

Consequently,

- $A \geq 0 \rightarrow \det[A] \geq 0$,
- $A > 0 \rightarrow \det[A] > 0$,
- $A \leq 0 \rightarrow \det[A] \leq 0$,
- $A < 0 \rightarrow \det[A] < 0$.

If A is positive semi-definite but not positive definite, then $\det[A] = 0$. Stated another way, a positive semi-definite matrix is positive definite if and only if it is invertible. An analogous statement holds for a negative semi-definite matrix.

Example B.15 (Example B.14 continued)   The 2×2 matrix (B.10) is p.d. when both $a > 0$ and $b > 0$. In that case, function (B.11) is a bowl lying in the upper region $(f(x) > 0)$ with vertex in the origin (of plane $x_1 - x_2$). By cutting the bowl with an horizontal plane, say $f(x) = 1$, we have

$$ax_1^2 + bx_2^2 = 1,$$

which is an ellipse with semi-axis of length $1/\sqrt{a}$ and $1/\sqrt{b}$. This is illustrated in Figure B.2.

If instead $a > 0$ and $b = 0$, then (B.10) is p.s.d. and not p.d.; indeed, the bowl degenerates into a valley with axis x_2 at its bottom. So, $x'Ax = 0$ in infinitely many points.

In general, for a matrix of any dimension, when $A > 0$, the function $y = x'Ax$ is a paraboloid with vertex in the origin. Its intersection with the plane $y = 1$ is an ellipsoid whose semi-axis has lengths given by the inverse of the square roots of the (real and positive) eigenvalues.

It is straightforward to verify that the sum of positive semi-definite matrices if positive semi-definite, while the sum of a positive semi-definite matrix and a positive definite matrix is positive definite.

Example B.16 (Positive semi-definiteness of a diad)   Given a vector $y \neq 0$ with $n > 1$ entries, the diad $A = yy'$ is positive semi-definite but not positive definite. Indeed, this matrix has $n - 1$ eigenvalues, which are null and the only non-null eigenvalue is positive (see Example B.6).

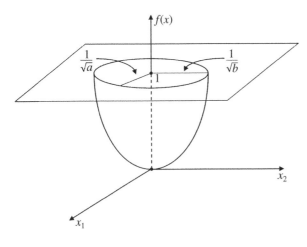

Figure B.2 The quadratic form of Example B.14 with non-null a and b.

We can order square matrices of the same dimension by using the symbols

$$A \geq B \quad [A > B]$$

to mean

$$A - B \geq 0 \quad [A - B > 0],$$

respectively.

Example B.17 (Ordering sum of diads) Consider two diads $A_1 = yy'$ and $A_2 = zz'$. Their sum $A = A_1 + A_2$ is a positive semi-definite matrix such that both $A \geq A_1$ and $A \geq A_2$ hold.

B.9 Derivative of a Scalar Function with Respect to a Vector

Consider a real scalar function $f(x)$ depending on a vector x with n entries

$$x = [x_1 \ x_2 \ \cdots \ x_n].$$

The derivative of f with respect to x is defined as

$$\frac{d}{dx}J(x) = \left[\frac{\partial}{\partial x_1} f(x) \ \frac{\partial}{\partial x_2} f(x) \ \cdots \ \frac{\partial}{\partial x_n} J(x) \right],$$

where $\frac{\partial}{\partial x_i} f(x)$ is the partial derivative of $f(x)$ with respect to the ith entry x_i of x.

The second derivative of J with respect to vector x is defined as

$$\frac{d^2}{dx^2} f(x) = \left[\frac{\partial^2}{\partial x_i \, x_j} f(x) \right]_{i=1,2,\ldots,n; \ j=1,2,\ldots n},$$

where $\frac{\partial^2}{\partial x_i x_j} J(x)$ is the second derivative of $f(x)$ with respect to x_i and x_j.

Hence, the first derivative is a row vector with n entries, while the second derivative is an $n \times n$ square matrix.

Example B.18 The scalar function $f(x) = 2x_1 + 5x_2^2$ has derivatives

$$\frac{d}{dx} f(x) = [2 \ 10x_2],$$

$$\frac{d^2}{dx^2} f(x) = \begin{bmatrix} 0 & 0 \\ 0 & 10 \end{bmatrix}.$$

B.10 Matrix Diagonalization via Similarity

As we have already seen, given an $n \times n$ invertible matrix T, the transformation leading from an $n \times n$ matrix A to the $n \times n$ matrix $\tilde{A}$

$$A \to \tilde{A} = T^{-1}AT$$

is named *similarity transformation*.

A main property of such transformation is that the eigenvalues of $\tilde{A}$ coincide with the eigenvalues of A.

A matrix is said to be *diagonalizable* if it can be transformed into a diagonal matrix by a similarity transformation. Any symmetric matrix is diagonalizable. However, not all matrices are diagonalizable. In general, with a similarity transformation, a matrix can be reduced to a *quasi-diagonal matrix* called *Jordan form*.

The Jordan form is – so to say – as "nearly diagonal as possible." To be precise, indicating with $\lambda_1, \lambda_2, \ldots, \lambda_q$ the distinct eigenvalues of the given matrix, this form is a block diagonal matrix of the type

$$\begin{bmatrix} J_1 & 0 & \cdots & 0 \\ 0 & J_2 & \cdots & 0 \\ 0 & 0 & \ddots & 0 \\ 0 & 0 & \cdots & J_q \end{bmatrix}.$$

The diagonal blocks J_k, named Jordan blocks, are square, possibly of different dimensions, with the same entry along the diagonal, eigenvalue λ_k.

In turn, each Jordan block is block diagonal of the type

$$J_k = \begin{bmatrix} J_k^{(1)} & 0 & \cdots & 0 \\ 0 & J_k^{(2)} & \cdots & 0 \\ 0 & 0 & \ddots & 0 \\ 0 & 0 & \cdots & J_k^{(r)} \end{bmatrix}.$$

Each $J_k^{(j)}$ is named Jordan mini-block. The mini-blocks may be of various dimensions, including the mono-dimensional case. When $J_k^{(j)}$ is 1×1, then $J_k^{(j)} = \lambda_k$. When $J_k^{(j)}$ is $s \times s$ with $s > 1$, then the Jordan mini-block is

$$J_k = \begin{bmatrix} \lambda_k & 1 & 0 & \cdots & 0 \\ 0 & \lambda_k & 1 & \cdots & 0 \\ \vdots & \vdots & \vdots & \ddots & \vdots \\ 0 & 0 & 0 & \cdots & 1 \\ 0 & 0 & 0 & \cdots & \lambda_k \end{bmatrix}. \tag{B.12}$$

The dimension of the Jordan block J_k is the algebraic multiplicity of eigenvalue λ_k, whereas the number of mini-blocks constituting J_k is the geometric multiplicity of eigenvalue λ_k.

B.11 Matrix Diagonalization via Singular Value Decomposition

An alternative diagonalization procedure is provided by the *singular value decomposition* (SVD). Note that such diagonalization method can be applied to non-square matrices as well.

Proposition B.2 (SVD theorem) *Given a matrix A, of dimension m × n, it is possible to factorize it as follows:*

$$A = U\ S\ V',$$

where

- *U is a square orthogonal matrix of dimension m × m,*
- *V is a square orthogonal matrix of dimension n × n,*
- *S is a matrix of the same dimension m × n of A, with all off-diagonal elements null; the diagonal entries, denoted as $\delta_1, \delta_2, \dots, \delta_k$, where $k = \min(m, n)$, are real and non-negative. They are set in decreasing order, i.e. $\delta_1 \geq \delta_2 \geq \delta_3 \cdots \geq \delta_k \geq 0$.*

The entries $\delta_1, \delta_2, \dots, \delta_k$ are named *singular values* of matrix A, and the above factorization is called SVD of A.

Note that, matrices U and V are not unique; however, the singular values are unique.

To visualize the SVD, we distinguish three cases: $m < n$, $m = n$, and $m > n$. A matrix with $m < n$ is sometimes referred to as *fat* matrix, while a matrix with $m > n$ is called *tall* matrix. The associated SVDs are depicted in Figure B.3.

The singular values provide the most useful information on a matrix, as stated in the following proposition.

Proposition B.3 (Singular values and rank) *The rank of a matrix coincides with the number of its non-null singular values.*

All off-diagonal entries of S being null, the above statement is equivalent to saying that any square or rectangular matrix can be written as

$$A = \sum_1^k \delta_i u_i v_i', \tag{B.13}$$

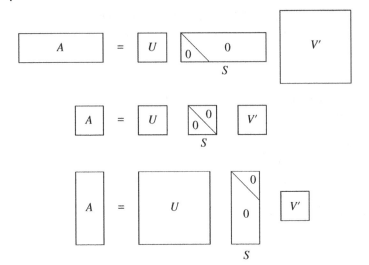

Figure B.3 SVD decomposition for a fat, square, and tall matrix.

where $u_1, u_2, \ldots, u_r$ are the columns of U and $v_1, v_2, \ldots, v_s$ are the columns of V.

Expression (B.13) shows that the SVD is just a way to split a matrix into a sum elementary matrices of rank 1.

The columns u_i and v_i are drawn from orthogonal matrices, so they are independent of each other. Therefore, the rank of A is maximum (i.e. rank$[A] = \min[r, s]$) unless some among the coefficients δ_i are null. In other words, the rank of A is equal to the number of non-null coefficients δ_i in the above linear combination.

Remark B.3 (Singular values and eigenvalues) By means of the SVD, the products AA' and $A'A$ can be written as

$$AA' = (USV')((USV'))' = USV'VS'U',$$

$$A'A = ((USV'))'(USV') = VS'U'USV'.$$

On the other hand, being V and U orthogonal, $V'V = I$ and $U'U = I$. Hence,

$$AA' = USS'U',$$

$$A'A = VS'SV'.$$

If A is square, then S is square as well. Also being diagonal with the singular values along the diagonal, we have

$$SS' = S'S = S^2 = \text{diag}[\delta_1^2, \delta_2^2, \ldots].$$

Moreover, U and V being orthogonal, $U' = U^{-1}$ and $V' = V^{-1}$. Hence,

$$AA' = US^2U^{-1}$$

and

$$A'A = VS^2V^{-1}.$$

These expressions show that the relationship between AA' (or $A'A$) and S^2 is a similarity transformation. On the other hand, we know that such transformation preserves eigenvalues. Therefore, the eigenvalues of AA' (or $A'A$) must coincide with the eigenvalues of S^2, which are obviously $\delta_1^2, \delta_2^2, \dots$.

Summing up, the singular values of a matrix A are the (non-negative) square roots of the eigenvalues of AA' (or AA'):

$$\delta_i[A] = |\sqrt{\lambda_i[AA']}| = |\sqrt{\lambda_i[AA']}|.$$

B.12 Matrix Norm and Condition Number

As is well known, given a vector x, its "length" (or, more properly, its *Euclidean norm* or *two-norm*) is defined as

$$\|x\|_2 = (x_1^2 + x_2^2 + \cdots + x_n^2)^{1/2}.$$

Thanks to such notion, we can define the two-norm for a square matrix as

$$\|A\|_2 = \sup_{\|x\|_2=1} \|Ax\|_2.$$

Note that such definition can be equivalently given in the form

$$\|A\|_2 = \sup_{x \neq 0} \frac{x'Ax}{\|x\|^2}.$$

Intuitively, this matrix norm describes the effect of "amplification or contraction" of the length of a vector when passed through transformation A: If $\|A\|_2 < 1$, then the transformed vector $y = Ax$ is shorter than x, namely $\|y\|_2 < \|x\|_2$, whereas if $\|A\|_2 > 1$, then $y = Ax$ is longer than x, namely $\|y\|_2 > \|x\|_2$.

The matrix norm just defined is related to the singular values of the matrix. To explain this point, consider first the simple case when A is a real diagonal matrix, $A = \text{diag}[a_{11}, a_{22}, \dots]$. Then $Ax = a_{11}x_1 + a_{22}x_2 + \cdots$. Thus, if we take as x the unit norm vector $x = [1\ 0\ 0\ \cdots]$, we obtain $Ax = a_{11}$. If instead we take $x = [0\ 1\ 0\ \cdots]$, then $Ax = a_{22}$, and so on. Thus, from the very definition of matrix norm, it is apparent that the norm is given by the absolute value of the largest element on the diagonal.

In particular, if $A = I$, then $\|I\|_2 = 1$. On the top of that, when A is any orthogonal matrix, again $\|A\|_2 = 1$, as it is easy to verify. A most important consequence of this last observation is that an orthogonal matrix does not affect the norm of a vector or the norm of a matrix, namely

$$\|Qx\|_2 = \|x\|_2, \text{ if } Q \text{ orthogonal,}$$

$$\|QA\|_2 = \|A\|_2, \text{ if } Q \text{ orthogonal.}$$

This fact has an important consequence in terms of SVD; indeed, with reference to any square real matrix A, from the expression of the SVD, we can straightforwardly conclude that

$$\|A\|_2 = \|S\|_2.$$

Moreover, S being diagonal, we have

$$\|A\|_2 = \delta_{\max},$$

where $\delta_{\max}$ is the maximum singular values.

Thus, we come to the conclusion that the norm of a matrix coincides with its maximum singular value.

Remark B.4 (Condition number) Given a square non-singular matrix A, the number

$$K(A) = \|A\|_2 \|A^{-1}\|_2 = \frac{\delta_{\max}}{\delta_{\min}}$$

is named *condition number* of the matrix.

Obviously, $K(A) \geq 1$ for any A. If $A = I$ or A is orthogonal, then the condition number is minimum, $K(A) = 1$. As $K(A)$ increases, the matrix becomes closer and closer to singularity. Indeed, $K(A)$ can be seen as a measure of the *distance to singularity* of A. To be precise, suppose that A is non-singular and add to it a *perturbation matrix* E such that $A + E$ becomes singular. Then, by referring to the SVD of A,

$$V'(A + E)U = V'AU + V'EU = S + F,$$

where

$$F = V'EU.$$

U and V being orthogonal, the norm of E coincides with the norm of F, so that perturbations E in A correspond to equal perturbations F to S.

We can then define the *distance to singularity*, d_{sing} as the norm of the smallest E such that $A + E$ is singular:

$$d_{\text{sing}} = \min \|E\|_2 \text{ s. t. } A + E \text{ singular.}$$

In view of the discussion above, this is equivalent to

$$d_{sing} = \min\|F\|_2 \text{ s. t. } S + F \text{ singular.}$$

S being diagonal, $S + F$ can easily be made singular by taking F diagonal with one entry equal to the opposite of a singular value; obviously, the minimum norm is achieved by taking

$$\hat{F} = \text{diag}[0, \ 0 \ \cdots, \ 0, -\delta_{min}],$$

with

$$\min\|F\|_2 = \delta_{min}.$$

In conclusion, the distance to singularity is provided by the smaller singular value

$$d_{sing} = \delta_{min}.$$

Consequently, the relative size to singularity is

$$\frac{d_{sing}}{\|A\|_2} = \frac{\delta_{min}}{\delta_{max}} = K(A)^{-1}.$$

Appendix C

Problems and Solutions

Problem 1 (Linear estimator)

Given a random variable v_1, with $E[v_1] = 0$, consider the variable

$$v_2 = \alpha v_1 + \epsilon,$$

where $\alpha \neq 0$ is a real number and ϵ a zero-mean random variable with variance μ^2, uncorrelated with v_1 (i.e. $E[v_1 \epsilon] = 0$).

Find the covariance coefficient between v_2 and v_1, and study its dependence upon variance μ^2.

Solution:

We first determine mean values and variance of v_2.

The mean value can be easily computed as $E[v_2] = \alpha E[v_1] + E[\epsilon] = 0$.

The variance is given by

$$
\begin{aligned}
\lambda_{22} &= E[(v_2 - E[v_2])^2] \\
&= E[v_2^2] \\
&= E[(\alpha v_1 + \epsilon)^2] \\
&= E[\alpha^2 v_1^2 + \epsilon^2 + 2\alpha\epsilon] \\
&= \alpha^2 E[v_1^2] + \mu^2 \\
&= \alpha^2 \lambda_{11} + \mu^2.
\end{aligned}
$$

We now compute the cross-variance λ_{12} between v_1 and v_2:

$$
\begin{aligned}
\lambda_{12} &= \lambda_{21} \\
&= E[(v_1 - E[v_1])(v_2 - E[v_2])] \\
&= E[v_1 v_2] \\
&= E[v_1(\alpha v_1 + \epsilon)]
\end{aligned}
$$

Model Identification and Data Analysis, First Edition. Sergio Bittanti.
© 2019 John Wiley & Sons, Inc. Published 2019 by John Wiley & Sons, Inc.

$$= E[\alpha v_1^2 + v_1 \epsilon]$$

$$= \alpha E[v_1^2]$$

$$= \alpha \lambda_{11}.$$

Therefore,

$$\rho = \frac{\alpha \lambda_{11}}{\sqrt{\lambda_{11}} \sqrt{\alpha^2 \lambda_{11} + \mu^2}} = \frac{\alpha \lambda_{11}}{\sqrt{\alpha^2 \lambda_{11}^2 + \mu^2}}.$$

If $\mu^2 = 0$, we return to the case studied in Example 1.1, where $v_2 = \alpha v_1$. When μ^2 is non-null and becomes larger and larger, then ρ decreases monotonically, tending to 0 when $\mu^2 \to \infty$. This corresponds to the intuition: With the increasing of μ^2, the relationship between variables v_1 and v_2 becomes more and more blurred so that the information contained in v_2 on v_1 is lower and lower. If μ^2 becomes very large, one is led to estimate v_2 with the *a priori* information only, namely its mean value (0 in this example).

Problem 2 (BLUE estimator)

Consider two uncorrelated random variables v and w, with the same mean value m, and variances $\text{Var}[v] = E[v^2] = \lambda^2$ and $\text{Var}[w] = E[w^2] = \mu^2$:

$$v \sim (m, \mu^2), \quad w \sim (m, \lambda^2).$$

To determine m from the measurements of v and w, consider the linear estimator

$$\hat{m} = \alpha v + \beta w.$$

Find the parameters α and β such that

- the estimator $\hat{m}$ is unbiased, i.e. $E[\hat{m}] = m$.
- the estimator $\hat{m}$ is unbiased and the variance of the estimation error $E[(\hat{m} - m)^2]$ is minimal. Such estimator is known as BLUE estimator (BLUE = best linear unbiased estimator).

Solution:
The mean value of $\hat{m}$ is

$$E[\hat{m}] = \alpha E[v] + \beta E[w] = (\alpha + \beta)m.$$

Therefore, the linear estimator is unbiased if and only if

$$\alpha + \beta = 1.$$

By setting $\beta = 1 - \alpha$, we can write the unbiased estimator as

$$E[\hat{m}] = \alpha v + (1 - \alpha)w.$$

Then, the variance of the error is

$$E[(\hat{m} - m)^2] = \alpha^2 \lambda^2 + (1 - \alpha)^2 \mu^2.$$

By computing the derivative of such expression with respect to α and setting it to 0, one can conclude that the minimum is achieved by choosing

$$\alpha = \frac{\mu^2}{\lambda^2 + \mu^2}, \quad \beta = \frac{\lambda^2}{\lambda^2 + \mu^2}.$$

Therefore, the BLUE is

$$\hat{m} = \frac{\mu^2}{\lambda^2 + \mu^2} v + \frac{\lambda^2}{\lambda^2 + \mu^2} w.$$

Problem 3 (Spectrum of an AR(2) process)

Consider the AR(2) process

$$v(t) = -0,64v(t - 2) + \eta(t), \quad \eta(t) \sim \text{WN}(0, 1)$$

and the process

$$y(t) = v(t) + w(t), \quad w(t) \sim \text{WN}(0, 1)$$

with $w(\cdot)$ and $\eta(\cdot)$ uncorrelated to each other, namely $E[w(t_1)\eta(t_2)] = 0, \ \forall t_1, t_2$. Draw the spectral diagram of process $y(t)$.

Solution:
The given AR(2) process has

$$W(z) = \frac{1}{1 + 0.64z^{-2}} = \frac{z^2}{z^2 + 0.64}$$

as transfer function. Hence, we have two zeros in the origin of the complex plane and two poles on the imaginary axis, in $\pm 0.8j$. The model being stable, it generates a stationary process.

The spectrum of process $v(t)$ can be computed with the *fundamental theorem of spectral analysis* (see Section 1.13)as

$$\Gamma_{vv}(\omega) = |W(e^{j\omega})|^2.$$

Introducing the complex numbers

$$v_1(\omega) = e^{j\omega},$$
$$v_2(\omega) = e^{j\omega} - 0.8,$$
$$v_3(\omega) = e^{j\omega} + 0.8,$$

we can express the spectrum as

$$\Gamma_{vv}(\omega) = \frac{|v_1(\omega)|^2}{|v_2(\omega)|^2 |v_3(\omega)|^2}.$$

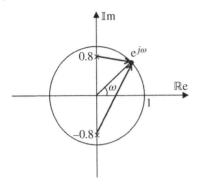

Figure C.1 Vectors in the complex plane for spectrum drawing – Problem 3.

Note that, in the complex plane, $e^{j\omega}$ is a point P on the unit circle; therefore, as shown in Figure C.1,

1. $v_1(\omega)$ is the vector connecting the zero in the origin to P,
2. $v_2(\omega)$ is the vector connecting the pole in +0.8 to P,
3. $v_3(\omega)$ is the vector connecting the pole in −0.8 to P.

Vector $v_1(\omega)$ has unitary length for any ω. When ω increases passing from $\omega = 0$ to $\omega = \pi$, vector $v_2(\omega)$ has minimum length for $\omega = \pi/2$. Analogously, when ω decreases from $\omega = 0$ to $\omega = -\pi$, vector $v_3(\omega)$ has a minimum length at $\omega = -\pi/2$. Hence, we expect that the spectrum exhibits two peaks around $\omega = \pm\pi/2$ and takes a minimum value at $\omega = 0$ and $\omega = \pi$. Easy computations lead to the values $\Gamma_{vv}(0) = \Gamma_{vv}(\pm\pi) = 0.37$ and $\Gamma_{vv}(\pm\pi/2) = 7.71$. The diagram is depicted in Figure C.2.

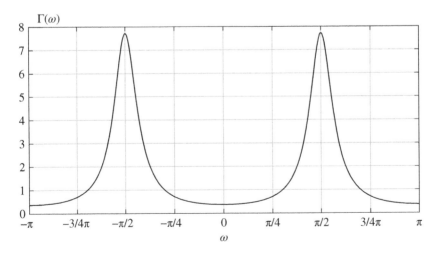

Figure C.2 Spectrum of process $v(t)$ – Problem 3.

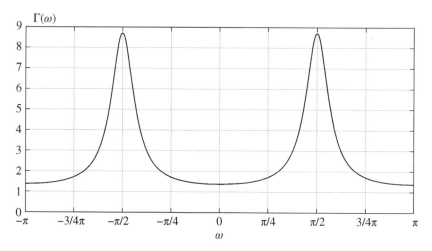

Figure C.3 Spectrum of process $y(t)$ – Problem 3.

Finally, process $y(t)$ is given by $v(t)$ plus a unit variance white noise $w(t)$ uncorrelated with noise $\eta(t)$ generating $v(t)$. Hence,

$$\Gamma_{yy}(\omega) = \Gamma_{vv}(\omega) + 1,$$

see Figure C.3.

Problem 4 (Spectrum of an AR(n) process)

An autoregressive process is described by the difference equation

$$v(t) = a_1 v(t-1) + \cdots + a_n v(t-n) + \eta(t), \quad \eta(t) \sim (0, \lambda^2). \tag{C.1}$$

In Figure C.4, four possible spectral plots are represented. In the same drawings, the spectrum of white noise η is given in dotted line.

Which diagram among the four may be the process spectrum?

Solution:
We denote by $W(z)$ the transfer function of the AR model (C.1). The first diagram in the above figure cannot be a spectrum, since any spectrum must be non-negative at any angular frequency. The fourth diagram is null at $\omega = \pm\pi$, so that the corresponding $W(z)$ should have a zero on the unit circle. This is impossible for an AR process, all zeros of which are located in the origin of the complex plane.

The third diagram is characterized by the fact that it is below the white noise spectrum, so that $\mathrm{Var}[v(t)] < \mathrm{Var}[\eta(t)]$. This is impossible. Indeed, consider the MA(∞) expansion of the given AR process:

$$W(z) = w_0 + w_1 z - 1 + \cdots ;$$

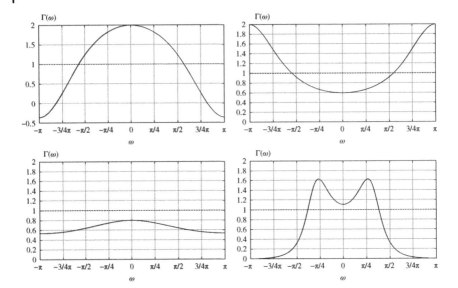

Figure C.4 Possible spectrum diagrams.

correspondingly, we have

$$\frac{\text{Var}[v(t)]}{\text{Var}[\eta(t)]} = w_0^2 + w_1^2 + \cdots .$$

The numerator and denominator of $W(z)$ being monic, the first term of such expansion must be unitary:

$$w_0 = 1.$$

Hence,

$$\frac{\text{Var}[v(t)]}{\text{Var}[\eta(t)]} = 1 + w_1^2 + \cdots \geq 1,$$

so that $\text{Var}[v(t)] < \text{Var}[\eta(t)]$ is impossible.
The only feasible diagram is the second one.

Problem 5 (Computing the variance from the spectrum)

An ARMA(1, 1) stationary process $v(t)$ has a the following spectrum:

$$\Gamma_v(\omega) = \frac{1.25 + \cos(\omega)}{1.64 + 1.6 \cos(\omega)}.$$

(i) Write the process in the time domain.
(ii) Compute the variance of the process.

Solution:

(i) The ARMA(1, 1) process has transfer function

$$W(z) = \frac{1 + cz^{-1}}{1 - az^{-1}}.$$

Assuming that the white noise at the input has unit variance, according to the fundamental theorem of spectral analysis (see Section 1.13), its spectrum is

$$\Gamma_v(\omega) = \frac{1 + c^2 + 2c \, \cos(\omega)}{1 + a^2 - 2a \, \cos(\omega)}.$$

Thus,

$$\frac{1 + c^2 + 2c \, \cos(\omega)}{1 + a^2 - 2a \, \cos(\omega)} = \frac{1.25 + \cos(\omega)}{1.64 + 1.6 \cos(\omega)}.$$

By imposing the identity of the numerators and denominators,

$$a = -\frac{4}{5}, \quad c = \frac{1}{2}.$$

Then, in the time domain, the process difference equation is

$$v(t) = -\frac{4}{5} v(t-1) + \eta(t) + \frac{1}{2}\eta(t-1), \quad \eta(t) \sim \mathrm{WN}(0, 1).$$

(ii) As for the process variance, we can proceed by computing it as either

$$\mathrm{Var}[v(t)] = E[v(t)^2] = E\left[\left(-\frac{4}{5} v(t-1) + \eta(t) + \frac{1}{2}\eta(t-1)\right)^2\right]$$

or

$$\mathrm{Var}[v(t)] = E[v(t)^2] = E\left[v(t)\left(\frac{4}{5} v(t-1) + \eta(t) + \frac{1}{2}\eta(t-1)\right)\right].$$

In any case, the conclusion is

$$\mathrm{Var}[v(t)] = \frac{5}{4}.$$

Problem 6 (Prediction of an AR(2) process)

For the AR(2) process

$$v(t) = \frac{1}{2} v(t-1) + \frac{1}{6} v(t-2) + \eta(t), \quad \eta(t) \sim \mathrm{WN}(0, 1),$$

find the optimal one-step- and two-steps-ahead predictors.

Solution:
The optimal predictors can be found with the usual method of the long division seen in Section 3.5. In the simple case of AR processes, one can adopt

an intuitive reasoning as follows. Consider the expression of $v(\cdot)$ at time $t+1$:

$$v(t+1) = \frac{1}{2}v(t) + \frac{1}{6}v(t-1) + \eta(t+1).$$

Knowing the past of $v(\cdot)$, we can say that, at time t, $v(t)$ and $v(t-1)$ are known; on the opposite, $\eta(t+1)$ is fully unpredictable since $\sim$ WN(0, 1). Therefore, the optimal one-step-ahead predictor is

$$\hat{v}(t+1|t) = \frac{1}{2}v(t) + \frac{1}{6}v(t-1).$$

The two-steps-ahead predictor can be worked out as follows. Consider the expression of $v(\cdot)$ at time $t+2$:

$$v(t+2) = \frac{1}{2}v(t+1) + \frac{1}{6}v(t) + \eta(t+2), \quad \eta(t) \sim \text{WN}(0,1).$$

Here, by replacing $v(t+1)$ with

$$v(t+1) = \frac{1}{2}v(t) + \frac{1}{6}v(t-1) + \eta(t+1),$$

one obtains

$$v(t+2) = \frac{5}{12}v(t) + \frac{1}{12}v(t-1) + \eta(t+2) + \frac{1}{2}\eta(t+1).$$

At time t, we know $v(t)$ and $v(t-1)$, but we cannot make an estimate of the future samples of the white noise $\eta(t+2) + \frac{1}{2}\eta(t+1)$ different form the trivial one (the mean value). Hence, the predictor is

$$\hat{v}(t+2|t) = \frac{5}{12}v(t) + \frac{1}{12}v(t-1).$$

The associated variances of the prediction errors are

$$\text{Var}[v(t+1) - \hat{v}(t+1|t)] = \text{Var}[\eta(t+1)] = 1,$$

$$\text{Var}[v(t+2) - \hat{v}(t+2|t)] = \text{Var}[\eta(t+2) + \frac{1}{2}\eta(t+1)] = \frac{5}{4}.$$

Problem 7 (Prediction for an ARMA(2, 1) process)

Find the one-step-ahead and the two-steps-ahead predictors for the ARMA (2, 1) process

$$v(t) + 0.9v(t-1) + 0.2v(t-2) = \eta(t) + 0.8\eta(t-1), \quad \eta \sim \text{WN}(0, \lambda^2).$$

Solution:
The transfer function from $\eta(t)$ to $v(t)$ is

$$W(z) = \frac{1 + 0.8z^{-1}}{1 + 0.9z^{-1} + 0.2z^{-2}} = \frac{z^2 + 0.8z}{z^2 + 0.9z + 0.2} = \frac{z^2 + 0.8z}{(z+0.5)(z+0.4)}.$$

Here, the numerator and denominator are monic and have the same degree. Moreover, poles and zeros are inside the unit disk. Hence, the given model is in canonical representation.

We perform two steps of the long division algorithm:

$$
\begin{array}{ll|l}
z^2 \quad +0.8z & & z^2 \quad +0.9z \quad +0.2 \\
\underline{z^2 \quad +0.9z \quad +0.2} & & \overline{1 \quad -0.1z^{-1}} \\
/ \quad -0.1z \quad -0.2 & \\
\underline{-0.1z \quad -0.09 \quad -0.02z^{-1}} & \\
/ \quad -0.11 \quad +0.02z^{-1} &
\end{array}
$$

Consequently,

$$
W(z) = 1 + \frac{-0.1z - 0.2}{z^2 + 0.9z + 0.2} = 1 + z^{-1}\frac{(-0.1z - 0.2)z}{z^2 + 0.9z + 0.2},
$$

so that the optimal one-step-ahead predictor from $\eta(\cdot)$ (fake predictor) has transfer function

$$
\hat{W}_1(z) = -\frac{0.1z^2 + 0.2z}{z^2 + 0.9z + 0.2}.
$$

The transfer function of the optimal predictor from data $v(\cdot)$ has the same numerator, while the denominator is given by the numerator of $W(z)$, so that

$$
W_1(z) = -\frac{0.1z^2 + 0.2z}{z^2 + 0.8z} = -\frac{0.1z + 0.2}{z + 0.8} = -\frac{0.1 + 0.2z^{-1}}{1 + 0.8z^{-1}}.
$$

In the time domain, the predictor equation is

$$
(1 + 0.8z^{-1})\hat{v}(t + 1|t) = -(0.1 + 0.2z^{-1})v(t),
$$

namely

$$
\hat{v}(t + 1|t) = -0.8\hat{v}(t|t - 1) - 0.1v(t) - 0.2v(t - 1).
$$

Turning to the two-steps-ahead predictor, from the long division, we have

$$
W(z) = 1 - 0.1z^{-1} + \frac{-0.11 + 0.02z^{-1}}{z^2 + 0.9z + 0.2} = 1 - 0.1z^{-1} + z^{-2}\frac{-0.11z^2 + 0.02z}{z^2 + 0.9z + 0.2}.
$$

The optimal two-step-ahead predictor from $\eta(\cdot)$ is

$$
\hat{W}_2(z) = -\frac{z(0.11z - 0.02)}{z^2 + 0.9z + 0.2},
$$

while the optimal two-step-ahead predictor from $v(\cdot)$ is

$$
W_2(z) = -\frac{z(0.11z - 0.02)}{z(z + 0.8)} = -\frac{0.11z - 0.02}{z + 0.8},
$$

with associated difference equation

$$\hat{v}(t+2|t) = -0.8\hat{v}(t+1|t-1) - 0.11v(t) + 0.02v(t-1).$$

Note that both the one-step- and the two-step-ahead predictors are stable.

The variance of the prediction error is λ^2 for the one-step-ahead and $(1 + 0.1^2)\lambda^2$ for the two-steps-ahead predictor.

Problem 8 (Prediction of an ARMA(2, 1) with missing term)

Consider the stationary process generated by the ARMA(2, 1) model

$$v(t) = \frac{1}{4}v(t-2) + \eta(t) + \frac{1}{3}\eta(t-1), \quad \eta(\cdot) \sim \text{WN}(0, 1). \tag{C.2}$$

Letting

$$A(z) = z^2 - \frac{1}{4}, \quad C(z) = z^2 + \frac{1}{3}z,$$

the transfer function from $\eta(t)$ to $v(t)$ is

$$W(z) = \frac{C(z)}{A(z)}.$$

Notice that $A(z)$ is a second-order polynomial with missing first power of z. Find the variance of process $v(t)$ and then determine the optimal one-step-ahead and two-steps-ahead predictors.

Solution:
It is straightforward to verify that the model of the process is a canonical factor. In particular, the system being stable, it generates a stationary process at the output, with zero mean value.

We compute first the process variance:

$$E[v(t)^2] = \frac{1}{16}E[v(t-2)^2] + E[\eta(t)^2] + \frac{1}{9}E[\eta(t-1)^2]$$

$$+ \frac{1}{4}E[v(t-2)\eta(t)] + \frac{1}{12}E[v(t-2)\eta(t-1)]$$

$$+ \frac{1}{3}E[\eta(t)\eta(t-1)].$$

Since $\eta(\cdot)$ is white, the random variables $\eta(t)$ and $\eta(t-1)$ are uncorrelated, so that the last term at the right-hand side is null. As for the terms $E[v(t-2)\eta(t)]$ and $E[v(t-2)\eta(t-1)]$, from the difference equation of the data generation mechanism, we see that $v(t-2)$ depends upon $\eta(\cdot)$ up to time $t-2$. Therefore, $v(t-2)$ is uncorrelated with both $\eta(t-1)$ and $\eta(t)$. Hence, $E[v(t-2)\eta(t)] = 0$ and $E[v(t-2)\eta(t-1)] = 0$, so that

$$\text{Var}[v] = \frac{1}{16}\text{Var}[v] + 1 + \frac{1}{9} \longrightarrow \text{Var}[v] = \frac{32}{27}.$$

To find the optimal one-step-ahead and two-steps-ahead predictors, we divide $C(z)$ by $A(z)$

$$
\begin{array}{ll|l}
z^2 & +\frac{1}{3}z & z^2 \quad +0z \quad -\frac{1}{4} \\
z^2 & +0z \quad -\frac{1}{4} & 1 \quad +\frac{1}{3}z^{-1} \\
\hline
/ & +\frac{1}{3}z \quad +\frac{1}{4} & \\
& +\frac{1}{3}z \quad +0 \quad -\frac{1}{12}z^{-1} & \\
\hline
& / \quad +\frac{1}{4} \quad +\frac{1}{12}z^{-1} &
\end{array}
$$

Therefore,

$$
W(z) = 1 + z^{-1}\frac{\frac{1}{3}z^2 + \frac{1}{4}z}{z^2 - \frac{1}{4}}
$$

and

$$
W(z) = 1 + \frac{1}{3}z^{-1} + z^{-2}\frac{\frac{1}{4}z^2 + \frac{1}{12}z}{z^2 - \frac{1}{4}}.
$$

The transfer function of the optimal one-step-ahead and two-steps-ahead predictors from v are therefore

$$
W_1(z) = \frac{\frac{1}{3}z^2 + \frac{1}{4}z}{z^2 + \frac{1}{3}z} = \frac{\frac{1}{3}z + \frac{1}{4}}{z + \frac{1}{3}},
$$

$$
W_2(z) = \frac{\frac{1}{4}z^2 + \frac{1}{12}z}{z^2 + \frac{1}{3}z} = \frac{1}{4}.
$$

In the time domain, we have

$$
\hat{v}(t+1|t) = -\frac{1}{3}\hat{v}(t|t-1) + \frac{1}{3}v(t) + \frac{1}{4}v(t-1),
$$

$$
\hat{v}(t+2|t) = \frac{1}{4}v(t).
$$

The variances of the prediction errors are

$$
\mathrm{Var}[v(t) - \hat{v}(t-1|t)] = 1,
$$

$$
\mathrm{Var}[v(t) - \hat{v}(t|t-2)] = 1^2 + \left(\frac{1}{3}\right)^2 = \frac{10}{9}.
$$

The expression for the two-steps-ahead predictor, $\hat{v}(t+2|t) = \frac{1}{4}v(t)$, is easily interpretable. Indeed, the data generation mechanism is $v(t+2) = \frac{1}{4}v(t) + \eta(t+2) + \frac{1}{3}\eta(t+1)$. We see that $v(t+2)$ is constituted by the sum of two terms,

$\frac{1}{4}v(t)$ and $\eta(t+2) + \frac{1}{3}\eta(t+1)$. If $v(\cdot)$ is known up to t, the first addendum can be computed with no error. However, the knowledge of $v(\cdot)$ up to t is of no help to predict the values of $\eta(t+2)$ and $\eta(t+1)$, since these are samples of process $\eta(\cdot)$ related to the future with respect to instant t. In other words, the knowledge of $v(t), v(t-1), \dots$ does not convey information on $\eta(t+2)$ and $\eta(t+1)$, so that the only possible prediction of these variables is their mean value, zero. It is therefore natural that $\hat{v}(t+2|t) = \frac{1}{4}v(t)$ is the best prediction rule.

Problem 9 (Prediction of an ARMA process with an exo term)

We reconsider the previous exercise with an exogenous constant term:

$$v(t) = \frac{1}{4}v(t-2) + \eta(t) + \frac{1}{3}\eta(t-1) + \bar{u}, \quad \bar{u} = 2, \quad \eta(t) \sim \text{WN}(0,1).$$

$$(C.3)$$

Find the optimal predictor.

Solution:
The system being stable, it generates a stationary process. However, the mean value of the process, say m, is now non-null. By applying the expectation operator to both sides of the above difference equation, we have

$$m = \frac{1}{4}m + 2$$

from which

$$m = \frac{8}{3}.$$

Observe that the same computation can be performed as follows. The given system has two inputs, $\eta(t)$ and $u(t)$ and a single output, $v(t)$, as indicated in Figure C.5. Denoting by $W(z)$ the transfer function from $\eta(t)$ to $v(t)$ and by $\Delta(z)$ the transfer function from $\bar{u}$ to $v(t)$, we have

$$W(z) = \frac{C(z)}{A(z)}, \quad \Delta(z) = \frac{1}{A(z)}$$

with

$$C(z) = z^2 + \frac{1}{3}z, \quad A(z) = z^2 - \frac{1}{4}.$$

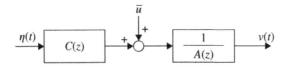

Figure C.5 Block scheme of the system of Problem 9.

The gains of these transfer functions are

$$\frac{C(1)}{A(1)} = \frac{16}{9}, \quad \frac{1}{A(1)} = \frac{4}{3}$$

(we refer to point A.1.6 of Appendix A for the notion of gain). The mean value of the output can be obtained by multiplying the gains of the transfer functions by the mean values of the respective inputs and summing them, i.e.

$$E[v(t)] = \frac{C(1)}{A(1)}E[\eta(t)] + \frac{1}{A(1)}\bar{u} = \frac{16}{9}0 + \frac{4}{3}2 = \frac{8}{3}.$$

We come now to the prediction problem. To solve it, we introduce process

$$\tilde{v}(t) = v(t) - m.$$

$\tilde{v}$, the mean value of which is null, is named *debiased process*. Since $\tilde{v}(t)$ and $v(t)$ differ from each other for the mean value only, their variances coincide, $\mathrm{Var}[v(t)] = \mathrm{Var}[\tilde{v}(t)]$. The equation governing $\tilde{v}$ can be easily deduced from that of v. Indeed,

$$\tilde{v}(t) = v(t) - m$$

$$= \frac{1}{4}v(t-2) + \eta(t) + \frac{1}{3}\eta(t-1) + 2 - m$$

$$= \frac{1}{4}(\tilde{v}(t-2) + m) + \eta(t) + \frac{1}{3}\eta(t-1) + 2 - m$$

$$= \frac{1}{4}\tilde{v}(t-2) + \eta(t) + \frac{1}{3}\eta(t-1) + 2 - \frac{3}{4}m.$$

Since $m = 8/3$, the quantity $2 - (3/4)m$ is null. Hence,

$$\tilde{v}(t) = \frac{1}{4}\tilde{v}(t-2) + \eta(t) + \frac{1}{3}\eta(t-1).$$

We see that $\tilde{v}$ is a zero-mean stationary process governed by the same difference equation studied in the previous Problem . Thus, we can say that its variance is given by $\mathrm{Var}[\tilde{v}(t)] = 32/27$, and its two-steps-ahead prediction rule is $\hat{\tilde{v}}(t+2|t) = (1/4)\tilde{v}(t)$.

We now return to the original process. Obviously, the predictor of v coincides with that of $\tilde{v}$ up to the translation of the mean value, i.e.

$$\hat{v}(t+2|t) = \hat{\tilde{v}}(t+2|t) + m.$$

Therefore,

$$\hat{v}(t+2|t) = \frac{1}{4}\tilde{v}(t) + m$$

$$= \frac{1}{4}(v(t) - m) + m$$

$$= \frac{1}{4}v(t) + \frac{3}{4}m$$

$$= \frac{1}{4}v(t) + 2$$

$$= \frac{1}{4}v(t) + \bar{u}.$$

From this last expression, we see that the presence of an additional constant $\bar{u} = 2$ at the right-hand side of expression (C.3) simply leads to the replication of that term in the predictor rule.

Problem 10 (Identification of an AR model with various delays)

Consider the stationary process generated by

$$\mathcal{S}: \quad y(t) = a^\circ y(t-1) + \eta(t), \quad \eta(t) \sim \text{WN}(0, \lambda^2)$$

and the predictive model

$$\hat{\mathcal{S}}: \quad \hat{y}(t, a) = ay(t-k).$$

Find the value $\hat{a}$ minimizing the variance of the prediction error

$$J(a) = E[(y(t) - \hat{y}(t, a))^2]$$

for $k = 1$ and $k = 2$. Is the prediction error of the estimated model white?

Solution:
For $k = 1$,

$$y(t) - \hat{y}(t, a) = (a^\circ - a)y(t-1) + \eta(t).$$

Hence,

$$J(a) = E[(a^\circ - a)^2 y(t-1)^2 + \eta(t)^2 + 2(a^\circ - a)y(t-1)\eta(t)].$$

$y(t-1)$ depends upon $\eta(\cdot)$ up to time $t-1$. Hence, $E[y(t-1)\eta(t)] = 0$, so that

$$J(a) = E[(a^\circ - a)^2 y(t-1)^2] + \lambda^2,$$

the minimum of which is obtained for

$$\hat{a} = a^\circ.$$

This result is expected, since, for $k = 1$, the system belongs to the class of models, so that the estimated parameter coincides with the true one.

Pass now to the case $k = 2$. In the expression $y(t) = a^\circ y(t-1) + \eta(t)$, by replacing $y(t-1)$ with $y(t-1) = a^\circ y(t-2) + \eta(t-1)$, we have

$$y(t) = (a^\circ)^2 y(t-2) + \eta(t) + a^\circ \eta(t-1).$$

Correspondingly,

$$J(a) = E[((a^\circ)^2 - a)^2 y(t-2)^2] + E[(\eta(t) + a^\circ \eta(t-1))^2]$$
$$+ E[y(t-2)(\eta(t) + a^\circ \eta(t-1))].$$

$y(t - 2)$ depends upon $\eta(\cdot)$ up to time $t - 2$. Hence, $E[y(t - 2)(\eta(t) + a^\circ\eta(t - 1))] = 0$, so that the minimum of $J(a)$ is achieved for

$$\hat{a} = (a^\circ)^2.$$

The prediction error of the identified model $y(t) - \hat{y}(t, \hat{a})$ is $\eta(t)$ for $k = 1$ and $\eta(t) + a^\circ\eta(t - 1)$ for $k = 2$. Hence, it is white for $k = 1$ only.

Problem 11 (Identification of an AR(2) model)

Consider the process

$$y(t) = v(t) + w(t),$$

where

$$v(t) = 0.5v(t - 1) + \eta(t), \quad \eta(t) \sim \text{WN}(0, 1),$$

$$w(t) \sim \text{WN}(0, 1)$$

and $w(t)$ and $\eta(t)$ are uncorrelated, namely

$$E[w(t_1)\eta(t_2)] = 0, \quad \forall t_1, t_2.$$

Determine

(i) the mean value of $y(t)$,
(ii) the covariance function of $y(t)$,
(iii) with a prediction error minimization criterion, identify an AR(2) model to fit process $y(t)$.

Solution:

(i) Process $v(t)$ has zero mean value; therefore, $E[y(t)] = E[v(t)] + E[w(t)] = 0$.
(ii) Processes $w(t)$ and $\eta(t)$ are uncorrelated; hence, $v(t)$ and $\eta(t)$ are uncorrelated as well. Therefore, the stationary process $y(t)$ has a covariance function $\gamma_{yy}(\tau)$ given by the sum of the two covariance functions

$$\gamma_{yy}(\tau) = \gamma_{vv}(\tau) + \gamma_{ww}(\tau).$$

$v(t)$ is an AR(1) process, so that its covariance function is given by

$$\begin{cases} \gamma_{vv}(0) = 4/3, \\ \gamma_{vv}(\tau) = (1/2)^\tau 4/3, \quad \tau > 0. \end{cases}$$

$w(t)$ being a white noise of unit variance,

$$\begin{cases} \gamma_{yy}(0) = 7/3, \\ \gamma_{vv}(\tau) = (1/2)^\tau 4/3, \quad \tau > 0. \end{cases}$$

In particular,

$$\gamma_{yy}(\tau) = \begin{cases} 7/3, & \tau = 0, \\ 2/3, & \tau = 1, \\ 1/3, & \tau = 2. \end{cases}$$

(iii) An AR(2) model in predictive form is written as

$$\hat{y}(t|t-1) = a_1 y(t-1) + a_2 y(t-2).$$

The predictive identification criterion is

$$J = E[(y(t) - a_1 y(t-1) - a_2 y(t-2))^2].$$

The minimization with respect to the parameters has been discussed in Section 1.9.2. The optimal parameters are given by expressions (1.14) and (1.15):

$$\hat{a}_1 = \frac{\gamma_{yy}(0)\gamma_{yy}(1) - \gamma_{yy}(1)\gamma_{yy}(2)}{\gamma_{yy}(0)^2 - \gamma_{yy}(1)^2},$$

$$\hat{a}_2 = \frac{\gamma_{yy}(0)\gamma_{yy}(2) - \gamma_{yy}(1)^2}{\gamma_{yy}(0)^2 - \gamma_{yy}(1)^2}.$$

Therefore, in the present case,

$$\hat{a}_1 = \frac{14/9 - 2/9}{49/9 - 4/9} = 4/15,$$

$$\hat{a}_2 = \frac{7/9 - 4/9}{49/9 - 4/9} = 1/15.$$

Problem 12 (Identification and prediction from numerical data)

For a zero-mean stationary process, the following samples are measured:

$$y(0) = 0.5, \quad y(1) = 0.25, \quad y(2) = -0.5, \quad y(3) = 0,$$
$$y(4) = 1, \quad y(5) = -0.25.$$

Estimate the value of $y(6)$.

Solution:
We estimate $y(6)$ by finding its best prediction, given the previous data: $\hat{y}(6|5)$. To this purpose, we first identify a model from the data and then we derive the predictor.

A simple model is the autoregressive model of order 1. With reference to such a type of model, we know that

$$\hat{y}(t|t-1) = ay(t-1)$$

is the corresponding predictor. Then, the prediction error is

$$\epsilon(t) = y(t) - \hat{y}(t|t-1).$$

We can estimate parameter a by minimizing

$$J(a) = \frac{1}{5} \sum_{t=1}^{5} \epsilon(t)^2$$

with respect to a.

Being $y(0) = 0.5$ and $\hat{y}(1|0) = ay(0) = 0.5a$, we have $\epsilon(1) = y(1) - \hat{y}(1|0) = 0.25 - 0.5a$. Analogously, we can proceed for the subsequent time points as summarized in the table below.

t	$\epsilon(t)$
1	$0.25 - 0.5a$
2	$-0.5 - 0.25a$
3	$0.5a$
4	1
5	$-0.25 - a$

In this way,

$$J(a) = \frac{1}{5}\left(\frac{25}{16}a^2 + \frac{1}{2}a + \beta\right),$$

where β does not depend upon a. The minimum is achieved for

$$\hat{a} = -\frac{4}{25}.$$

Therefore, we formulate the following estimate

$$\hat{y}(6|5) = \hat{a}y(5) = -\frac{4}{25}\left(-\frac{1}{4}\right) = \frac{1}{25}.$$

Problem 13 (Minimum variance control for an ARX system)

Consider the ARX system

$$\mathcal{S} : y(t+2) = 0.5y(t+1) + u(t) + 0.9u(t-1) + \eta(t+2). \tag{C.4}$$

In order to impose that the output signal $y(\cdot)$ be as close as possible to a given reference signal $\bar{y}^\circ$, adopt the minimum variance control strategy, requiring the minimization of

$$J = E[(y(t+2) - \bar{y}^\circ)^2].$$

Design the optimal controller and study the corresponding feedback system.

Solution:
The given system is described by a model of the type

$$\mathcal{S} : A(z)y(t) = B(z)u(t-k) + C(z)\eta(t), \quad \eta(\cdot) \sim WN(0, \lambda^2),$$

with delay $k = 2$ and

$$A(z) = 1 - 0.5z^{-1}, \quad B(z) = 1 + 0.9z^{-1}, \quad C(z) = 1.$$

As seen in Section 8.5, the two-steps-ahead predictor can be found by the long division of $C(z)$ by $A(z)$ for two steps, to obtain polynomials $E(z)$ and $F(z)$ according to diophantine equation

$$C(z) = A(z)E(z) + z^{-k}\tilde{F}(z).$$

The optimal predictor is then

$$\hat{\mathcal{S}} : C(z)\hat{y}(t+k|t) = \tilde{F}(z)y(t) + B(z)E(z)u(t).$$

In our problem, by performing the long division for $k = 2$ steps, we have

$$E(z) = 1 + 0.5z^{-1}, \quad \tilde{F}(z) = 0.25.$$

Hence,

$$B(z)E(z) = 1 + 1.4z^{-1} + 0.45z^{-2},$$

so that

$$\hat{y}(t+2|t) = 0.25y(t) + u(t) + 1.4u(t-1) + 0.45u(t-2). \tag{C.5}$$

In the simple case of ARX models, the predictor can be also easily derived from Eq. (C.4), by replacing $y(t+1)$ at the right-hand side with the expression provided by the model itself, i.e. $\mathcal{S} : y(t+1) = 0.5y(t) + u(t-1) + 0.9u(t-2) + \eta(t+1)$. In this way, we can express $y(t+2)$ as

$$\mathcal{S} : y(t+2) = 0.25y(t) + u(t) + 1.4u(t-1) + 0.45u(t-2)$$
$$+ 0.5\eta(t+1) + \eta(t+2).$$

In this formula, we can distinguish the term $0.25y(t) + u(t) + 1.4u(t-1) + 0.45u(t-2)$ from the term $0.5\eta(t+1) + \eta(t+2)$. The former one is computable from past input–output data up to time t. On the contrary, $\eta(\cdot)$ being a white noise, $\eta(t+1)$ and $\eta(t+2)$ are unpredictable from the past up to time t. The

Figure C.6 System of
Problem 14.

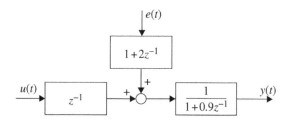

only estimate of $0.5\eta(t+1) + \eta(t+2)$ we can formulate is its mean value, 0. So, we come again to predictor (C.5).

The optimal MV controller, obtained by imposing $\hat{y}(t+k|t) = \bar{y}°$, is therefore

$$\mathscr{S} : u(t) = -1.4u(t-1) - 0.45u(t-2) + \bar{y}° - 0.25y(t). \tag{C.6}$$

This expression indicates that $u(t)$ is constructed by means of a feedback action from $y(t)$, taking also into account the reference signal $\bar{y}°$ and two past samples of the input, $u(t-1)$ and $u(t-2)$. This feedback action leads to a control loop, illustrated in Figure 8.1, with a set of blocks in cascade. The open-loop transfer function $L(z)$ is the product of the transfer functions of the various blocks:

$$L(z) = C(z)\frac{1}{B(z)E(z)}z^{-k}B(z)\frac{1}{A(z)}\tilde{F}(z) = \frac{0.25(z+0.9)}{(z+0.9)(z+0.5)(z-0.5)}.$$

As can be seen, there is a common factor at numerator and denominator, as polynomial $B(z)$ appears at both numerator and denominator.

The characteristic polynomial of the loop is therefore

$$\Delta(z) = (z+0.9)[(z+0.5)(z-0.5) + 0.25] = (z+0.9)z^2.$$

Hence, the closed-loop system has three poles, two located in $z = 0$ and one in -0.9. This implies that the feedback system is stable.

Problem 14 (Minimum variance control for an ARMAX system)

Consider the system depicted in the block scheme of Figure C.6, where $e(\cdot) \sim \text{WN}(0, 1)$.

(i) Design a minimum variance control law for a reference signal $y°(t) = \bar{y}° = 2$,

(ii) Draw the block diagram of the control system.

(iii) Find the mean value and the variance of the output $y(t)$ of the control system.

(iv) In Figure C.7, the realizations of four stationary processes are represented (the dotted line represents the mean value). Which one among them can correspond to the output of the designed control system?

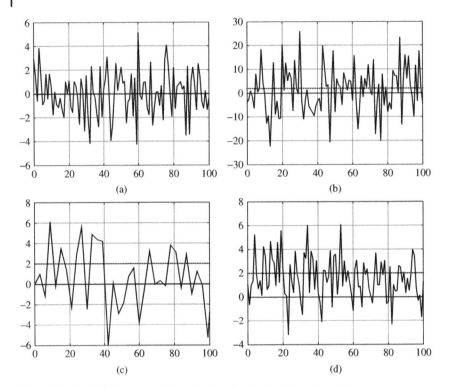

Figure C.7 Possible diagrams of the output of the designed control system.

Solution:

(i) The given block diagram corresponds to the ARMAX system

$$\mathcal{S} : A(z)y(t) = B(z)u(t - k) + C(z)e(t),$$

with delay $k = 1$ and

$$A(z) = 1 + 0.9z^{-1}, \quad B(z) = 1, \quad C(z) = 1 + 2z^{-1}, \quad e(\cdot) \sim \text{WN}(0, 1).$$

The MA process $C(z)e(t)$, however, is not in the canonical representation. Therefore, we first rewrite the system as

$$\mathcal{S} : A(z)y(t) = B(z)u(t - k) + \hat{C}(z)\eta(t),$$

with delay $k = 1$ and

$$A(z) = 1 + 0.9z^{-1}, \quad B(z) = 1, \quad \hat{C}(z) = 1 + 0.5z^{-1}, \quad \eta(\cdot) \sim \text{WN}(0, 4).$$

In the time domain, this means

$$\mathcal{S} : y(t + 1) = -0.9y(t) + u(t) + \eta(t + 1) + 0.5\eta(t).$$

The one-step-ahead predictor is

$$\hat{y}(t+1|t) = \frac{(C(z) - A(z))}{C(z)} y(t+1) + \frac{B(z)}{C(z)} u(t),$$

namely

$$\hat{y}(t+1|t) = \frac{-0.4}{1+0.5z^{-1}} y(t) + \frac{1}{1+0.5z^{-1}} u(t).$$

By imposing $\hat{y}(t+1|t) = y^\circ(t)$, we obtain the control law

$$y^\circ(t) = \frac{-0.4}{1+0.5z^{-1}} y(t) + \frac{1}{1+0.5z^{-1}} u(t).$$

(ii) The block diagram of the control system is represented in Figure C.8.

(iii) According to the control scheme above, the output $y(t)$ is the sum of a deterministic signal $y_d(t)$, the effect of $y^\circ(t)$, and a stochastic signal $y_s(t)$, the effect of noise $e(t)$. It is easy to compute the transfer function from $y^\circ(t)$ to $y(t)$ and see that it is given by z^{-1}. Therefore, $y_d(t) = y^\circ(t-1) = \bar{y}^\circ = 2$. As for the stochastic effect, the transfer function from $e(t)$ to $y(t)$ is

$$\frac{1+2z^{-1}}{1+0.5z^{-1}}.$$

This is an all-pass filter generating a white noise $\eta(t)$ of 0 mean value and variance 4. In conclusion,

$$y(t) = 2 + \eta(t), \quad \eta(t) \sim WN(0, 4),$$

which is equivalent to

$$y(t) = \eta(t), \quad \xi(t) \sim WN(2, 4).$$

In conclusion, the output is white noise with mean value 2 and variance 4.

(iv) The mean value of the output being 2, we discharge diagram (i) where the mean value is 0. The variance is 4; hence, the output should take most

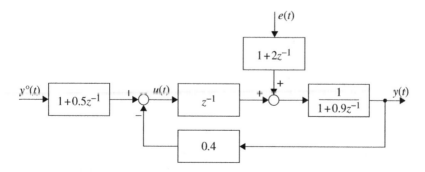

Figure C.8 Control system for Problem 14.

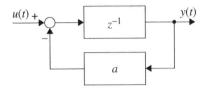

$u(t)$ $y(t)$

z^{-1}

a

Figure C.9 A feedback system – Problem 15.

values in the range $2 \pm 3\sqrt{4}$, namely in the range $(-4, 8)$. This excludes diagram (ii) since the depicted signal oscillates in an oversized range of values. Between the two remaining cases, we exclude (iii) as the diagram moves too slowly to be a white noise.

Problem 15 (Parametric control)

Consider the feedback system depicted in Figure C.9, where input $u(t)$ is given by

$$u(t) = \bar{u} + \eta(t), \quad \bar{u} = 1 \quad \eta(t) \sim \text{WN}(0, 1).$$

(i) Find the range of values of parameter a for which the system is stable. Correspondingly, output $y(t)$ is a stationary process. Determine the mean value and variance of such process.
(ii) Find the value of parameter a in the range determined at point (i) leading to the minimization of $\text{Var}[y(t)]$.
(iii) Find the value of parameter a in the range determined at point (i) leading to the minimization of $E[y(t)^2]$.

Solution:

(i) The transfer function from $u(t)$ to $y(t)$ is

$$G(z) = \frac{z^{-1}}{1 + az^{-1}} = \frac{1}{z + a}. \tag{C.7}$$

Hence, the system is stable when $|a| < 1$.
Note that, in the time domain, relation (C.7) corresponds to

$$(1 + az^{-1})y(t) = z^{-1}u(t), \tag{C.8}$$

i.e.

$$y(t) = -ay(t - 1) + u(t - 1). \tag{C.9}$$

Therefore, $y(t)$ is an autoregressive process.
Its mean value $\bar{y} = E[y(t)]$ can be found by applying the expected value operator to both sides of this last equation, so obtaining $\bar{y} = \bar{u}/(1 + a)$.

Alternatively, one can derive the gain of (C.7) from $G(z)$ by posing $z = 1$: $G(1) = 1/(1+a)$. Hence, $\bar{y} = G(1)\bar{u} = \bar{u}/(1+a)$.

As for the variance, we observe that it coincides with the variance of the zero-mean autoregressive process:

$$y(t) = -ay(t-1) + \eta(t-1). \tag{C.10}$$

As seen Section 1.8, such variance is $1/(1-a^2)$.

In conclusion,

$$E[y(t)] = \frac{1}{1+a}, \tag{C.11}$$

$$\mathrm{Var}[y(t)] = \frac{1}{1-a^2}. \tag{C.12}$$

(ii) The variance of $y(t)$ being given by expression (C.12), the minimum is achieved for $a = 0$.

(iii) To compute $E[y(t)^2]$, we observe that

$$E[y(t)^2] = E[(y(t) - \bar{y} + \bar{y})^2] = \mathrm{Var}[y(t)] + E[\bar{y}^2] + 2\bar{y}E[y(t)^2 - \bar{y}].$$

The third term at the right-hand side is null. Hence, from Eqs. (C.11) and (C.12),

$$E[y(t)^2] = \frac{1}{1-a^2} + \frac{1}{(1+a)^2} = \frac{2}{(1-a^2)(1+a)}$$

$$= \frac{2}{(1-a)(1+a)^2} = \frac{2}{f(a)},$$

where

$$f(a) = (1-a)(1+a)^2. \tag{C.13}$$

In the range $-1 < a < +1$, this function has the following properties: $f(-1) = 0$, $f(+1) = 0$, and $f(a) > 0$, $\forall a$. An easy computation shows that the maximum of (C.13), and therefore the minimum of $E[y(t)^2]$, is achieved for $a = 1/3$.

Problem 16 (State space identification via Hankel matrix)

Consider a deterministic system with transfer function

$$W(z) = \frac{z^2 - 3/5z}{z^2 - (1/10)z - 3/10}.$$

(i) Construct the associated Hankel matrix of dimension 3×3.
(ii) Determine a state space representation.

Solution:

(i) Denoting by $h(\cdot)$ the impulse response of the given system, the 3×3 Hankel matrix is

$$\mathcal{H}^3 = \begin{bmatrix} h(1) & h(2) & h(3) \\ h(2) & h(3) & h(4) \\ h(3) & h(4) & h(5) \end{bmatrix}.$$

To construct it, we need the first samples of the impulse response up to $h(5)$. These samples are easily found by the long division of the numerator and denominator of the given transfer function:

$$
\begin{array}{ll|ll}
z^2 & -\dfrac{3}{5}z & z^2 & -\dfrac{1}{10}z \qquad -\dfrac{3}{10} \\[2mm]
\cline{3-4}
z^2 & -\dfrac{1}{10}z \quad -\dfrac{3}{10} & 1 & -\dfrac{1}{2}z^{-1} \quad +\dfrac{1}{4}z^{-2} \\[2mm]
\cline{1-2}
/ & -\dfrac{1}{2}z \quad +\dfrac{3}{10} & & \\[2mm]
& -\dfrac{1}{2}z \quad +\dfrac{1}{20} \quad +\dfrac{3}{20}z^{-1} & & \\[2mm]
\cline{1-2}
/ & +\dfrac{1}{4} \quad -\dfrac{3}{20}z^{-1} & & \\
\end{array}
$$

By further steps of the long division, the following expansion of $W(z)$ in negative powers of z is worked out

$$W(z) = 1 - \frac{1}{2}z^{-1} + \frac{1}{4}z^{-2} - \frac{1}{8}z^{-3} + \frac{1}{16}z^{-4} - \frac{1}{32}z^{-5} \cdots .$$

Therefore, the first coefficients of the impulse response are

$$h(0) = 1, \quad h(1) = -\frac{1}{2}, \quad h(2) = \frac{1}{4}, \quad h(3) = -\frac{1}{8},$$

$$h(4) = \frac{1}{16}, \quad h(5) = -\frac{1}{32}.$$

Hence, the 3×3 Hankel matrix is

$$\mathcal{H}^3 = \begin{bmatrix} -\dfrac{1}{2} & +\dfrac{1}{4} & -\dfrac{1}{8} \\[2mm] +\dfrac{1}{4} & -\dfrac{1}{8} & +\dfrac{1}{16} \\[2mm] -\dfrac{1}{8} & +\dfrac{1}{16} & -\dfrac{1}{32} \end{bmatrix}.$$

(ii) To find a state space realization of the system, we start by observing that, from the result of the long division above, $W(z)$ can be written as

$$W(z) = W_a(z) + W_b(z),$$

with

$$W_a(z) = -\frac{1}{2}z^{-1} + \frac{1}{4}z^{-2} - \frac{1}{8}z^{-3} + \frac{1}{16}z^{-4} - \frac{1}{32}z^{-5} + \cdots,$$

$$W_b(z) = 1.$$

Denoting with $u(t)$ the input of the system and with $y(t)$ its output, we have

$$y(t) = y_a(t) + y_b(t),$$

with

$$y_a(t) = W_a(z)u(t) = \left(-\frac{1}{2}z^{-1} + \frac{1}{4}z^{-2} - \frac{1}{8}z^{-3} + \frac{1}{16}z^{-4} - \frac{1}{32}z^{-5} + \cdots\right)u(t),$$

$$y_b(t) = W_b(z)u(t) = u(t).$$

A state space representation (F_a, G_a, H_a) of $W_a(z)$ can be obtained from the Hankel matrix $\mathcal{H}^3$ as discussed in Chapter 7.

In particular, we know that the dimension of the state space realization is the rank of such matrix. In our problem, by analyzing the Hankel matrix, it is apparent that

$$\text{rank}[\mathcal{H}^3] = 1.$$

So a first-order state space system suffices. This is not surprising since the given transfer function can be written as

$$W(z) = \frac{z(z - 3/5)}{(z + 1/2)(z - 3/5)},$$

so that there is a pole-zero cancellation leading to

$$W(z) = \frac{z}{z + 1/2}.$$

Second, we know that the impulse response coefficients are related to the system matrices by $h(k) = H_a F_a^{k-1} G_a$, $k \geq 1$. Hence, in our problem, $H_a G_a = -1/2$, $H_a F_a G_a = 1/4$, $H_a F_a^2 G_a = -1/8$, and so on. A triple (F_a, G_a, H_a) satisfying these conditions is $F_a = -1/2$, $G_a = 1$, and $H_a = -1/2$, so that

$$x(t + 1) = -\frac{1}{2}x(t) + u(t)$$

$$y_a(t) = -\frac{1}{2}x(t)$$

is a realization of $W_a(z)$.

Taking into account the contribution of W_b, the conclusion is that

$$x(t + 1) = -\frac{1}{2}x(t) + u(t)$$

$$y(t) = -\frac{1}{2}x(t) + u(t)$$

is a state space realization of the given system (with transfer function $W(z) = W_a(z) + W_b(z)$).

Problem 17 (Kalman predictor for an unstable first order system)

Consider the scalar system

$$x(t + 1) = 2x(t) + v_1(t), \qquad\qquad\qquad\qquad (c.14a)$$

$$y(t) = x(t) + v_2(t), \qquad\qquad\qquad\qquad\qquad (c.14c)$$

where $v_1(t) \sim \text{WN}(0, 1)$ and $v_2(t) \sim \text{WN}(0, 1)$. Moreover, $v_1(t)$ and $v_2(t)$ are uncorrelated to each other and uncorrelated with the initial condition.

(i) Show that the solution $P(t)$ of the corresponding DRE (difference Riccati equation) converges to a limit matrix $\bar{P}$ and determine $\bar{P}$.

(ii) Determine the optimal steady-state one-step-ahead Kalman predictor associated with $\bar{P}$.

(iii) Verify that the state prediction error is a stationary process and compute its variance.

Solution:

(i) The system is scalar ($n = 1$) with dynamic matrix $F = 2$ and output matrix $H = 1$. The variance of the state equation disturbance is $V_1 = 1$, the variance of the output equation disturbance is $V_2 = 1$ and the cross variance is $V_{12} = 0$. The square root of $V_1 = 1$ is $G_v = 1$.
The DRE (difference Riccati equation) is given by

$$P(t + 1) = FP(t)F' + V_1 - FP(t)H'(HP(t)H' + V_2)^{-1}HP(t)F'$$

$$= 4P(t) + 1 - \frac{4P(t)^2}{P(t) + 1}$$

$$= \frac{5P(t) + 1}{P(t) + 1}.$$

The system is not stable since $|F| > 1$; however, it is reachable and observable. Hence, from the second convergence theorem, we know that $P(t)$ converges as $t \to \infty$. The limit matrix $\bar{P}$ can be found by solving the ARE (algebraic Riccati equation), given by

$$P = \frac{5P + 1}{P + 1}.$$

This leads to

$$P^2 - 4P - 1 = 0,$$

which has the two solutions

$$P = 2 \pm \sqrt{5}.$$

One of them is negative, so the only feasible solution is

$$\bar{P} = 2 + \sqrt{5} = 4.236.$$

Summing up, all solutions of the DRE associated with a non-negative initial condition tend to such $\bar{P}$ as $t \to \infty$.

(ii) The steady-state optimal one-step-ahead Kalman predictor is

$$\mathcal{S}_1 : \begin{cases} \hat{x}(t+1|t) = F\hat{x}(t|t-1) + Ke(t), \\ \hat{y}(t+1|t) = H\hat{x}(t+1|t). \end{cases}$$

where the innovation is

$$e(t) = y(t) - \hat{y}(t|t-1)$$

and the predictor gain is

$$\bar{K} = F\bar{P}H'(H\bar{P}H' + V_2)^{-1}. \tag{C.15}$$

Being

$$\bar{K} = \frac{2\bar{P}}{\bar{P}+1} = 1.618, \tag{C.16}$$

we have

$$\mathcal{S}_1 : \begin{cases} \hat{x}(t+1|t) = 2\hat{x}(t|t-1) + 1.618(y(t) - \hat{y}(t|t-1)), \\ \hat{y}(t+1|t) = \hat{x}(t+1|t). \end{cases}$$

(iii) We now compute the state prediction error:

$$\begin{aligned}
\epsilon(t+1) &= x(t+1) - \hat{x}(t+1|t) \\
&= 2x(t) + v_1(t) - 2\hat{x}(t|t-1) - 1.618y(t) + 1.618\hat{x}(t|t-1) \\
&= 2\epsilon(t) + v_1(t) - 1.618y(t) + 1.618\hat{x}(t|t-1) \\
&= 2\epsilon(t) + v_1(t) - 1.618x(t) - 1.618v_2(t) + 1.618\hat{x}(t|t-1) \\
&= 2\epsilon(t) + v_1(t) - 1.618\epsilon(t) - 1.618v_2(t) \\
&= 0.382\epsilon(t) + v_1(t) - 1.618v_2(t).
\end{aligned}$$

Define now

$$\eta(t) = v_1(t) - 1.618v_2(t).$$

$\eta(t)$ is a zero-mean white noise with variance $1^2 + 1.618^2 = 3.618$. Therefore,

$$\epsilon(t+1) = 0.382\epsilon(t) + \eta(t), \quad \eta(t) \sim \text{WN}(0, 3.618).$$

In conclusion, even if the given system is unstable (so that it generates a diverging stochastic process), the Kalman predictor can track the state with an finite variance error; indeed, the state prediction error is governed by a *stable* difference equation. To be precise, $\epsilon(t)$ is an AR(1) process. Its variance is given by

$$\text{Var}[\epsilon(t)] = \frac{1}{1 - 0.382^2} 3.618 = 4.236.$$

As expected $\text{Var}[\epsilon(t)]$ coincides with $\bar{P}$.

Problem 18 (Kalman predictor for a second-order system)

In the second-order system

$$x_1(t + 1) = ax_1(t) + v_{11}(t)$$
$$x_2(t + 1) = \beta x_2(t) + v_{12}(t)$$
$$y(t) = x_2(t) + v_2(t).$$

$v_{11}(t)$, $v_{12}(t)$, and $v_2(t)$ are white noises of zero mean value and unit variance, uncorrelated to each other. For the following cases,

(a) $|\alpha| < 1$ and $|\beta| < 1$,
(b) $|\alpha| > 1$ and $|\beta| > 1$,

discuss

(i) the asymptotic characteristics of the state prediction error provided by the Kalman predictor,
(ii) the asymptotic characteristics of the prediction error of the output when the state is estimated with the Kalman predictor.

Solution:
The system has two state variables. The covariance matrix of the state prediction error is therefore a 2×2 matrix of the type

$$P(t) = \begin{bmatrix} P_{11}(t) & P_{12}(t) \\ P_{21}(t) & P_{22}(t) \end{bmatrix},$$

where $P_{11}(t)$ and $P_{22}(t)$ are the variances of the prediction errors of the first state and the second state variables, respectively.

We observe that the given system is constituted by two noninteracting subsystems. Subsystem 1 is

$$x_1(t + 1) = ax_1(t) + v_{11}(t)$$

and subsystem 2 is

$$x_2(t + 1) = \beta x_2(t) + v_{12}(t)$$
$$y(t) = x_2(t) + v_2(t).$$

In case (a), both such subsystems are stable. Therefore, in view of the first convergence theorem for the Riccati equation, $P(t)$ asymptotically converges to a finite matrix. This entails that

case (a) $P_{11}(t) \to \bar{P}_{11}, \quad P_{22}(t) \to \bar{P}_{22}.$

In case (b), both subsystems are not asymptotically stable, so the first convergence theorem of Kalman theory is not applicable. The second convergence theorem is also not applicable since the observability condition is not satisfied. On the other hand, we see that the subsystem 2 is observable, so that the variance of its error estimate converges to a finite value. On the contrary, subsystem 1 is unstable and the measured output variable does not bring any information on its state. Therefore,

case (b) $P_{11}(t) \to \infty, \quad P_{22}(t) \to \bar{P}_{22}.$

Pass now to the output. From the system equations, we see that $y(t)$ does not depend upon $x_1(t)$. Therefore, the variance of its estimation is asymptotically limited in both cases (a) and (b). This example shows that the error in the estimation of the output may be finite even if state estimation error matrix $P(t)$ is divergent.

Problem 19 (Sensor choice)

In the system

$$x_1(t + 1) = x_1(t) + v_{11}(t)$$
$$x_2(t + 1) = x_1(t) + 2x_2(t) + v_{12}(t),$$

where $v_{11}(t)$ and $v_{12}(t)$ are white noises of zero mean value and unit variances, uncorrelated to each other, it is possible to measure either $x_1(t)$ or $x_2(t)$ with a sensor characterized by a disturbance $v_2(t)$ of zero mean and unit variance, uncorrelated with $v_{11}(t)$ and $v_{12}(t)$. What is the appropriate choice? For that choice, study the convergence of the one-step-ahead predictor and find the steady-state predictor.

Solution:
The given system can be written in the matrix form

$$x(t + 1) = Fx(t) + v_1(t),$$

with

$$F = \begin{bmatrix} 1 & 0 \\ 1 & 2 \end{bmatrix},$$

and state noise vector

$$v_1(t) = \begin{bmatrix} v_{11}(t) \\ v_{11}(t) \end{bmatrix}.$$

The covariance matrix of $v_1(t)$ is

$$V_1 = \begin{bmatrix} 1 & 0 \\ 0 & 1 \end{bmatrix}.$$

The state variables can be seen as

$$x_1(t) = \frac{1}{z-1} v_{11}(t),$$

$$x_2(t) = \frac{1}{(z-1)(z-2)} v_{11}(t) + \frac{1}{z-2} v_{12}(t).$$

So, they are generated from $v_{11}(t)$ and $v_{12}(t)$ via non-stable transfer functions. Therefore, they are not stationary processes. The same conclusion can be drawn by noting that the eigenvalues of matrix F are not in the stability region of the complex plane.

We can complement the state equation with the following output equation:

$$y(t) = \alpha x_1(t) + \beta x_2(t) + v_2(t), \quad v_2(t) \sim WN(0, 1).$$

Then, according to the problem statement, we distinguish two cases:

(i) $\alpha = 1$ and $\beta = 0$ (measuring $x_1(t)$).
(ii) $\alpha = 0$ and $\beta = 1$ (measuring $x_2(t)$).

The output matrix is

$$H = \begin{bmatrix} \alpha & \beta \end{bmatrix},$$

leading to an observability matrix

$$\mathcal{O} = \begin{bmatrix} H \\ HF \end{bmatrix} = \begin{bmatrix} \alpha & \beta \\ \alpha + \beta & 2\beta \end{bmatrix}.$$

Hence,

$$\text{case (i)} : \mathcal{O} = \begin{bmatrix} 1 & 0 \\ 1 & 0 \end{bmatrix}, \quad \text{case (ii)} : \mathcal{O} = \begin{bmatrix} 0 & 1 \\ 1 & 2 \end{bmatrix}.$$

This means that the observability condition is satisfied in case (ii) only, so that the appropriate sensor choice is measuring $x_2(t)$. This is also in agreement with the state equations, from which we see that, measuring $x_1(t)$, no information on $x_2(t)$ would be available.

Focus then on case (ii). The system is not stable, so the first convergence theorem is not applicable. However, the observability conditions are satisfied. Moreover, as it is easy to see, the reachability condition from noise $v_1(t)$ is also satisfied. Hence, one can resort to the second convergence theorem and conclude that the difference Riccati equation (DRE) has a solution $P(t)$, which is asymptotically convergent. Correspondingly, the Kalman gain will converge too and the optimal predictor will tend to a steady-state predictor.

The asymptotic value $\bar{P}$ of $P(t)$ (from which the asymptotic value $\bar{K}$ of the gain can be computed) is the positive semi-definite solution of the algebraic Riccati equation (ARE):

$$P = FPF' + V_1 - (FPH')(HPH' + V_2)(FPH')'. \tag{C.17}$$

By writing the 2×2 matrix P as

$$P = \begin{bmatrix} P_{11} & P_{12} \\ P_{21} & P_{22} \end{bmatrix},$$

with $P_{21} = P_{21}$, from the ARE, it is easy to see that

$$P_{12}^2 - (P_{22} + 1) = 0,$$

$$P_{11} + P_{12} - \frac{P_{12}(P_{12} + 2P_{22})}{P_{22} + 1} = 0,$$

$$P_{11} + 4P_{12} + 3P_{22} + 1 - \frac{(P_{12} + 2P_{22})^2}{P_{22} + 1} = 0,$$

so that

$$\bar{P} = \begin{bmatrix} \bar{P}_{11} & \bar{P}_{12} \\ \bar{P}_{21} & \bar{P}_{22} \end{bmatrix} = \begin{bmatrix} 3.39 & 3.04 \\ 3.04 & 8.27 \end{bmatrix}.$$

The gain of the steady-state Kalman predictor, given by expression

$$\bar{K} = F\bar{P}H'(H\bar{P}H' + 1)^{-1},$$

takes the value

$$\bar{K} = \begin{bmatrix} 0.328 \\ 2.112 \end{bmatrix}.$$

$$\bar{K} = \begin{bmatrix} 0.328 \\ 2.112 \end{bmatrix}.$$

The steady-state Kalman predictor

$$\bar{x}(t + 1) = F\bar{x}(t) + \bar{K}e(t) = F\bar{x}(t) + \bar{K}(y(t) - H\bar{x}(t))$$

has dynamic matrix

$$F - \bar{K}H = \begin{bmatrix} 1 & -0.328 \\ 1 & -2.112 \end{bmatrix},$$

whose eigenvalues are

$$0.444 \pm j0.137.$$

Their modulus is 0.464, so that the steady-state Kalman predictor is stable.

Problem 20 (Kalman predictor with correlated noises)

Consider the first-order system

$$x(t+1) = 0.5x(t) + e(t)$$
$$y(t) = 1.5x(t) + e(t),$$

where $e(t) \sim WN(0, 1)$.

(i) Write the equations of the optimal one-step-ahead Kalman predictor.
(ii) Find the asymptotic value of the variance of the state prediction error:

$$\bar{P} = \lim_{t\to\infty} \text{Var}[x(t+1) - \hat{x}(t+1|t)].$$

(iii) Find the asymptotic value of the variance of the output prediction error:

$$\bar{Q} = \lim_{t\to\infty} \text{Var}[y(t+1) - \hat{y}(t+1|t)].$$

Solution:

(i) In this problem, the state noise $v_1(t)$ and the output noise $v_2(t)$ coincide, so that

$$V_1 = 1, \quad V_2 = 1, \quad V_{12} = 1.$$

The Kalman gain is given by

$$K(t) = \frac{0.75P(t) + 1}{2.25P(t) + 1}.$$

Correspondingly, the difference Riccati equation is

$$P(t+1) = 0.25P(t) + 1 - \frac{(0.75P(t) + 1)^2}{2.25P(t) + 1}.$$

Easy computations lead to

$$P(t+1) = \frac{P(t)}{2.25P(t) + 1}.$$

(ii) From the Riccati equation above, we see that, if $P(1) = 0$, then $P(t)$ is constant and equal to 0 for each time point. If instead $P(1) > 0$, then $P(2) < P(1)$, $P(3) < P(2)$, and so on:

$$\lim_{t\to\infty} P(t) = 0, \quad \forall P(1) \geq 0.$$

Hence,

$$\bar{P} = 0.$$

By the way, this is the only positive semi-definite solution of the algebraic Riccati equation (ARE).

(iii) Let $Q(t) = \mathrm{Var}[y(t) - \hat{y}(t|t-1)]$ be the prediction error variance of the output. To compute it, we recall that the optimal predictor of the output can be obtained from the optimal predictor of the state as

$$\hat{y}(t|t-1) = 1.5\hat{x}(t|t-1).$$

Consequently,

$$y(t) - \hat{y}(t|t-1) = 1.5[x(t) - \hat{x}(t|t-1)] + e(t).$$

Therefore,

$$Q(t) = \mathrm{Var}[1.5[x(t) - \hat{x}(t|t-1)]] + \mathrm{Var}[e(t)] + 3E[[x(t) - \hat{x}(t|t-1)]e(t)].$$

From the state equation $x(t+1) = 0.5x(t) + e(t)$, we see that $x(t)$ depends upon $e(\cdot)$ up to time $t-1$. The predictor $\hat{x}(t|t-1)$ also must depend on past data. Hence, $x(t) - \hat{x}(t|t-1)$ and $e(t)$ are uncorrelated. Therefore,

$$Q(t) = 1.5^2 P(t) + 1.$$

For $t \to \infty$, this matrix tends to

$$\bar{Q} = 1.$$

Bibliography

Further reading

Time series analysis

Bisgaard, S. and Kulahci, M. (2011). *Time Series Analysis and Forecasting by Example*. Wiley.

Bittanti, S. (ed.) (1986). *Time Series and Linear Systems*. Springer-Verlag.

Box, G.E.P. and Jenkins, G.M. (1970). *Time Series Analysis: Forecasting and Control*. Wiley. A modernized edition with G.C. Reinsel as third author has been published in 2013.

Brockwell, P.J. and Davis, E.A. (1991). *Time Series: Theory and Methods*. Springer.

Shumway, R.H. and Stoffer, D.S. (2000). *Time Series Analysis and its Applications*. Springer.

Stochastic systems

Hannan, E.J. and Deistler, M. (1988). *The Statistical Theory of Linear Systems*. Wiley.

Lindquist, A. and Picci, G. (2015). *Linear Stochastic Systems*. Springer.

Speyer, J.L. and Chung, W. (2008). *Stochastic Processes, Estimation and Control*. SIAM.

System identification

Bittanti, S. and Picci, G. (eds) (1996). *Identification, Adaptation, Learning - The Science of Learning Models from Data*. Springer.

Chen, H.F. and Zhao, W. (2017). *Recursive Identification and Parameter Estimation*. CRC Press.

Guidorzi, R. (2003). *Multi-Variable System Identification. From Observations to Models*. Bononia University Press.

Model Identification and Data Analysis, First Edition. Sergio Bittanti.
© 2019 John Wiley & Sons, Inc. Published 2019 by John Wiley & Sons, Inc.

Kailath, T., Sayed, A.H., and Hassibi, B. (2000). *Linear Estimation*. Prentice-Hall.

Katayama, T. (2005). *Subspace Methods for System Identification*. Springer.

Ljung, L. (1995). *System Identification Toolbox for Use with MATLAB*. The MathWorks.

Ljung, L. (1999). *System Identification: Theory for the User*. Prentice-Hall.

Pintelon, R. and Schoukens, J. (2012). *System Identification: A Frequency Domain Approach*. Wiley.

Söderström, T. and Stoica, P. (1989). *System Identification*. Prentice-Hall.

Tangirala, A.K. (2015). *Principles of System Identification*. CRC Press.

Van Overschee, P. and De Moor, B. (1996). *Subspace Identification for Linear Systems*. Kluwer Academic Publishers.

Digital control

Åström, K.J. (2006). *Introduction to Stochastic Control Theory*. Reprinted Dover.

Åström, K.J. and Wittenmark, B. (2011). *Computer-Controlled Systems: Theory and Design*. Courier Corporation.

Kučera, V. (1991). *Analysis and Design of Discrete Linear Control Systems*. Prentice-Hall.

Filtering

Anderson, B.D.O. and Moore, J.B. (2005). *Optimal Filtering*. Reprinted Dover.

Lewis, F., Xie, L., and Popa, D. (2008). *Optimal and Robust Estimation*. CRC Press.

Simon, D. (2006). *Optimal State Estimation: Kalman, H_∞, and Nonlinear Approaches*. Wiley Inter-sciences.

Systems and control - basics

Antsaklis, P.J. and Michel, A.N. (2006). *Linear Systems*. Birkhåuser.

Åström, K.J. and Murray, R.M. (2010). *Feedback Systems: An Introduction for Scientists and Engineers*. Princeton University Press.

Matrices

Golub, G.H. and Van Loan, C.F. (2013). *Matrix Computations*. Baltimore, MD: The John Hopkins University Press.

Strang, G. (2016). *Introduction to Linear Algebra*. Wellesley-Cambridge Press.

Chapters references

Akaike, H. (1974). A new look to statistical model identification. *IEEE Transactions on Automatic Control* 19: (6): 716–723.

Albertos, P. and Mareels, I. (2010). *Feedback and Control for Everyone*. Springer.

Allgöwer, F. and Zheng, A. (eds.) (2000). *Nonlinear Model Predictive Control*. Birkhäuser.

Åstrom, K. J. and Wittemark, B. (1989). *Adaptive Control*. Addison-Wesley.

Bellhouse, D. (2005). *Historia Mathematica* 32 (2): 180–202.

Bemporad, A. and Morari, M. (2017). *Predictive Control*. Cambridge University Press.

Basar, T. and Bernhard, P. (1991). H_∞ *Optimal Control and Related Minimax Design Problems*. Birkhaüser.

Bitmead, R.R., Gevers, M., and Wertz, V. (1990) *Adaptive Optimal Control: The Thinking Man Approach*. Prentice Hall.

Bittanti, S. and Campi, M. (1994). Bounded error identification of time-varying parameters by RLS techniques. *IEEE Transactions on Automatic Control* 39 (5): 1106–1110.

Bittanti, S. and Piroddi, L. (1994). GMV techniques for nonlinear control with neural networks. *IEE Proceedings – Control Theory and Applications* 141 (2): 57–69.

Bittanti, S. and Piroddi, L. (1997). Nonlinear identification and control of a heat exchanger: a neural network approach. *Journal of Franklin Institute* 334 (1): 135–153.

Bittanti, S. and Savaresi, S.M. (2000). On the parametrization and design of an extended Kalman filter frequency tracker. *IEEE Transactions on Automatic Control* 45 (9): 1718–1724.

Bittanti, S., Bolzern, P., and Campi, M. (1990). Convergence and exponential convergence of identification algorithms with directional forgetting factor. *Automatica* 26 (5): 929–932.

Bittanti, S., Campi, M., and Savaresi, S.M. (1997). Unbiased estimation of a sinusoid in colored noise via adapted notch filters. *Automatica* 33 (2): 209–215.

Camacho, E. and Bordons Alba, C. (2007). *Model Predictive Control*. Springer.

Campi, M.C. and Garatti, S. (2018). *Introduction to the Scenario Approach*. SIAM.

Campi, M.C., Lecchini, A., and Savaresi, S.M. (2002). Virtual reference feedback tuning: a direct method for the design of feedback controllers. *Automatica* 38 (8): 1337–1346.

Clarke, D.W., Mohtadi, C., and Tufts, P.S. (1987a). Generalized predictive control Part I. The basic algorithm. *Automatica* 23 (2): 137–148.

Clarke, D.W., Mohtadi, C., and Tufts, P.S. (1987b). Generalized predictive control Part II. Extensions and interpretations. *Automatica* 23 (2): 149–160.

Crisan, D. and Doucet, A. (2002). A survey of convergence results on particle filtering methods for practitioners. *IEEE Transactions on Signal Processing* 50 (3): 736–746.

Ding, F. and Chen, T. (2005). Performance bounds of forgetting factor least-squares algorithms for time-varying systems with finite measurement data. *IEEE Transactions on Circuits and Systems* 52 (3): 555–566.

Garatti, S. and Bittanti, S. (2008). Parameter estimation via artificial data generation with the *Paso Doble* approach. 17th IFAC World Congress of 2008 in Seoul.

Garatti, S. and Bittanti, S. (2012). A new paradigm for parameter estimation in system modeling. *International Journal of Adaptive Control and Signal Processing* 27 (8): 667–687.

Garatti, S. and Bittanti, S. (2014). A model identification approach to the analysis of the Kobe earthquake time series. International Work-Conference on Time Series Analysis (ITISE2014), held in Granada (Spain) in 2014.

Goodwin, G.C. and Sin, K.S. (1984). *Adaptive Filtering, Prediction and Control*. Englewood Cliffs, NJ: Prentice-Hall.

Guo, L., Ljung, L., and Priouret, P. (1993). Performance analysis of the forgetting factor RLS algorithm. *International Journal Adaptive Control and Signal Processing* 7 (6): 525–527.

Ho, B.L. and Kalman, R.E. (1965). Effective construction of linear state-variable models from input-output functions. *Regelungstechnik* (12): 545–548.

Kung, S.Y. (1978). A new identification method and model reduction algorithm via singular value decomposition. In: *Proceedings of the 12th Asilomar Conference on Circuits, Systems and Computers*, 705–714,

Landau, I.D., Lozano, R., and M'Saad, M. (1998). *Adaptive Control*. Springer.

Ljung, L., Soderstrom, T., and Gustavson, I. (1975). Counterexamples to general convergence of a commonly used recursive identification method. *IEEE Transactions on Automatic Control* AC-20 (5): 643–652.

Maciejowski, J.M. (2002). *Predictive Control*. Pearson.

Magni, L. and Scattolini, R. (2014). *Advanced and Multivariable Control*. Pitagora Editrice.

Rawlings, J.B., Mayne, D.Q., and Diehl, M.M. (2017). *Model Predictive Control: Theory, Computation, and Design*. Nob Hill Publishing.

Reif, K. and Unbehauen, R. (1999). The extended Kalman filter as an exponential observer for nonlinear systems. *IEEE Transaction on Signal Processing* 47 (8): 2324–2328.

Rissanen, J. (1978). Modeling by shortest data description. *Automatica* 14 (5): 467–471.

Savaresi, S.M., Bittanti, S., and So, H.C. (2003). Closed-form unbiased frequency estimation of a noisy sinusoid using notch filters. *IEEE Transactions on Automatic Control* 48 (7): 1285–1292.

Strang, G. (2016). *Linear Algebra*. Wellesley-Cambridge Press.

Wilkinson, J.H. (1988). *The Algebraic Eigenvalue Problem*. Oxford University Press, Inc.

Zeiger, H.P. and McEwen, A.J. (1974). Approximate linear realizations of given dimensions via Ho's algorithm. *IEEE Transactions on Automatic Control* 19 (2): 153.

Index

Model Identification and Data Analysis, First Edition. Sergio Bittanti.
© 2019 John Wiley & Sons, Inc. Published 2019 by John Wiley & Sons, Inc.

1 M. Lovera 2 F. Algöwer 3 A. Brankovic 4 M. Prandini 5 G.O. Guardabassi 6 R. Guidorzi 7 S. Formentin 8 A. Isidori 9 G. Panzani 10 T. Katayama 11 I.K. Craig 12 J. Speyer 13 G. Goodwin 14 M. Rapizza 15 T. Başar 16 P. Rocco 17 M. Fanna 18 P. Bolzern 19 K.J. Åström 20 S. Garatti 21 M. Gevers 22 S. Bittanti 23 F. Dabbene 24 R. Scattolini 25 M.C. Campi 26 D. Pareschi 27 P. Lluka 28 P. Albertos 29 V. Kučera 30 P. Colaneri

A set of participants in the workshop *Perspectives on System Identification and Control Science* nicknamed *Bit Fest*, held in Como-Italy, 2017. The picture, devised and set up by Marco Campi, was taken on 17-7-2017.

Printed and bound by CPI Group (UK) Ltd, Croydon, CR0 4YY

16/04/2025

14658418-0003